BASIC MATHEMATICS

Derek I. Bloomfield, Ph.D.

Orange County Community College
Middletown, New York

West Publishing Company

Minneapolis/St. Paul ▌ New York
Los Angeles ▌ San Francisco

DEDICATED TO MY WONDERFUL CHILDREN JENNIFER, MAX, DEREK JR, AND DAVID

Copyeditor: Susan Gerstein

Cover and Text Design: Susan Guillory, TECHarts
Boulder, Colorado

Proofreader: Robert Moody

Art: Christine Bentley, Visual Graphic Systems,

Chapter Openers and Cover Image Art: Kristopher K. Hill, FinalCopy

Compositor: Clarinda Company

Index: E. Virginia Hobbs

Production, Prepress, Printing and Binding by West Publishing Company.

By WEST PUBLISHING COMPANY
610 Opperman Drive
P.O. Box 64526
St. Paul, MN 55164-0526

Printed in the United States of America
01 00 99 98 97 96 95 94 8 7 6 5 4 3 2 1 0
Library of Congress Cataloging–in–Publication Data

Bloomfield, Derek I., 1940–
Basic mathematics / Derek I. Bloomfield.
p. cm.
Includes index.
ISBN 0-314-02783-1 (soft—student version)
ISBN 0-314-02881-1 (soft—annotated teacher's edition)
1. Mathematics. I. Title.
QA39.2.B588 1994
513'.14—dc20 93-42968
CIP

West's commitment to the environment

In 1906, West Publishing Company began recycling materials left over from the production of books. This began a tradition of efficient and responsible use of resources. Today, up to 95 percent of our legal books and 70 percent of our college and school texts are printed on recycled, acid-free stock. West also recycles nearly 22 million pounds of scrap paper annually—the equivalent of 181,717 trees. Since the 1960s, West has devised ways to capture and recycle waste inks, solvents, oils, and vapers created in the printing process. We also recycle plastics of all kinds, wood, glass, corrugated cardboard, and batteries, and have eliminated the use of styrofoam book packaging. We at West are proud of the longevity and the scope of our commitment to the environment.

About the cover and chapter opener images

More than eye-catching computer graphics, these images are known as fractals. Fractals are mathematical characterizations of the complexities found in nature. Jagged coastlines, fluffy clouds, the bark of trees, and snowflakes all exhibit these intricate patterns. These fractals are from the Mandelbrot and Julia sets. They are named for Benoit B. Mandelbrot, who introduced the concept of fractals in the early 1970s, and French mathematician Gaston Julia, who studied self-similar boundaries with Pierre Fatou during the first world war. An interesting property of fractals is their iterative nature. No matter how much you enlarge the fractal, it looks uncannily similar to lesser enlargements. The simple yet complex beauty of nature and our universe is represented in the colorful rings, whorls, and spirals.

CONTENTS

CHAPTER 7 Descriptive Statistics 235

CHAPTER 8 Signed Numbers—The Language of Algebra 265

CHAPTER 9 Algebra 313

PREFACE

Purpose and Style

This book is intended for anyone who needs to learn the fundamental skills of mathematics. This audience consists of students who didn't do as well in high school as they wish they had, adults interested in returning to school to better themselves or to change careers, people wanting to take high school equivalency examinations, people needing more mathematics to advance in their work, and a long list of others. This book is written with these people in mind. Its sole purpose is to bring such students up to a level of competence in mathematics that will enable them to take additional mathematics courses or to take courses in other areas in which a knowledge of basic mathematical skills is required.

This book is intended for use in any of three instructional modes:

1. a conventional lecture-type class;
2. a mathematics laboratory;
3. a self-study program in which the student works at his or her own pace.

The books consists of 12 chapters, and the material in each chapter is presented in the following manner: A concise explanation of the fundamental concepts for the particular topic is given; examples illustrating these fundamental concepts are worked out step-by-step; and an ample number of similar problems are given in the exercise sets enabling the student to master the concepts involved. The exercises progress from simple to more difficult in an effort to give the student confidence in the ability to solve problems. The answers to odd-numbered exercises are given at the end of the text.

The major goal of this method of presentation is a correct understanding of the topics and maximum skill in performing the mathematical operations.

The material is presented in measured amounts so that the student can complete a topic before moving on to the next concept. An achievement test is given at the end of each chapter with answers to all exercises given in the answer section.

Outstanding Features

- **Number Knowledge:** At the beginning of each chapter is a "human interest" mathematical interlude.
- **How to be Successful at Mathematics:** Hints to help students are scattered throughout the first two chapters.
- STOP **Warning Signs:** These help students avoid frequently committed errors.
- **Critical Thinking Exercises:** Students are required to find errors in worked-out solutions to problems.
- **Mental Mathematics Exercises:** Practice exercises that are to be done without paper and pencil; these help students to master certain important fundamental skills as well as help to learn the lost art of doing mathematics "in their head."
-  **In-Your-Own-Words Exercises:** Students are required to think through

and to write out definitions and procedures in their own words. This lets them know whether they really understand a concept or not.

- **Living-in-the-Real-World Exercises:** These represent real-world applications of mathematics in everyday situations.
- **Calculators:** Chapter 12, the last chapter in the book is about how to use a calculator. The instruction and exercises stand alone and may be included at the beginning of the semester, at the end, or not at all, at the individual instructor's discretion.
- **Chapter Summary:** At the end of each chapter is a list of definitions, rules, and procedures, each of which is accompanied by a worked-out example illustrating that procedure.
- **Chapter Review Exercises:** A set of exercises, keyed to the appropriate section in the chapter, is included at the end of each chapter.
- **Achievement Test:** A test is included at the end of each chapter with answers to all exercises to show the student whether or not he or she has mastered the concepts of the chapter.
- **Cumulative Reviews:** At the end of every third chapter is a cumulative set of exercises to ensure that students remember previous material.
- **Pedagogical Use of Color:** It's not just decoration.
- **Workbook Style with Perforated Pages:** These may be handed in if the instructor so desires.
- Over 800 worked-out examples with step-by-step explanations are presented, with important steps highlighted in color.
- Over 5,000 exercises have been carefully chosen to clarify explanations and to provide necessary drill.
- Sufficient space is provided for working out the exercises right in the text. This also provides a good reference when it comes time to review.

Ancillaries

- **Westest 3.0:** A computer-generated testing program. Problems may be selected and mixed and matched to suit the instructor's needs. There are versions for both MacIntosh and IBM PCs or compatibles running DOS or Windows 3.0.
- **Instructor's Manual with Test Bank** containing a test bank, review sheets, and chapter tests.
- **Instructor's Solutions Manual** containing worked-out solutions for all even-numbered problems in the test.
- **Student's Solutions Manual** containing worked-out solutions for all odd-numbered problems in the text.
- **Annotated Instructor's Edition** containing answers to all exercises written adjacent to the problem for easy instructor reference.
- **Mathens Tutorial Software** containing multiple examples of the basic problem types in *Basic Mathematics*. The program is easy to use, interactive with the student, keyed to each section in the text, and versions are available for both the MacIntosh and IBM PCs and compatibles.
- **Video Tapes** keyed to the text section-by-section.

I would like to thank the reviewers for their many helpful and constructive suggestions. These include:

- Jeff Mock—*Diablo Valley College*
- Elaine Miller—*University of Toledo*

- Sally Copeland—*Johnson County Community College*
- Deann Christianson—*University of the Pacific*
- Steven S. Terry—*Ricks College*
- Elaine M. Klett—*Brookdale Community College*
- Arleen J. Watkins—*University of Arizona*
- Sharon C. Ross—*DeKalb College*
- Ned W. Shillow—*Lehigh County Community College*
- Marion Glasby—*Anne Arundel Community College*
- Wendell L. Motter—*Florida A&M*
- Debra Madrid-Doyle—*Santa Fe Community College*
- Nancy G. Hyde—*Broward Community College*
- Chris Burditt—*Napa Valley College*
- James Marcotte—*Cincinnati Technical College*
- James W. Newsom—*Tidewater Community College*
- James R. Fryxell—*College of Lake County*
- John R. Garlow—*Tarrant County Junior College*
- Sandra Evans—*Cumberland County College*
- Michael Payne—*College of Alameda*
- Ann Thorne—*College of DuPage*
- Harold S. Engelsohn—*Kingsborough Community College*
- Richard C. Spangler—*Tacoma Community College*
- Lloyd G. Roeling—*University of Southwestern Louisiana*
- Karen Sue Cain—*Eastern Kentucky University*
- Bill Orr—*Crafton Hills College*
- Wayne Milloy—*Crafton Hills College*
- Linda J. Murphy—*Northern Essex Community College*
- Nancy K. Nickerson—*Northern Essex Community College*
- Bill Elliott—*Keene State College*
- Sylvia Kennedy—*Broome County Community College*
- Julie R. Monte—*Daytona Beach Community College*
- James R. Brasel—*Phillips County Community College*
- Sharon Edgmon—*Bakersfield College*

Acknowledgments

I want to thank the people at West Publishing Company and everyone who assisted with this book. Special thanks go to my editors Ron Pullins and Denise Bayko for their friendship and for their guidance and encouragement through many steps along the way; Laura Mezner Nelson and Poh Lin Khoo, my production editors, who took special care to ensure that every detail in the finished product would look just right; Doug Abbot for his marketing efforts; and Susan Gerstein, my copy editor, who does her work as well as anyone can. Maria Blon prepared the Instructor's Manual and Robert Moody proofread the final manuscript. My most sincere thanks to all of them.

I am grateful to the administration, my colleagues, and especially to the students at Orange County Community College for providing a stimulating and positive environment. It's a great place to work. My wife, Marcella, typed the manuscript, prepared both the Instructor and Student Solutions Manuals, and made innumerable suggestions that improved the book greatly. Marcella and my children, Jennifer, Max, Derek, Jr., and David, gave up many Sunday afternoons. I thank them for their incentive, their encouragement, their understanding, and their love.

A Note from the Author to the Student

There is a lot of truth in the old adage, "There's no success like success." I learned a long time ago that the first problem on a math exam should be one that *everyone* can do. It leads to confidence and success. Rearranging the same problems on the same exam so that the first problem is an especially difficult one can so destroy a student that you can't even spell your own name. Your anxiety level goes up, your confidence level plummets.

So, in my writing, I have attempted to alleviate anxiety, build confidence, and ensure mathematical success in the following ways.

1. I speak personally to you, the student, letting each of you know that I care about you and about your success.
2. I explain how to solve problems in a way that you can understand.
3. I very, very carefully make sure that the exercise sets progress slowly from simple to the more difficult ones. If you start at the beginning and work through the set in sequence, you will learn a tremendous amount, sometimes subtly, without even knowing what is happening to you. And, as on my exams, *everybody* can do the first few problems.
4. Nothing in the book is incidental. There are no fillers. In that regard, it's like a Mozart symphony—every note means something. In my book, every word counts.
5. My style is convincing and encouraging. In general I have more confidence in you as a student than perhaps you have in yourself. I speak to that throughout.

This book is written for you and I'm genuinely interested in what you think about it. Please write to me and let me know what you like about it or how you think it can be improved.

Thank you.

Derek Bloomfield
R. D. #4, Box 834
Middletown, NY 10940

BASIC MATHEMATICS

CHAPTER 1

Whole Numbers

INTRODUCTION

I asked my daughter, Jennifer, what her favorite word was. She answered "candy," which came as no great surprise to me. What does this have to do with mathematics? you ask. As far as Jennifer is concerned, the only thing mathematical about candy is, "How many pieces?" As she grows a little older, she will probably become more concerned with, "How much does it cost?" These two questions, *how many* and *how much,* account for much of what mathematics deals with. Of course, these two questions can become much more complicated if we consider all the processes involved in a large candy-manufacturing business: recipes, costs, profits, payrolls, sales statistics, and many, many more.

Everyone who uses this book would like to understand the mathematical concepts involved in these processes. The reasons why people don't understand them are many. At this point, however, it is too late to worry about why you didn't learn many of these skills. Therefore, your immediate concern must be how to learn them now. One point must not be overlooked: all of you are capable of learning these skills no matter how bad you think you are at math.

This knowledge will open up the doors to new educational goals. Work hard, be persistent, and success will follow.

In this first chapter we will take a look at the numbers we are most familiar with: 0, 1, 2, 3, 4, ... We will learn how to read and write them properly and also how to combine them in various ways.

CHAPTER I—NUMBER KNOWLEDGE

It is difficult to imagine how we could get along without using numbers. For example, how could we make the following comparisons of different countries if we could not use numbers?

Country	Area (square miles)	Population	Literacy Rate	Annual Income per Person	Life Expectancy	
					Males	Females
United States	3,540,939	246,800,000	96%	$16,444	71	78
Japan	143,572	123,200,000	99%	$21,820	75	81
India	1,229,737	853,000,000	36%	$ 250	53	52

1.1 Naming, Reading, and Writing Whole Numbers

The numbers that we all know the most about are 0, 1, 2, 3, 4, 5, 6, 7, 8, 9, and so on. These numbers are called the *whole numbers*. They form perhaps the most important set of numbers in mathematics because all other numbers used in mathematics can be defined in terms of the whole numbers. For this reason, the student should master all the operations of the whole numbers before going on to study other, more complicated, sets of numbers.

Technically speaking, a *number* is an abstract concept and a *numeral* is a symbol or group of symbols that is used to represent or name a number. One rather well known group of numerals used to represent numbers is the system of *Roman numerals*—I, II, III, IV, V, VI, and so on. The most common group of numerals used to represent numbers is 0, 1, 2, 3, 4, 5, 6, 7, 8, 9, which is used today throughout much of the world. This group is called the *Hindu–Arabic system* because it was invented in India and transmitted to the Western world by the Arabs.

To write any whole number, we use only the digits 0, 1, 2, 3, 4, 5, 6, 7, 8, and 9. If you want to write a whole number larger than 9, you must combine two or more of these digits.

For the number eight hundred fifty-two, we write the numeral 852. For the number eighty-seven, we write the numeral 87. Note that the 8 means something different in each numeral. In 852, the 8 means 8 hundreds. In 87 the 8 means 8 tens. For this reason, our system of numeration is called a *place-value* system. A digit takes on a specific value depending on what place it occupies in a given numeral.

In the following figure, look at the place value of each digit in the numeral 7,304,526,398:

billions		hundred millions	ten millions	millions		hundred thousands	ten thousands	thousands		hundreds	tens	units or ones
7	,	3	0	4	,	5	2	6	,	3	9	8

In this number, the 8 represents 8 units or 8 ones. A digit in this far-right position is in the *units place*. The 9 immediately to the left of the 8 is in the *tens place* and represents 9 tens. Moving to the left, we have 3 in the hundreds place, meaning 3 hundreds; 6 in the thousands place, meaning 6 thousands, and so on.

How many hundred thousands are represented in the figure? What does the 4 represent? How many 10 millions do we have?

EXAMPLE 1 Name the place value of the underlined digit: $\underline{5}26{,}342$.

SOLUTION Hundred thousands ▮▮

EXAMPLE 2 Name the place value of the underlined digit: $3{,}862{,}1\underline{2}4$.

SOLUTION Tens ▮▮

You will note that as we go from right to left we separate the digits into groups of three digits each by using commas. With the use of these groups let us now read our numeral. We read 7,304,526,398 as seven billion, three hundred four million, five hundred twenty-six thousand, three hundred ninety-eight. Note that in the last group on the right we say three hundred ninety-eight instead of three hundred ninety-eight ones.

EXAMPLE 3 Write the numeral 826,217 in words.

SOLUTION Eight hundred twenty-six thousand, two hundred seventeen. ▮▮

EXAMPLE 4 Write the numeral 5,006,308 in words.

SOLUTION Five million, six thousand, three hundred eight. ▮▮

EXAMPLE 5 Write the numeral 60,000,066 in words.

SOLUTION Sixty million, sixty-six. ▮▮

Not only must we be able to write numerals in words, but we must also be able to go the other way and write words as numerals.

EXAMPLE 6 Write six hundred seventy-eight as a numeral.

SOLUTION 678 ▮▮

EXAMPLE 7 Write forty-two million, six hundred eighteen thousand, seven as a numeral.

SOLUTION 42,618,007 ❙❙

Expanded Notation

The number 428 consists of the digit 4, the digit 2, and the digit 8. As we know, each of these digits has a specific value, depending on which place it occupies. The 4 represents 4 hundreds, the 2 represents 2 tens, and the 8 represents 8 ones. We could write this as

$$428 = 4 \text{ hundreds} + 2 \text{ tens} + 8 \text{ ones}$$

or, more simply,

$$428 = 400 + 20 + 8$$

This way of writing a numeral according to the place value of each digit is called writing the numeral in **expanded form** or **expanded notation.** We shall find this notation extremely useful when we perform operations with whole numbers, so it is important to learn the process well.

EXAMPLE 1 Write 5,382 in expanded notation.

SOLUTION $5{,}382 = 5{,}000 + 300 + 80 + 2$ ❙❙

EXAMPLE 2 Write 26,318,524 in expanded notation.

SOLUTION $20{,}000{,}000 + 6{,}000{,}000 + 300{,}000 + 10{,}000 + 8{,}000 + 500 + 20 + 4$ ❙❙

EXAMPLE 3 Write 20,006 in expanded notation.

SOLUTION $20{,}000 + 6$ ❙❙

1.1 Exercises

▼ Name the place value of the underlined digit in each of the following

1. $3{,}8\underline{2}4{,}127$
2. $5{,}1\underline{8}2$
3. $23{,}612{,}89\underline{4}$
4. $52{,}86\underline{0}$
5. $\underline{3}{,}821$
6. $5\underline{1}6$
7. $30{,}\underline{0}04$
8. $38\underline{2}{,}114$
9. $\underline{6}82{,}148{,}362$
10. $5\underline{0}8{,}362{,}785{,}510$
11. $701{,}40\underline{0}{,}216$
12. $30\underline{0}{,}000{,}000{,}000$

▼ Name the place value of the 3 in each of the following:

13. 1,037
14. 3,127
15. 7,003
16. 43

17. 30

18. 30,576

19. 1,073,522

20. 3,516,285

▼ Write the following numerals in words, then read them.

21. 52

22. 316

23. 5,328

24. 62,531,218

25. 702

26. 50,500

27. 308,000

28. 500,000,005

29. 60,606,060,606

30. 21,300,096

31. 4,001

32. 246,820

▼ Write each of the following written statements as a numeral.

33. Thirty-six

34. Five hundred sixteen

35. Seventy-four thousand, nine hundred ninety-seven

36. Twenty-five million, twenty-five thousand, twenty-five

37. Seventy million

38. Seventy million, seven

39. Seventy million, seven thousand

40. Five hundred twenty-six thousand, four hundred eight-five

41. Seventy-four billion, sixty million, four hundred sixty

42. Three hundred thousand, six

43. Six hundred thousand

44. Six hundred thousand, six hundred six

▼ Write each of the following numerals in expanded notation.

45. 26

46. 518

47. 508

48. 3,186

49. 20,038

50. 500,005

51. 26,308,124

52. 5,180,002

53. 30,031,118

54. 21,482,679

LIVING IN THE REAL WORLD

55. Columbus arrived in America in 1492. What is the place value of the 4 in this number?

56. Light travels at the rate of 186,282 miles per second. What is the place value of the 6?

57. One mile is the same as 5,280 feet. What place value does the 2 have?

58. There are 31,536,000 seconds in one year. What place value does the 1 have in this number?

59. How big is a trillion? It is a million millions, or 1,000,000,000,000. A trillion seconds is almost 31,710 years. What place value does the 3 represent in the number 31,710?

60. The Library of Congress in Washington, D.C., contains over 67,400,000 volumes. What place value does the 4 have in this number?

61. What are the whole numbers?

62. There are two 3s in the number 7,313. Why do they represent different amounts?

1.2 Rounding Whole Numbers

If the population of the city where you live is 18,824, you would probably remember this number as "about 19,000." You would be rounding the number 18,914 to the nearest thousand. It is closer to 19,000 than 18,000, so it is rounded to 19,000. If we wanted to round 18,824 to the nearest hundred, we would get 18,800 since 18,824 is closer to 18,800 than it is to 18,900.

Whenever you round a whole number, you must first decide what place you want it rounded to. The rules for rounding whole numbers are given next. The examples that follow these rules will then help you learn the procedure well.

To round a whole number

1. Identify the place to which you are rounding.
2. The digit in the desired place is
   a. not changed if the digit immediately to the right is less than 5;
   b. increased by 1 if the digit immediately to the right is 5 or greater.
3. Replace all digits to the right of the rounded place by zeros.

EXAMPLE 1 Round 7,285 to the nearest hundred.

SOLUTION First locate the hundreds place: it is the 2. The digit immediately to the right of the 2 is 8, which is greater than 5, so we increase the 2 to a 3. All digits to the right of the 3 are then changed to zeros.

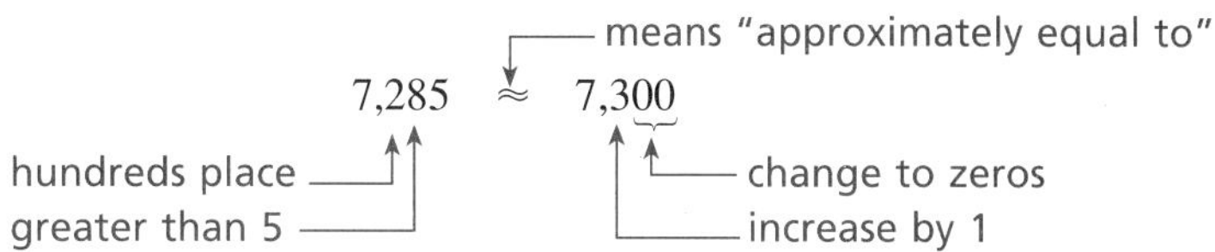

Notice that 7,285 is closer to 7,300 than it is to 7,200. ▮▮

EXAMPLE 2 Round 714 to the nearest hundred.

SOLUTION

714 ≈ 700

hundreds place
less than 5
change to zeros
remains unchanged ▮▮

EXAMPLE 3 Round 962 to the nearest hundred.

SOLUTION The hundreds digit is a 9, and the digit to its right is a 6, which is greater than 5, so we increase the 9 by 1, giving us a 10.

962 ≈ 1,000

hundreds place
greater than 5
changed to zeros
increased by 1 ▮▮

The following table gives more examples of rounding whole numbers. Study it carefully.

Given Number	Rounded to the nearest: Ten	Hundred	Thousand
7,216	7,220	7,200	7,000
5,841	5,840	5,800	6,000
7,993	7,990	8,000	8,000
9,844	9,840	9,800	10,000
320,814	320,810	320,800	321,000

1.2 Exercises

Round to the nearest hundred.

1. 741
2. 836
3. 402
4. 766
5. 914
6. 923
7. 987
8. 973
9. 4,386
10. 5,273
11. 6,184
12. 8,302
13. 4,912
14. 7,943
15. 3,988
16. 4,972

Round to the indicated place.

17. 5,280 to the nearest hundred
18. 24,785 to the nearest thousand
19. 7,821,312 to the nearest thousand
20. 43,299 to the nearest ten
21. 7,899 to the nearest ten
22. 27,896 to the nearest ten

Complete the table

	Given Number	Round to the nearest: Ten	Hundred	Thousand
23.	3,852			
24.	14,812			
25.	23,676			
26.	9,047			
27.	3,006	3010	3000	3000
28.	67,044			
29.	7,111			
30.	3,999	4000	4000	4000
31.	42,999			
32.	67,099			

LIVING IN THE REAL WORLD

33. The population of Australia in 1990 was 16,794,622. Round this number to the nearest hundred thousand.

34. In World War II, United States armed forces lost 291,557 members. Round this number to the nearest thousand.

35. The movie *E.T.* earned $228,618,939. How large is this number, rounded to the nearest million dollars?

36. The population of Portland, Oregon, was 429,415 in 1990. Express this number rounded to the nearest ten thousand.

37. In 1992 a total of 208,423 vehicles were reported stolen in the United States. Round this number to the nearest ten thousand.

38. In a recent year, 48,198,208 passengers passed through the Dallas–Ft. Worth Airport. Round this number to the nearest million.

39. The circulation of the *Detroit Free Press* newspaper is 598,418. Round this number to the nearest hundred thousand.

40. Mount Kilimanjaro, which is 19,340 feet high, is the highest mountain in Africa. Round this number to the nearest thousand.

▼ The following pie chart represents one semester's expenses for a college student. Use this information to answer Exercises 41–43.

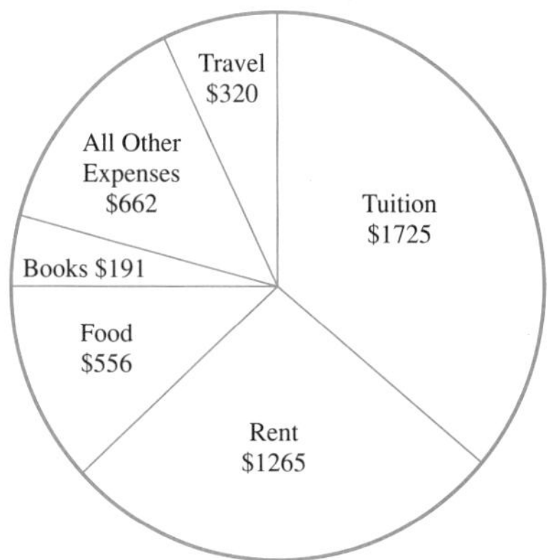

41. To the nearest hundred dollars, how much was tuition?

42. How much was spent for books, rounded to the nearest ten dollars?

43. Round the amount paid for rent to the nearest hundred dollars.

1.3 Addition of Whole Numbers

Suppose that you have four $1 bills and also suppose that, through some good fortune, someone gives you three more $1 bills. How much money do you have? Well, if you count them all up—1, 2, 3, 4, 5, 6, 7—you have $7. Or you could just say $4 + 3 = 7$. This is what addition of whole numbers is all about. First, let's construct what's called a **number line.** We just take a straight line and put the whole numbers evenly spaced along the line, like this:

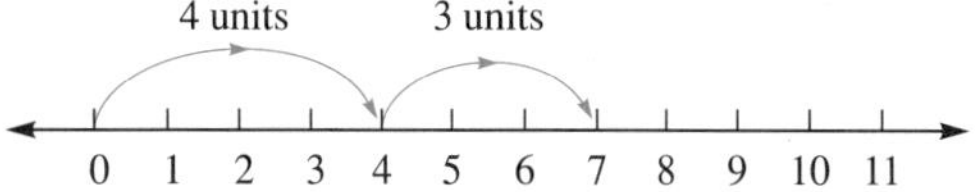

The arrow on the right indicates that the line keeps going as far as we want it to. Our number line is quite useful for showing the idea behind addition of whole numbers. Let's illustrate. In our original example, we wanted to add $4 + 3$. Let's start at zero on the number line and mark off four spaces or **units.** Since we are adding 3, we mark off three more spaces. We end up at 7, which is our answer: $4 + 3 = 7$. This simple example shows exactly what it means to add two whole numbers. It means to find the total or the **sum** of the two numbers.

I'm sure you'll agree that we don't want to draw a number line every time we are asked to add two whole numbers. To avoid this problem, we must learn what are called the **addition facts.** These are the sums of all pairs of single-digit whole numbers and are given in this table:

+	0	1	2	3	4	5	6	7	8	9
0	0	1	2	3	4	5	6	7	8	9
1	1	2	3	4	5	6	7	8	9	10
2	2	3	4	5	6	7	8	9	10	11
3	3	4	5	6	7	8	9	10	11	12
4	4	5	6	7	8	9	10	11	12	13
5	5	6	7	8	9	10	11	12	13	14
6	6	7	8	9	10	11	12	13	14	15
7	7	8	9	10	11	12	13	14	15	16
8	8	9	10	11	12	13	14	15	16	17
9	9	10	11	12	13	14	15	16	17	18

Before you continue on in mathematics, it is necessary for you to know these addition facts thoroughly. You should know them so well that the answers come automatically.

The following table is similar to the preceding one, but it is only partially filled in. You are to complete the table. There are also some timed exercises that you should practice to be sure that you have learned the addition facts well.

Practice Problems

▼ Complete the following addition table:

+	0	1	2	3	4	5	6	7	8	9
0			2							
1										
2		3					8			11
3										
4					8					
5										
6		7								
7									15	
8										
9			11							

▼ Time yourself on these problems. Try to answer them in 3 minutes or less and not get any wrong. Good luck.

7 + 9	9 + 3	8 + 3	9 + 2	9 + 1	4 + 2	6 + 9	5 + 5	3 + 6
7 + 3	5 + 8	7 + 2	4 + 7	9 + 4	9 + 6	6 + 4	3 + 5	8 + 5
9 + 7	4 + 8	6 + 9	1 + 1	5 + 3	3 + 9	5 + 9	6 + 8	8 + 8
8 + 4	2 + 2	3 + 1	8 + 6	6 + 7	8 + 8	6 + 5	6 + 1	1 + 5
6 + 2	8 + 1	6 + 3	2 + 1	5 + 6	8 + 2	7 + 4	7 + 9	5 + 7
1 + 4	9 + 5	5 + 4	4 + 5	2 + 3	5 + 2	6 + 6	4 + 4	3 + 7
4 + 3	7 + 5	4 + 6	2 + 5	4 + 1	7 + 8	3 + 4	2 + 4	3 + 7
8 + 7	7 + 6	9 + 8	3 + 3	7 + 1	2 + 6	3 + 2	1 + 3	2 + 9
4 + 6	7 + 7	5 + 1	9 + 9	2 + 7	1 + 2	3 + 8	8 + 9	2 + 8

▼ If you need more practice, try the next set of problems.

Time: about 3 minutes

2 + 8	2 + 4	2 + 9	5 + 7	4 + 4	1 + 5	3 + 5	5 + 5	6 + 8
1 + 3	8 + 9	3 + 7	6 + 6	6 + 5	8 + 8	7 + 9	6 + 4	6 + 9
3 + 8	7 + 8	3 + 2	8 + 2	6 + 1	9 + 6	5 + 9	3 + 6	3 + 7
1 + 2	3 + 4	7 + 1	2 + 3	5 + 6	8 + 8	8 + 5	5 + 3	9 + 1
9 + 9	2 + 5	3 + 3	5 + 2	8 + 6	7 + 4	4 + 7	3 + 9	4 + 2
2 + 7	4 + 1	2 + 6	4 + 5	6 + 7	6 + 3	1 + 1	9 + 4	9 + 2

7 + 6   7 + 7   9 + 5   4 + 6   2 + 2   2 + 1   4 + 8   9 + 3   7 + 2

5 + 1   4 + 3   9 + 8   5 + 4   6 + 2   9 + 7   7 + 3   8 + 3   3 + 1

4 + 6   7 + 5   8 + 7   1 + 4   8 + 4   8 + 1   6 + 9   5 + 8   7 + 9

▼ If you need still more practice, try the next set of problems.

Time: About 3 minutes

5 + 6   7 + 9   8 + 2   5 + 7   8 + 4   2 + 1   6 + 3   6 + 2   7 + 1

2 + 3   6 + 6   3 + 7   4 + 4   5 + 4   1 + 4   4 + 5   9 + 5   5 + 2

3 + 4   3 + 7   7 + 8   2 + 4   4 + 3   7 + 5   4 + 6   4 + 1   2 + 5

7 + 1   3 + 2   2 + 9   1 + 3   8 + 7   9 + 8   2 + 6   7 + 6   3 + 3

1 + 2   3 + 8   2 + 8   8 + 9   4 + 6   7 + 7   5 + 1   2 + 7   9 + 9

3 + 6   9 + 1   5 + 5   6 + 9   8 + 3   7 + 9   9 + 3   8 + 3   4 + 2

9 + 6   8 + 5   6 + 4   5 + 8   3 + 5   7 + 3   7 + 2   9 + 4   4 + 7

5 + 3   8 + 8   5 + 9   6 + 9   9 + 7   4 + 8   1 + 1   3 + 9   6 + 8

8 + 8   6 + 1   1 + 5   6 + 5   8 + 4   3 + 1   2 + 2   6 + 7   8 + 6

Adding Larger Numbers

The next kind of whole numbers that we shall add are those that have more than one digit.

EXAMPLE 1 Add 675 and 247.

SOLUTION First, we write each of the numbers in expanded notation:

$$\begin{aligned} 675 &= 6 \text{ hundreds} + 7 \text{ tens} + 5 \text{ ones} \\ \underline{+247} &= \underline{2 \text{ hundreds} + 4 \text{ tens} + 7 \text{ ones}} \end{aligned}$$

We start with the ones first. 5 ones plus 7 ones is 12 ones. But 12 ones is 1 ten with 2 ones left over. Se we put down the 2 ones and put the 1 ten in the tens column. We usually call it **carrying** the 1 ten.

$$
\begin{array}{r}
\text{1 ten}\quad\quad\quad\quad\\
6 \text{ hundreds} + 7 \text{ tens} + 5 \text{ ones}\\
+\underline{2 \text{ hundreds} + 4 \text{ tens} + 7 \text{ ones}}\\
2 \text{ ones}
\end{array}
$$

Next, we add all the tens: 1 + 7 is 8, plus 4 more is 12 tens altogether. Remember that 10 tens is 100, so we have 1 hundred with 2 tens left over. We put down the 2 tens and carry the 1 hundred to the hundreds column.

$$
\begin{array}{r}
\text{1 hundred}\quad\quad\text{1 ten}\quad\quad\quad\quad\\
6 \text{ hundreds} + 7 \text{ tens} + 5 \text{ ones}\\
+\underline{2 \text{ hundreds} + 4 \text{ tens} + 7 \text{ ones}}\\
9 \text{ hundreds} + 2 \text{ tens} + 2 \text{ ones}
\end{array}
$$

Now we add all the hundreds: 1 + 6 is 7, plus 2 more is 9. So our answer is

$$9 \text{ hundreds} + 2 \text{ tens} + 2 \text{ ones} \qquad \text{or} \qquad 922$$

This method explains very nicely what goes on when we add whole numbers, but we can shorten it somewhat and make our work easier. Let's look again at Example 1:

$$
\begin{array}{r}
675\\
+\underline{247}
\end{array}
$$

We start at the right in the ones column. We add 5 + 7 and get 12. This is 12 ones, which can also be thought of as 1 ten and 2 ones, so we put down the 2 and carry the 1 ten to the tens column:

$$
\begin{array}{r}
\\
675\\
+\underline{247}\\
?
\end{array}
\qquad
\begin{array}{r}
1\\
675\\
+\underline{247}\\
2
\end{array}
\qquad
\begin{array}{r}
1\\
675\\
+\underline{247}\\
22
\end{array}
\qquad
\begin{array}{r}
\\
675\\
+\underline{247}\\
922
\end{array}
$$

Next, adding in the tens column gives us 1 + 7 = 8 and 8 + 4 = 12 tens. Okay, now 12 tens is 1 hundred + 2 tens left over. We put down the 2 tens and carry the 1 hundred to the hundreds place. Finally, we add all the hundreds: 1 + 6 is 7, plus 2 is a total of 9, and we're all finished. Our answer is 922, which, happily, is the same answer that we obtained earlier. ■

EXAMPLE 2 Add:
$$
\begin{array}{r}
5{,}707\\
+\underline{6{,}873}
\end{array}
$$

SOLUTION Again, we start in the ones column at the far right: 7 + 3 equals 10 ones, [illegible] is how many tens? 10 ones is exactly 10, with no ones left over. Another saying that there are no ones left over is that there are *zero* ones left over. So we put down the 0 and carry the 1 to the tens position:

$$\begin{array}{rrrr} {}^{1} & {}^{1} & {}^{1\ \ 1} & {}^{1\ \ 1} \\ 5{,}707 & 5{,}707 & 5{,}707 & 5{,}707 \\ +\underline{6{,}873} & +\underline{6{,}873} & +\underline{6{,}873} & +\ \underline{6{,}873} \\ 0 & 80 & 580 & 12{,}580 \end{array}$$

Now we add the tens: $1 + 0$ is 1, plus 7 more is 8 tens, which is 80. This is less than 100, so we have no hundreds to carry and we just record the 8 in the tens column and proceed to the hundreds column. Here $7 + 8$ is 15 hundreds, and remembering that 10 hundreds is a thousand, we see that our 15 hundreds is equal to 1 thousand with 5 hundreds left over. Again, we put down the 5 and carry the 1. Now, adding in the thousands column, we have $1 + 5$ is 6, plus 6 is 12. Since we don't have any more numbers to add, we can just put down the 12, and our answer is 12,580.

In the next problem of this type we'll shorten our thinking even more.

EXAMPLE 3 Add 8,379 and 7,553.

SOLUTION

$$\begin{array}{rrrr} {}^{1} & {}^{1\,1} & {}^{1\,1} & \\ 8{,}379 & 8{,}379 & 8{,}379 & 8{,}379 \\ +\underline{7{,}553} & +\underline{7{,}553} & +\underline{7{,}553} & +\ \underline{7{,}553} \\ 2 & 32 & 932 & 15{,}932 \end{array}$$

We add $9 + 3$, which is 12. We put down the 2 and carry the 1. Since $1 + 7$ is 8 and $8 + 5$ is 13, we put down the 3 and carry the 1. Next, $1 + 3$ is 4, and $4 + 5$ is 9. We have nothing to carry, so we just put down the 9. Now we have $8 + 7 = 15$, which we put right down since we don't have any more numbers to add. Our answer is 15,932.

Next we'll put our newfound skills to work on a couple of practical problems.

EXAMPLE 4 Suppose that you took a nice long weekend trip. You drove 188 miles on Saturday and then you took the scenic route home on Sunday and drove 274 miles. How far did you drive altogether?

SOLUTION The answer is obtained simply by adding the distance you drove on Saturday to the distance you drive on Sunday.

$$\begin{array}{rrrr} & {}^{1} & {}^{1\,1} & {}^{1\,1}\phantom{0\ \text{miles}} \\ 188\text{ miles} & 188 & 188 & 188\text{ miles} \\ +\underline{274\text{ miles}} & +\underline{274} & +\underline{274} & +\underline{274\text{ miles}} \\ ? & 2 & 62 & 462\text{ miles} \end{array}$$

Since $8 + 4$ is 12, we put down the 2 and carry the 1. Next, $1 + 8$ is 9, and $9 + 7$ is 16, so we put down the 6 and carry the 1. Finally, $1 + 1$ is 2, plus more is 4. You drove a total of 462 miles.

You may have already wondered: Can you add more than two numbers at once? Our next example answers that question. Let's temporarily become a movie-theater manager.

EXAMPLE 5 We're interested in knowing the total attendance at your theater for the weekend. Suppose that attendance was 1,514 on Friday, 1,854 on Saturday, and 753 on Sunday. To find the total attendance for the three days, we have to add up the three numbers.

SOLUTION

	1	11	211	2 11
1,514	1,514	1,514	1514	1,514
1,854	1,854	1,854	1,854	1,854
+ 753	+ 753	+ 753	+ 753	+ 753
	1	21	121	4,121

Be careful to line up the 753 correctly; it is shorter than the other two numbers. A good rule to remember is to line up all the units digits, that is, the right column. If you do that, all the other digits will line up correctly. Now let's add. We follow the same rules here that we followed when we added two numbers. We start at the right: 4 + 4 is 8, plus 3 is 11. We put down 1 and carry 1. Next, 1 + 1 is 2, plus 5 is 7, plus 5 is 12. We put down the 2 and carry the 1. Next, 1 + 5 is 6, plus 8 is 14, and 14 + 7 is 21. That last one, 14 + 7, is not one of our addition facts, but it shouldn't give you any trouble. Okay, so we have 21. Put down the 1 and carry the 2. This is the first time we've ever carried a 2, but it's all right; let's keep going. Since 2 + 1 is 3, and 1 more is 4, our total attendance is 4,121. ❚❚

We'll do one last problem together and then you can do some for yourself in the exercises.

EXAMPLE 6 Mr. Easytake took his car into Honest Charley's Garage for a tune-up. Later that afternoon he went to pick up his car and found that the bill had the following items: 1 engine, \$921; 1 transmission, \$832; 1 radiator, \$94; labor, \$806; and sales tax, \$160. How much is it going to cost Mr. Easytake to get his car back?

SOLUTION Let's add everything up. We must be careful to keep things even in the right column.

\$921
832
94
806
+ 160
\$2,813

Notice how the units digits are lined up on the right. Incidentally, when you carry you don't have to write the number down. It's probably quicker if you can carry it mentally and add it in to the next column. Do whatever you feel is easiest for you. The answer is \$2,813. Not bad for a tune-up. Good old Honest Charley. ❚❚

Try to do some of the exercises as soon as possible, while the ideas we have just covered are still fresh in your mind. If you need help, don't forget to ask.

1.3 Exercises

Add the following:

1. $53 + 25$
2. $61 + 28$
3. $15 + 22$
4. $84 + 14$
5. $322 + 147$
6. $381 + 615$
7. $166 + 723$
8. $424 + 365$
9. $67 + 24$
10. $38 + 56$
11. $37 + 88$
12. $97 + 24$
13. $426 + 588$
14. $927 + 149$
15. $765 + 389$
16. $428 + 775$
17. $4{,}316 + 7{,}859$
18. $7{,}512 + 3{,}098$
19. $7{,}428 + 3{,}944$
20. $5{,}820 + 7{,}989$
21. $67{,}328 + 41{,}576$
22. $38{,}242 + 78{,}599$
23. $21{,}384 + 76{,}785$
24. $91{,}234 + 41{,}788$
25. 43, 19, 87, +32
26. 77, 81, 13, +41
27. 56, 88, 23, +91
28. 54, 81, 44, +76
29. 689, 425, +388
30. 776, 36, +518
31. 66, 846, + 39
32. 558, 18, + 79
33. 9,645, 889, 18, +4,313
34. 46, 3,811, 958, + 12
35. 7,884, 375, 900, +8,414
36. 12, 9,832, 65, + 841
37. 4,326, 7,584, 1,881, +34,515
38. 74,381, 6,842, 71,114, + 89
39. 46,442, 77, 5,433, + 712
40. 5,280, 312, 6,161, + 98

In Exercises 41–45, arrange the numbers in a vertical column and add.

41. 68,421 + 985 + 6,711
42. 38 + 7,842 + 9,844
43. 7,850 + 3,214 + 6
44. 77 + 886 + 32,785 + 8
45. 5,361 + 918 + 39,424 + 67,521

LIVING IN THE REAL WORLD

46. A couple bought a refrigerator for $939, a stove for $656, and a dishwasher for $340. The sales tax was $116. How much did they spend for everything?

47. A family's heating-oil tank was filled three times this year. They received 416 gallons, 382 gallons, and 371 gallons. How much oil did they receive altogether?

48. While on vacation, a family drove 387 miles the first day, 401 miles the second day, and 288 miles the third day. How far did they travel altogether?

49. On a trip, Maryann spent $56 the first day, $61 the second day, $38 the third day, and $66 the fourth day. How much did she spend in the four days?

50. The first of four numbers is 3,125, the second is greater than the first by 5,108, the third is equal to the sum of the first and the second, and the fourth is equal to the sum of the third and first. What is the sum of the four numbers?

51. Rachael estimates her college expenses for the fall semester as follows:

Tuition	$2,750
Room	1,175
Books	210
Food	1,120
Travel	225
Miscellaneous	350

What is the total of Rachael's estimated expenses for the fall semester?

52. On a business trip, Sandra spent $188 for lodging, $91 for food, $381 for travel, and $28 for miscellaneous expenses. How much did she spend altogether?

53. Sometimes it is necessary to "carry" when adding. Why does this work?

1.4 Subtraction of Whole Numbers

Let's construct a number line with the whole numbers evenly spaced along the line.

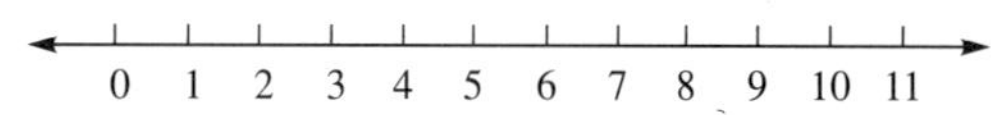

Then subtraction of one whole number from another is defined as fi
tance from one of the numbers to the other on our number line.

EXAMPLE 1 Subtract 5 from 8.

SOLUTION Look at our number line. We find 5 first, next we find 8, and then we count the

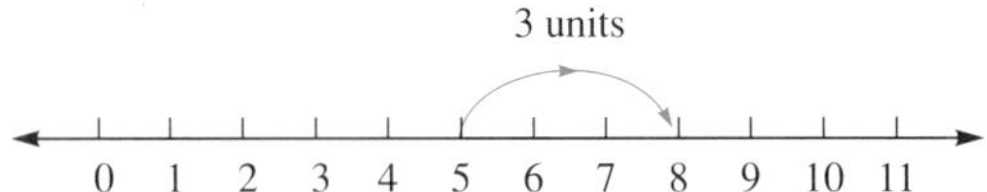

number of spaces between them. There are three spaces, so we conclude that $8 - 5 = 3$.

EXAMPLE 2 Subtract: $9 - 3$.

SOLUTION If we count the spaces on the number line from 3 to 9, we get 6, so $9 - 3 = 6$.

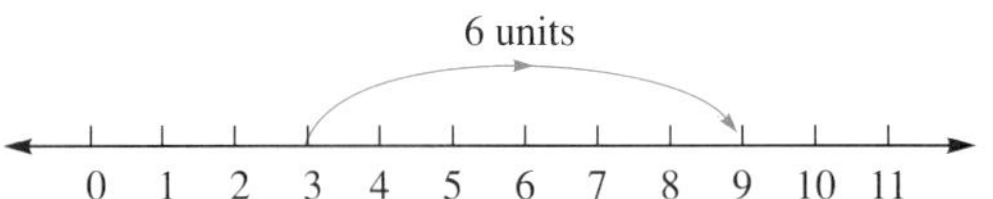

Another way of saying this is: What number must we add to 3 to get 9? From our knowledge of addition we know that 6 added to 3 gives 9. Therefore $9 - 3 = 6$, because $3 + 6 = 9$.

A further example of this type of thinking would be: $12 - 5 =$ what? Well, what distance do you have to add to 5 on the number line to get to 12? Or what number must you add to 5 to get 12? The answer is 7, so $12 - 5 = 7$, because $5 + 7 = 12$. It appears that in order to subtract, all we really have to know is how to add.

EXAMPLE 3 $15 - 6 = 9$, because $6 + 9 = 15$.

EXAMPLE 4 $17 - 5 =$ what? You should be thinking 12, because $5 + 12 = 17$.

EXAMPLE 5 $25 - 21 =$ what? The answer is 4, because $21 + 4 = 25$.

EXAMPLE 6 $13 - 8 =$ what? The answer is 5, because $8 + 5 = 13$.

As with the addition facts, you should become so familiar with simple subtraction problems that the answers come to you immediately. Practice the following timed exercises to be sure you know them well.

Practice Problems

▼ Allow yourself about $3\frac{1}{2}$ minutes to do the following problems.

9	15	5	7	8	2	11	12	13	3
− 6	− 7	− 3	− 1	− 6	− 1	− 3	− 7	− 9	− 1
10	4	4	9	9	11	12	13	9	4
− 2	− 2	− 3	− 9	− 4	− 7	− 8	− 8	− 5	− 4
4	13	8	12	8	12	3	9	10	7
− 1	− 7	− 2	− 4	− 8	− 5	− 3	− 8	− 9	− 2
14	8	9	10	9	7	18	8	10	6
− 8	− 1	− 7	− 2	− 4	− 4	− 9	− 5	− 4	− 3
3	5	7	17	11	10	10	1	17	9
− 2	− 5	− 7	− 8	− 5	− 6	− 7	− 1	− 9	− 3
13	11	10	14	14	12	10	16	12	11
− 4	− 9	− 8	− 5	− 7	− 9	− 5	− 7	− 3	− 4
2	14	15	9	6	8	14	7	8	5
− 2	− 9	− 6	− 1	− 6	− 3	− 6	− 6	− 7	− 1
10	12	11	6	13	7	7	5	16	11
− 1	− 6	− 6	− 2	− 6	− 5	− 3	− 4	− 8	− 2
5	16	15	8	6	11	10	6	15	13
− 2	− 9	− 8	− 4	− 5	− 8	− 3	− 1	− 9	− 5

▼ If you need more practice, try the next exercise set.

Time: About $3\frac{1}{2}$ minutes

13	10	14	11	12	14	16	12	11	10
− 4	− 8	− 5	− 9	− 9	− 7	− 7	− 3	− 4	− 5
8	14	5	7	8	14	15	2	6	9
− 3	− 6	− 1	− 6	− 7	− 9	− 6	− 2	− 6	− 1
7	7	5	16	11	10	6	12	11	13
− 5	− 3	− 4	− 8	− 2	− 1	− 2	− 6	− 6	− 6
6	11	10	15	6	5	16	15	13	8
− 5	− 8	− 3	− 9	− 1	− 2	− 9	− 8	− 5	− 4
2	11	12	3	13	15	9	7	5	8
− 1	− 3	− 7	− 1	− 9	− 7	− 6	− 1	− 3	− 6

12 − 8	11 − 7	4 − 4	9 − 5	13 − 8	10 − 2	4 − 3	4 − 2	9 − 9	9 − 4
12 − 5	3 − 3	10 − 9	9 − 8	7 − 2	4 − 1	8 − 2	13 − 7	8 − 8	12 − 4
7 − 4	6 − 3	8 − 5	18 − 9	10 − 4	14 − 8	9 − 7	8 − 1	10 − 2	9 − 4
11 − 5	10 − 7	1 − 1	9 − 3	17 − 9	3 − 2	5 − 5	7 − 7	17 − 8	10 − 6

▼ If you need still more practice, try the next exercise set.

Time: About 3 minutes

10 − 6	12 − 4	10 − 2	9 − 4	8 − 6	13 − 5	9 − 1	11 − 4	13 − 4	13 − 6
17 − 8	9 − 4	8 − 8	5 − 3	15 − 8	6 − 6	12 − 3	8 − 3	9 − 9	11 − 6
7 − 7	8 − 2	9 − 7	4 − 3	15 − 7	12 − 6	2 − 2	16 − 9	10 − 5	7 − 5
5 − 5	13 − 7	4 − 2	8 − 1	7 − 1	8 − 4	6 − 2	15 − 6	16 − 7	6 − 5
3 − 2	14 − 8	10 − 2	4 − 1	9 − 6	5 − 2	14 − 9	10 − 1	14 − 7	2 − 1
17 − 9	10 − 4	7 − 2	9 − 5	15 − 9	13 − 9	11 − 2	12 − 9	11 − 7	8 − 7
1 − 1	8 − 5	13 − 8	9 − 8	12 − 7	5 − 4	5 − 1	14 − 5	12 − 5	6 − 1
9 − 3	18 − 9	4 − 4	3 − 3	10 − 3	16 − 8	7 − 6	11 − 9	7 − 4	3 − 1
10 − 8	14 − 6	11 − 8	11 − 3	7 − 3	12 − 8	10 − 9	11 − 5	6 − 3	10 − 7

Subtracting Larger Numbers

Now let's consider a subtraction problem containing larger numbers.

EXAMPLE 7 Subtract 324 from 467.

SOLUTION As before, we write each number in expanded notation:

$$\begin{array}{r} 467 = 4 \text{ hundreds} + 6 \text{ tens} + 7 \text{ ones} \\ -\underline{324 = 3 \text{ hundreds} + 2 \text{ tens} + 4 \text{ ones}} \\ 1 \text{ hundred} \ + 4 \text{ tens} + 3 \text{ ones} \end{array}$$

Starting at the right, we subtract each column: 7 ones minus 4 ones is 3 ones; 6 tens minus 2 tens is 4 tens, and 4 hundreds minus 3 hundreds is 1 hundred. Our answer is 143.

Let's look at a slightly more complicated example

EXAMPLE 8 537 − 362 = what?

SOLUTION We approach the problem in the same way as we did before.

$$\begin{array}{r} 537 = 5 \text{ hundreds} + 3 \text{ tens} + 7 \text{ ones} \\ -\underline{362 = 3 \text{ hundreds} + 6 \text{ tens} + 2 \text{ ones}} \\ 5 \text{ ones} \end{array}$$

Subtracting 2 ones from 7 ones gives us 5 ones. Now, moving to the tens column, we notice a problem: we can't subtract 6 tens from 3 tens. We overcome this problem in the following way. Since 1 hundred = 10 tens, we borrow one of the hundreds and write it as 10 tens.

10 tens

$\overset{4}{\cancel{5}}$ hundreds + 3 tens + 7 ones = 4 hundreds + 13 tens + 7 ones

Now we can subtract:

$$\begin{array}{r} 4 \text{ hundreds} + 13 \text{ tens} + 7 \text{ ones} \\ -\underline{3 \text{ hundreds} + \ \ 6 \text{ tens} + 2 \text{ ones}} \\ 1 \text{ hundred} \ + \ \ 7 \text{ tens} + 5 \text{ ones} = 175 \end{array}$$

Let's do the same problem again, using a simpler, more common form.

$$\begin{array}{r} 537 \\ -\underline{362} \end{array}$$

First, we subtract 2 ones from 7 ones and record the answer in the ones place.

$$\begin{array}{r} 537 \\ -\underline{362} \\ 5 \end{array} \longrightarrow \begin{array}{r} \overset{4}{\not5}\ {}^{1}3\ \ 7 \\ -\underline{3\ \ 6\ \ 2} \\ 5 \end{array} \longrightarrow \begin{array}{r} \overset{4}{\not5}\ {}^{1}3\ \ 7 \\ -\underline{3\ \ 6\ \ 2} \\ 1\ \ 7\ \ 5 \end{array}$$

Now, 6 tens cannot be subtracted from 3 tens, so we borrow 1 hundred (that is, 10 tens) from the 5 hundreds and add them to the 3 tens that we already have, giving us 13 tens. Continuing to subtract gives us 13 tens minus 6 tens, or 7 tens; and in the hundreds column, 4 hundreds minus 3 hundreds equals 1 hundred, and we're finished.

Let's do another example of this type to be sure that you have the idea.

EXAMPLE 9

$$\begin{array}{r} 742 \\ -\underline{638} \end{array}$$

SOLUTION

$$\begin{array}{r} 742 \\ -\underline{638} \\ ? \end{array} \longrightarrow \begin{array}{r} 7\ \ \overset{3}{\not4}\ {}^{1}2 \\ -\underline{6\ \ 3\ \ 8} \end{array} \longrightarrow \begin{array}{r} 7\ \ \overset{3}{\not4}\ {}^{1}2 \\ -\underline{6\ \ 3\ \ 8} \\ 1\ \ 0\ \ 4 \end{array}$$

We can't subtract 8 ones from 2 ones, so we borrow 1 ten, which is equal to 10 ones, and add these 10 ones to the 2 ones that we already have. We now have 12 ones, from which we can subtract 8 ones and get 4. We continue to subtract: 3 tens minus 3 tens is 0 tens, and 7 hundreds minus 6 hundreds equals 1 hundred, and we're finished.

STOP The phrase "subtract 6 from 11" means 11 − 6, not 6 − 11. Be careful to keep the numbers in the correct order.

Special care must be taken when you are faced with a subtraction problem in which zeros appear in the top number.

EXAMPLE 10 Subtract 428 from 805.

SOLUTION

$$\begin{array}{r} 805 \\ -\underline{428} \\ ? \end{array} \longrightarrow \begin{array}{r} \overset{7}{\not8}\ {}^{1}\overset{9}{\not0}\ {}^{1}5 \\ -\underline{4\ \ 2\ \ 8} \end{array} \longrightarrow \begin{array}{r} \overset{7}{\not8}\ {}^{1}\overset{9}{\not0}\ {}^{1}5 \\ -\underline{4\ \ 2\ \ 8} \\ 3\ \ 7\ \ 7 \end{array}$$

Since we cannot subtract 8 ones from 5 ones, we must borrow a ten. But we find that we have no tens to borrow, so we must move on to the hundreds column and we borrow 1 hundred, leaving 7 hundreds. This hundred is equal to 10 tens. We now have tens to borrow from, so we borrow 1 ten, leaving 9. This ten that we borrowed is equal to 10 ones, and we add it to the 5 ones we already have. This gives us 15 ones, from which we can subtract the 8 ones, giving us 7 ones. Continuing, we have 9 tens minus 2 tens, which gives us 7 tens, and 7 hundreds minus 4 hundreds equals 3 hundreds. Our answer is 377.

One additional note: It is easy to check the answer to any subtraction problem, as illustrated by the following example:

EXAMPLE 11 Subtract 532 from 779 and check the answer:

SOLUTION First we do the subtraction:

$$\begin{array}{r} 779 \\ -\underline{532} \\ 247 \end{array}$$

To check, we add the answer, 247, to the number that we subtracted, 532, and if we worked the problem correctly, we should get 779.

$$\begin{array}{r} 532 \\ +\underline{247} \\ 779 \end{array}$$ and our problem checks

1.4 Exercises

▼ Subtract and check:

1. $\begin{array}{r} 819 \\ -\underline{318} \end{array}$ 2. $\begin{array}{r} 684 \\ -\underline{252} \end{array}$ 3. $\begin{array}{r} 687 \\ -\underline{115} \end{array}$

4. $\begin{array}{r} 733 \\ -\underline{523} \end{array}$ 5. $\begin{array}{r} 7{,}644 \\ -\underline{3{,}531} \end{array}$ 6. $\begin{array}{r} 8{,}844 \\ -\underline{2{,}533} \end{array}$

7. $\begin{array}{r} 642 \\ -\underline{534} \end{array}$ 8. $\begin{array}{r} 752 \\ -\underline{628} \end{array}$ 9. $\begin{array}{r} 637 \\ -\underline{628} \end{array}$

10. $\begin{array}{r} 7{,}006 \\ -\underline{3{,}258} \end{array}$ 11. $\begin{array}{r} 7{,}231 \\ -\underline{4{,}287} \end{array}$ 12. $\begin{array}{r} 6{,}104 \\ -\underline{3{,}507} \end{array}$

13. $\begin{array}{r} 5{,}238 \\ -\underline{4{,}279} \end{array}$ 14. $\begin{array}{r} 6{,}118 \\ -\underline{2{,}884} \end{array}$ 15. $\begin{array}{r} 5{,}223 \\ -\underline{4{,}867} \end{array}$

16. Subtract 205 from 504.

17. Subtract 617 from 5,119.

18. 5,820 minus 625 equals what number?

19. 6,723 minus 1,844 equals what number?

20. Find the difference between 7,042 and 6,134.

21. Find the difference between 60,123 and 5,074.

22. Find the difference between 6,000 and 3,841.

LIVING IN THE REAL WORLD

23. Mrs. Smith and Mr. Jones ran for Mayor. Mrs. Smith received 18,364 votes, and Mr. Jones received 15,588 votes. How many more votes did Mrs. Smith receive?

24. If you had $368 in cash and wanted to buy a used car that cost $500, how much more do you need?

25. If your checking account balance was $374, at the beginning of the month, and you wrote checks during the month for $62, $58, $36, $6, and $104, what is the new balance in your account at the end of the month?

26. The Schwartzes are driving from New York to Phoenix, a distance of 2,447 miles. If they drive 448 miles the first day, 502 miles the second day, and 492 miles the third day, how many more miles do they still have to drive to get to Phoenix?

27. The Shanleys paid $82,000 for their house six years ago. What will their profit be if they sell it for $118,000? What will their net profit be after deducting $4,000 for fix-up costs and a $7,000 real estate commission?

28. David has $436 in his checking account. On Monday he writes a check for rent for $275, on Tuesday he makes a deposit of $170, and on Wednesday he writes one check for $42 to pay the phone bill and another check for $78 for utilities. What is David's current balance?

29. Megan has $336 in her checking account. She deposits $225 and then writes checks for $161, $125, and $77. What is her balance?

30. Wendy is saving for a used car that costs $2800. If she started with $820 in her savings account and has made weekly deposits of $240, $210, $260, $275, and $300, how much more does she need to purchase the car?

1.5 Multiplication of Whole Numbers

The next mathematical operation that we will consider is multiplication. Multiplication can be thought of as **repeated addition.** For example, if you buy four 10-cent stamps, they cost 10 cents + 10 cents + 10 cents + 10 cents, which is equal to 40 cents. This is expressed mathematically as $4 \times 10 = 40$, which we read as "4 times 10 equals 40." The numbers that are being multiplied, in this case 4 and 10, are called **factors,** and the answer to any multiplication problem is called the **product.**

Several different symbols are used to indicate multiplication:

- $5 \times 2 = 10$ "×" means to multiply.
- $5 \cdot 2 = 10$ "·" means to multiply.
- $(5)(2) = 10$ Parentheses written right next to each other means to multiply.
- $5(2) = 10$
- $(5)2 = 10$

(for the last two) Frequently only one set of parentheses is used to mean multiplication.

As in the case of addition and subtraction, we have some essential facts that must be memorized before we can do further work in multiplication. It really can't be emphasized too much that you **must** know the facts in this table so well that they come to you automatically, without thinking.

×	0	1	2	3	4	5	6	7	8	9
0	0	0	0	0	0	0	0	0	0	0
1	0	1	2	3	4	5	6	7	8	9
2	0	2	4	6	8	10	12	14	16	18
3	0	3	6	9	12	15	18	21	24	27
4	0	4	8	12	16	20	24	28	32	36
5	0	5	10	15	20	25	30	35	40	45
6	0	6	12	18	24	30	36	42	48	54
7	0	7	14	21	28	35	42	49	56	63
8	0	8	16	24	32	40	48	56	64	72
9	0	9	18	27	36	45	54	63	72	81

Study the table of multiplication facts and do the practice exercises until you know them very well.

Practice Problems

▼ Complete the following multiplication table:

×	0	1	2	3	4	5	6	7	8	9
0			0							
1										
2										
3						15				
4				12						36
5										
6										
7								49		
8		8								
9										

▼ Allow about 4 minutes to do the following problems.

9 × 6	8 × 6	5 × 8	6 × 7	4 × 8	8 × 8	9 × 4	6 × 4	7 × 9	6 × 5
7 × 7	3 × 5	6 × 3	6 × 4	6 × 3	8 × 6	6 × 9	9 × 3	4 × 3	8 × 5
7 × 5	2 × 8	9 × 5	5 × 4	7 × 3	8 × 9	8 × 8	6 × 6	7 × 6	4 × 5
4 × 6	3 × 4	5 × 8	6 × 4	8 × 4	6 × 5	8 × 2	5 × 6	6 × 9	3 × 8
3 × 7	7 × 5	5 × 9	5 × 2	7 × 9	7 × 5	8 × 4	6 × 7	5 × 7	6 × 9
4 × 5	4 × 9	9 × 7	9 × 5	4 × 8	5 × 4	7 × 3	9 × 2	4 × 4	3 × 5
7 × 2	3 × 5	5 × 6	7 × 2	4 × 7	9 × 5	7 × 8	8 × 8	7 × 3	9 × 9
5 × 3	7 × 2	9 × 2	7 × 3	3 × 9	7 × 9	9 × 6	5 × 7	6 × 9	6 × 7

▼ If you need more practice, try the next exercise set.

Time: About 4 minutes

5 × 6	6 × 8	7 × 7	6 × 6	4 × 7	3 × 9	6 × 7	9 × 5	8 × 2	7 × 5
8 × 8	9 × 3	8 × 2	8 × 6	9 × 5	9 × 7	8 × 4	9 × 9	7 × 7	4 × 8
7 × 5	4 × 9	7 × 8	9 × 6	6 × 7	6 × 9	8 × 9	6 × 5	6 × 2	3 × 7
7 × 9	8 × 8	4 × 5	5 × 7	8 × 6	9 × 6	8 × 4	7 × 3	3 × 8	6 × 9
6 × 4	5 × 4	9 × 4	7 × 2	5 × 3	6 × 3	9 × 2	5 × 2	5 × 9	4 × 4
5 × 9	7 × 3	3 × 5	7 × 2	4 × 8	6 × 4	9 × 5	5 × 8	6 × 6	3 × 2
9 × 2	5 × 6	5 × 4	7 × 5	3 × 4	2 × 8	6 × 4	9 × 4	7 × 4	7 × 1
3 × 7	4 × 8	7 × 2	7 × 3	5 × 8	3 × 5	9 × 5	4 × 6	3 × 9	5 × 5

▼ If you need additional work, try the following exercise set.

Time: About 4 minutes

7×6 5×8 3×6 7×6 4×5 8×7 9×6 5×4 3×4 6×6

5×6 8×9 9×3 3×2 6×4 3×7 4×6 4×9 5×3 5×9

7×7 5×2 6×9 4×9 9×9 7×2 6×3 9×5 4×4 3×2

3×8 1×9 6×6 5×9 3×1 4×7 6×8 5×7 7×9 6×1

5×8 5×5 7×8 6×8 9×6 6×9 3×3 8×5 7×9 4×5

5×2 9×8 2×9 3×2 8×4 7×5 4×8 6×2 3×5 6×8

5×9 7×2 5×5 4×5 2×8 3×9 4×4 6×5 3×7 3×3

7×6 9×9 8×3 6×8 5×6 9×2 8×7 4×2 6×9 5×8

Multiplying Larger Numbers

Now let's consider a more complicated problem.

EXAMPLE 1 $4 \times 5{,}368$

SOLUTION This means adding four 5,368s. Let's add them.

$$\begin{array}{r} 5{,}368 \\ 5{,}368 \\ 5{,}368 \\ +5{,}368 \\ \hline 21{,}472 \end{array}$$

Since the sum is 21,472, it follows that $4 \times 5{,}368 = 21{,}472$. You're forced to admit that this is a tedious task. However, if we look carefully at what we've done, we'll see that our work can be shortened considerably. Let's look at how we got our answer. The first thing we did was add up the 8s in the units column. But since these were all 8s, the answer was merely the sum of the 8s, or 4×8, which is 32. We could have gotten the same answer by multiplying the units digit by 4, getting 32 directly. We put down the 2 and carry the 3. Now in the tens column, forgetting for a second the 3 that we carried, we have to add 4 sixes, which is the same as multiplying 4×6, which gives us 24. Now we add in the 3 that we carried, giving us 27 for our answer in the tens place. We record the 7 and carry the 2. In the hundreds column we have to add 4 threes and add the 2 that we carried. This gives us 4×3, which equals 12, plus the 2 that was carried, which yields 14 in the hundreds place. We put down the 4 and carry the 1. Finally, we have 4 × 5 = 20, plus the 1, or a total of 21, and we're all finished.

$$\begin{array}{r} {\scriptstyle 1\,23} \\ 5{,}368 \\ \underline{\times 4} \\ 21{,}472 \end{array}$$

EXAMPLE 2 Multiply 576 by 7.

SOLUTION

$$\begin{array}{r} {\scriptstyle 54\,} \\ 576 \\ \underline{\times 7} \\ 4{,}032 \end{array}$$

We start in the units column: 6 × 7 is 42. We enter the 2 in the units place and carry 4. Next, 7 × 7 is 49, plus the 4 that we carried gives us 53; we put down the 3 and carry the 5. Finally, 5 × 7 is 35, plus 5 is 40, and our answer is 4,032. This is certainly easier than adding 576 seven times.

Now let's look quickly at a special type of multiplication. Suppose that we want to multiply 324 by 10. We could add 324 ten times to get the correct answer; if we did, we'd get 3240. So 324 × 10 = 3,240. If you want, you can actually add them to be sure the answer is correct. This example illustrates a general rule: **to multiply any whole number by 10, simply put a zero after it.** Let's look at some examples.

EXAMPLE 3 35 × 10 = 350

EXAMPLE 4 7248 × 10 = 72,480

Now to apply a slight extension of this rule: let's multiply by numbers like 20. Since 20 is 2 × 10, to multiply by 20, we multiply by 2 and then attach a zero to get the answer.

EXAMPLE 5 To multiply 7 × 20, we multiply 7 × 2, which gives us 14, and then attach a zero, giving 140 as our answer.

Similarly, we can multiply by 30 or 40 or 50, and so forth.

EXAMPLE 6 To multiply 36 × 30, we multiply

$$\begin{array}{r} 36 \\ \underline{\times 3} \\ 108 \end{array}$$

and then add a zero to our result; we have 36 × 30 = 1,080.

Can you guess what we get if we multiply a number by 100? If you were thinking that we put two zeros after the number, you're correct.

EXAMPLE 7 $7 \times 100 =$ what? We put two zeros after the 7, and our answer is 700.

EXAMPLE 8 $326 \times 100 =$ what? We put two zeros after the 326 and we get $326 \times 100 = 32{,}600$.

EXAMPLE 9 Try 326×200. Again, since 200 is 2×100, we multiply 326×2 and attach two zeros.

$$\begin{array}{r} 326 \\ \underline{\times 2} \\ 652 \end{array}$$

Now we add two zeros, and our answer is 65,200.

EXAMPLE 10 What would 26×700 be? We multiply 26×7 and attach two zeros:

$$\begin{array}{r} 26 \\ \underline{\times 7} \\ 182 \end{array}$$

plus two zeros gives us $26 \times 700 = 18{,}200$.

Now we're ready to go on and try more complicated problems.

EXAMPLE 11 Multiply: 48×36.

SOLUTION Since $36 = 30 + 6$, what we really have is thirty 48s plus six 48s. Multiplying 48×36, then, should be the same as multiplying 48×30 and adding that result to 48 times 6.

$$\begin{array}{r} 48 \\ \underline{\times 36} \end{array} \quad \text{is the same as} \quad \begin{array}{r} 48 \\ \underline{\times (30 + 6)} \end{array}$$

which is the same as

$$\begin{array}{r} 48 \\ \underline{\times 30} \end{array} + \begin{array}{r} 48 \\ \underline{\times 6} \end{array}$$

We multiply 48×30 first. Remember how? Multiply 48×3 and attach a zero:

$$\begin{array}{r} 48 \\ \underline{\times 3} \\ 144 \end{array} \quad \text{plus a zero on the end gives us} \quad 48 \times 30 = 1{,}440$$

Next, we multiply 48×6:

$$\begin{array}{r} 48 \\ \underline{\times 6} \\ 288 \end{array}$$

Our final answer is the sum of these two products. We can organize our work like this:

$$\begin{array}{rl} 48 & \\ \times 36 & \\ \hline 288 & (6 \times 48) \\ +1440 & (30 \times 48) \\ \hline 1{,}728 & \end{array}$$

It should be clear that nothing is really gained by writing the zero in 1,440. The same result can be achieved by multiplying 48 × 3 and **indenting** the answer one place, like this:

$$\begin{array}{rl} 48 & \\ \times 36 & \\ \hline 288 & (6 \times 48) \\ +144 & (3 \times 48\text{, indented}) \\ \hline 1{,}728 & \end{array}$$

▮▮

To illustrate further, let's look at another problem.

EXAMPLE 12

$$\begin{array}{r} 629 \\ \times 218 \\ \hline \end{array}$$

SOLUTION First we multiply 629 × 8.

$$\begin{array}{r} 629 \\ \times 218 \\ \hline 5032 \end{array}$$

Next, we multiply 629 × 1, remembering to indent.

$$\begin{array}{r} 629 \\ \times 218 \\ \hline 5032 \\ 629 \end{array}$$

We keep going, multiplying 629 by 2, and this time we indent 2 spaces:

$$\begin{array}{r} 629 \\ \times 218 \\ \hline 5032 \\ 629 \\ 1258 \\ \hline \end{array}$$

Next, we add to get the final answer.

$$\begin{array}{r} 629 \\ \times 218 \\ \hline 5032 \\ 629 \\ +1258 \\ \hline 137{,}122 \end{array}$$

▮▮

One additional procedure to illustrate here is what to do if one of the digits in the second number is a zero.

EXAMPLE 13 Multiply 218 × 205.

SOLUTION

```
   218        218
  ×205       ×205
  1090       1090
  000       +4360
+436        44,690
 44,690
```

Following the procedure that we have just learned, we first multiply by 5, giving us 1,090. Next we come to zero. Since multiplication by zero yields zero, it contributes nothing to our product, and we can leave out the line of zeros, as shown on the right. However, if it is left out completely, we are apt to forget that the third line must be indented 2 places, so we put in one zero properly indented and then put the product of 218 times 2 to the left of this zero. As you can see, we get the same answer in both cases.

Here is an additional illustration.

EXAMPLE 14

```
 462
 ×30
   ?
```

SOLUTION

```
 462       462
 ×30       ×30
   0    13,860
```

Instead of multiplying 462 × zero, we put down just one zero and then multiply 462 × 3, writing the product immediately to the left of the zero.

Before we end, one final observation is in order. Carrying numbers in a multiplication problem **should** be done mentally. This is important, because often a number is multiplied by many digits, and writing in the carried digits leads to error and confusion.

1.5 Exercises

In Exercises 1–10, multiply without showing any work.

1. 6 × 10
2. 8 × 10
3. 4 × 100
4. 7 × 100
5. 3 × 1,000
6. 7 × 1,000
7. 35 × 10
8. 42 × 10
9. 75 × 100
10. 58 × 100
11. 738 × 100
12. 542 × 100

▼ Multiply:

13. 265×8

14. 318×7

15. 721×23

16. 342×32

17. 576×37

18. $3{,}816 \times 59$

19. $5{,}623 \times 78$

20. 725×359

21. $3{,}529 \times 7{,}005$

22. 718×58

23. $7{,}024 \times 352$

24. $9{,}241 \times 706$

25. 315×14

26. $7{,}432 \times 504$

27. $5{,}325 \times 750$

28. $2{,}564 \times 318$

29. $7{,}183 \times 560$

30. $3{,}261 \times 7{,}040$

LIVING IN THE REAL WORLD

31. A college offers 17 sections of psychology. There are 28 students registered in each section. How many students are enrolled for psychology?

32. A student deposits $86 in his savings account each week during summer vacation. If his vacation is 15 weeks long, how much does he save altogether?

33. If band uniforms cost $308 each, how much does it cost to supply uniforms for a 60-piece band?

34. If each year had 365 days, how many days are there in 28 years?

35. If one mile equals 5,280 feet, and it is 106 miles from New York to Philadelphia, how many feet is it from New York to Philadelphia?

36. How much will you earn if you make $6 per hour and work 40 hours?

37. How far will you travel in 16 hours if you average 46 mph?

38. Find the total cost of 30 ten-cent stamps, 24 fifteen-cent stamps, and 12 twenty-cent stamps.

39. If Mr. Johnson earns $307 per week, how much does he earn in one year (52 weeks)?

40. A corporation declares a special dividend of 42¢/share. There are a total of 589,214 shares of stock in the corporation. What is the total amount of the dividend?

41. A woman buys 5 pounds of pork chops and an 8-pound pork roast. How much does she spend if the pork chops cost $3.99 per pound and the roast costs $4.29 per pound?

42. If you make \$7.20 per hour for the first 40 hours and \$10.80 per hour for the time worked over 40 hours, how much will you earn if you work 46 hours?

43. A jogger starts from a certain place and runs at the rate of 550 feet per minute. Five minutes later a cyclist starts after the jogger, beginning at the same place. If the cyclist is traveling at a rate of 1,320 feet per minute, how far apart are the cyclist and the jogger after the jogger has been running for ten minutes? Who is ahead?

44. The diameter of Mercury is 2,967 miles; the diameter of Saturn is 24 times that of Mercury; and the diameter of the Sun is 12 times that of Saturn. What is the Sun's diameter?

45. What are factors?

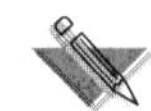

46. What is a product?

47. Why is multiplication sometimes called "repeated addition"?

1.6 Division of Whole Numbers

Suppose that you borrowed \$24 from a friend and agreed to pay him back at the rate of \$6 per week. How long will it take you to repay your friend? I think we can all figure out that it will take 4 weeks to get out of debt. The number of payments is found by *dividing* the debt, \$24, by the amount of each payment, \$6. We can check our result easily using multiplication. If you pay \$6 per week for 4 weeks, you will have paid back \$6 $\times$ 4, or \$24, to your friend. In other words,

$$24 \div 6 = 4$$

This can also be written

$$\frac{24}{6} = 4 \quad \text{or} \quad 6\overline{)24}^{\,4}$$

Somehow division has gained a reputation as being one of the most difficult operations in mathematics. With a little work, we will show that this is an unearned reputation. The first thing that you should do is review what we call the division facts. If you know the multiplication facts well, then you will have no trouble. Here are some timed practice problems to help you.

Practice Problems

Time Limit: About 5 minutes

1. $4\overline{)16}$ $4\overline{)28}$ $4\overline{)4}$ $7\overline{)56}$ $7\overline{)49}$ $7\overline{)7}$ $2\overline{)16}$
2. $4\overline{)36}$ $4\overline{)32}$ $7\overline{)35}$ $7\overline{)21}$ $7\overline{)28}$ $7\overline{)14}$ $9\overline{)45}$
3. $1\overline{)1}$ $8\overline{)72}$ $5\overline{)10}$ $3\overline{)18}$ $1\overline{)8}$ $8\overline{)16}$ $6\overline{)54}$
4. $5\overline{)10}$ $3\overline{)18}$ $5\overline{)30}$ $3\overline{)18}$ $1\overline{)5}$ $8\overline{)24}$ $2\overline{)2}$
5. $5\overline{)15}$ $3\overline{)15}$ $1\overline{)7}$ $8\overline{)48}$ $5\overline{)25}$ $3\overline{)24}$ $9\overline{)63}$
6. $4\overline{)20}$ $1\overline{)6}$ $8\overline{)56}$ $5\overline{)40}$ $3\overline{)21}$ $1\overline{)2}$ $3\overline{)27}$
7. $1\overline{)9}$ $7\overline{)63}$ $4\overline{)12}$ $8\overline{)8}$ $5\overline{)35}$ $5\overline{)5}$ $9\overline{)72}$
8. $3\overline{)3}$ $3\overline{)27}$ $4\overline{)24}$ $7\overline{)63}$ $1\overline{)3}$ $8\overline{)32}$ $2\overline{)8}$
9. $5\overline{)20}$ $3\overline{)12}$ $4\overline{)8}$ $7\overline{)42}$ $1\overline{)4}$ $8\overline{)64}$ $3\overline{)15}$
10. $5\overline{)45}$ $3\overline{)6}$ $6\overline{)30}$ $6\overline{)6}$ $2\overline{)14}$ $2\overline{)8}$ $4\overline{)24}$
11. $9\overline{)36}$ $9\overline{)27}$ $6\overline{)42}$ $2\overline{)4}$ $9\overline{)9}$ $6\overline{)12}$ $5\overline{)35}$
12. $2\overline{)18}$ $9\overline{)27}$ $6\overline{)36}$ $2\overline{)12}$ $9\overline{)54}$ $6\overline{)48}$ $8\overline{)40}$
13. $2\overline{)6}$ $9\overline{)18}$ $6\overline{)18}$ $2\overline{)10}$ $9\overline{)81}$ $6\overline{)24}$ $6\overline{)36}$

If you need additional practice, try the next exercise.

Time Limit: 5 minutes

1. $3\overline{)6}$ $6\overline{)54}$ $2\overline{)2}$ $9\overline{)63}$ $5\overline{)45}$ $8\overline{)64}$ $1\overline{)4}$
2. $3\overline{)12}$ $6\overline{)24}$ $2\overline{)16}$ $9\overline{)45}$ $5\overline{)20}$ $8\overline{)32}$ $7\overline{)42}$
3. $3\overline{)27}$ $6\overline{)18}$ $2\overline{)10}$ $9\overline{)81}$ $5\overline{)35}$ $8\overline{)8}$ $4\overline{)8}$
4. $3\overline{)21}$ $6\overline{)48}$ $2\overline{)6}$ $9\overline{)54}$ $5\overline{)40}$ $8\overline{)56}$ $7\overline{)63}$
5. $3\overline{)3}$ $6\overline{)36}$ $2\overline{)12}$ $9\overline{)18}$ $5\overline{)5}$ $8\overline{)40}$ $5\overline{)40}$
6. $3\overline{)15}$ $6\overline{)12}$ $2\overline{)18}$ $9\overline{)72}$ $5\overline{)25}$ $8\overline{)48}$ $4\overline{)16}$
7. $3\overline{)24}$ $6\overline{)42}$ $2\overline{)4}$ $9\overline{)9}$ $5\overline{)15}$ $8\overline{)24}$ $1\overline{)7}$
8. $3\overline{)9}$ $6\overline{)6}$ $2\overline{)8}$ $9\overline{)27}$ $5\overline{)30}$ $8\overline{)16}$ $7\overline{)49}$
9. $3\overline{)18}$ $6\overline{)30}$ $2\overline{)14}$ $9\overline{)36}$ $5\overline{)10}$ $8\overline{)72}$ $5\overline{)15}$

10.	$1\overline{)8}$	$4\overline{)16}$	$7\overline{)49}$	$1\overline{)1}$	$7\overline{)7}$	$4\overline{)28}$	$3\overline{)27}$
11.	$4\overline{)32}$	$4\overline{)4}$	$7\overline{)56}$	$7\overline{)21}$	$1\overline{)5}$	$1\overline{)6}$	$7\overline{)21}$
12.	$4\overline{)12}$	$4\overline{)36}$	$7\overline{)35}$	$1\overline{)7}$	$7\overline{)14}$	$1\overline{)9}$	$8\overline{)24}$
13.	$4\overline{)24}$	$1\overline{)2}$	$7\overline{)28}$	$4\overline{)20}$	$7\overline{)63}$	$1\overline{)3}$	$3\overline{)15}$

Short Division

Returning to the friend from whom you borrowed money, suppose that you borrow $25 and are to repay him at the rate of $6 per week. How many payments must you make? With a little thought, we find that 4 payments are not enough and 5 payments are too many. If we make 4 payments, we'll still owe a dollar. This can be thought of as a division problem where 25 divided by 6 equals 4 with a **remainder** of 1. The result, 4, is called the **quotient.** In fact, the answer to any division problem is called the quotient.

$$25 \div 6 = 4 + 1 \qquad \text{(remainder)}$$

The number 25 is called the **dividend** and the number we are dividing by, 6, is called the **divisor.**

We can easily check our problem by seeing that 4 payments of $6 each plus the $1 remainder equals the $25 that was borrowed.

Usually we don't have any trouble with division problems like 25 ÷ 6 or 9 ÷ 3. However, when we deal with larger numbers, we have to do a little more work. For example, suppose that we are asked to divide 512 by 3, that is, 512 ÷ 3. To find the answer, we must find the largest whole number that, when multiplied by 3, will be smaller than 512. Take a guess. Try 200. Well, 200 × 3 = 600, so 200 is too large. How about 100? 100 × 3 = 300, which is certainly less than 512; in fact, there is 212 left over. In other words, 512 = 100 × 3 + 212. Now do the same thing to 212. How many times does 3 divide 212? How about 60? 60 × 3 is 180. How much is left over? Let's see, 212 minus 180 is 32. That means that 212 = 60 × 3 + 32. Let's keep going: 32 ÷ 3 = 10 with 2 remainder. Well, then, how many 3s are there in 512?

$$\begin{aligned} 512 &= (100 \times 3) + (60 \times 3) + (10 \times 3) + (2) \\ &= \quad 300 \quad + \quad 180 \quad + \quad 30 \quad + 2 \\ &= \quad 512 \end{aligned}$$

so we have (100 + 60 + 10) threes or 170 threes with a remainder of 2. Therefore, 512 ÷ 3 = 170 with 2 remainder.

Hopefully this all seems to make sense, but the method is pretty long and involved. Perhaps we can shorten things up a bit. There is a mechanical method, called **short division,** that is much easier and will produce the same results. Let's take a look at it. We use the following symbols to indicate division.

$$3\overline{)512}$$

We start by dividing 5, the left digit of 512, by 3:

$$3\overline{)5\,{}^{2}1\;2}\quad\text{quotient: }1$$

Since 3 goes into 5 once with a remainder of 2, we write the quotient above the 5, "carry" the 2 forward, and combine it with the 1 in the tens position. Actually, the 2 is 2 hundreds and the 1 is 1 ten. Now we divide the 21 by 3; it goes in 7 times with no remainder. We put the 7 above the 1.

$$3\overline{)5\,{}^{2}1\;2}\quad\text{quotient: }1\;7$$

We have no remainder to carry forward, so we move to the right and divide the 2 by 3. But 3 won't divide into 2 or, another way of saying the same thing, 3 divides 2 zero times with a remainder of 2. We put the 0 above the 2 and write the remainder 2 like this:

$$3\overline{)5\,{}^{2}1\;2}\quad\text{quotient: }1\;7\;0\text{ R2}$$

You will note that we got the same answer as before.

We can check our answer by multiplying the quotient by the divisor and adding in the remainder. This result should be the dividend.

Check

170	⟵	quotient
× 3	⟵	divisor
510		
+ 2	⟵	remainder
512	⟵	dividend

EXAMPLE 1 32,171 ÷ 6

SOLUTION We use the short-division method, setting it up as follows. First, we divide 6 into 3:

$$6\overline{)3\;2\;1\;7\;1}$$

It goes 0 times with 3 remainder. We carry the 3 forward.

$$6\overline{)3\,{}^{3}2\;1\;7\;1}\quad\text{quotient: }0$$

Next, we divide 32 by 6; it goes in 5 times with a remainder of 2. We record the 5 and carry the 2 forward:

$$6\overline{)3\,{}^{3}2\,{}^{2}1\;7\;1}\quad\text{quotient: }0\;5$$

At this point you may have noticed that since 6 didn't divide the first 3, we could have divided 6 into 32 right away, which would eliminate the 0 at the beginning of our quotient. It has no value in this position anyway, so let's leave it off. Next, 21 divided by 6 is 3, with a remainder of 3. We record 3 above the 21 and carry the 3 remainder forward.

$$6\overline{)3\,{}^{3}2\,{}^{2}1\,{}^{3}7\,1}\quad \text{quotient: } 5\ 3$$

Next, 6 divides 37 six times with 1 left over. We record the 6 and carry the 1 forward. Finally, we divide 11 by 6. It goes in once with a remainder of 5, and we're finished.

$$6\overline{)3\,{}^{3}2\,{}^{2}1\,{}^{3}7\,{}^{1}1}\quad \text{quotient: } 5\ 3\ 6\ 1\ \text{R5}$$

Again, one nice thing about division is the fact that we can always check our answer by using multiplication to see whether we've done the problem correctly. Let's check the last problem to see how we did.

5,361	⟵	quotient
× 6	⟵	divisor
32,166		
+ 5	⟵	remainder
32,171	⟵	dividend

The method that we have just described is generally called **short division** and we can use it any time the number that we are dividing by is a single digit. Remember that the number that we divide by is called the **divisor,** and the number that we divide into is called the **dividend.** So, if the divisor is a single digit, use short division.

EXAMPLE 2 Divide: $\dfrac{23{,}021}{6}$.

SOLUTION

$$6\overline{)2\ 3\,{}^{5}0\,{}^{2}2\,{}^{4}1}\quad \text{quotient: } 3\ 8\ 3\ 6\ \text{R5}$$

First, 6 divides 2 zero times, but remember that a zero in this position is meaningless, so we divide 6 into 23 right away. It goes in 3 times, since 6 × 3 is 18, and we have a remainder of 5. We record the 3 and carry the 5 forward. Next, 50 divided by 6 is 8, because six 8s are 48 with 2 left over. Now, 6 divides 22 three times, and 22 minus 18 is 4, so we have a remainder of 4. We record the 3 and carry the 4 forward. Finally, 41 divided by 6 is 6, because 6 times 6 is 36, and that leaves us a remainder of 5.

STOP

15 ÷ 3 is read "15 divided by 3," which means $\dfrac{15}{3}$ or $3\overline{)15}$, *not* $\dfrac{3}{15}$ or $15\overline{)3}$.

Division Involving Zero

$$0 \div 3 = \frac{0}{3} = 0 \qquad \text{because} \qquad 0 \times 3 = 0$$

What about

$$\frac{3}{0} = \boxed{?}$$

What number can replace the question mark so that $\boxed{?} \times 0 = 3$? Since any number multiplied by zero gives zero, there is no possible replacement for the question mark. We conclude then that $\frac{3}{0}$ has no answer. Since the problem $\frac{3}{0}$ has no answer, we say that it is *undefined.*

Next we consider

$$\frac{0}{0} = \boxed{?}$$

What number can replace the question mark so that $\boxed{?} \times 0 = 0$. Since any number times zero is zero, we can replace the question mark by any number we like. This gives us $\frac{0}{0} = 0$, 6, 14, or any other number. This situation is certainly not allowable. We say that $\frac{0}{0}$ is *indeterminant* or *not uniquely defined.* We conclude that division by zero in all cases is not permitted, and we say that it is undefined.

Division by zero is undefined. Zero may never be used as a divisor. Never divide by zero.

EXAMPLE 3

(a) $\frac{16}{0}$ is undefined.

(b) $\frac{0}{12} = 0$ since $12 \times 0 = 0$.

(c) $21 \div 0$ is undefined.

(d) $0 \div 8 = 0$

(e) $0 \div 0$ is undefined.

1.6 Exercises

Use short division in the following problems.

1. $3\overline{)4,612}$
2. $8\overline{)1,926}$
3. $6\overline{)3,820}$
4. $5\overline{)321,846}$

5. $2\overline{)386{,}129}$

6. $5\overline{)21{,}893}$

7. $8\overline{)30{,}046}$

8. $6\overline{)300{,}019}$

9. $6\overline{)10{,}000}$

10. $7\overline{)583}$

11. $5\overline{)38{,}200}$

12. $4\overline{)38{,}123}$

13. $3\overline{)420{,}091}$

14. $9\overline{)837{,}112}$

15. $8\overline{)642{,}051}$

▼ Divide each of the following using short division.

16. $\dfrac{326}{4}$ 17. $\dfrac{928}{6}$

18. $\dfrac{3{,}289}{7}$ 19. $\dfrac{4{,}243}{7}$

20. $\dfrac{2{,}468}{3}$ 21. $\dfrac{8{,}251}{3}$

▼ Do the following division problems involving zero.

22. $\dfrac{0}{6}$ 23. $\dfrac{0}{13}$

24. $\dfrac{4}{0}$ 25. $\dfrac{9}{0}$

26. $8\overline{)0}$ 27. $3\overline{)0}$

28. $268 \div 0$ 29. $14 \div 0$

30. $0 \div 315$ 31. $\dfrac{0}{0}$

32. $0 \div 0$ 33. $0\overline{)0}$

34. $0\overline{)6}$ 35. $0\overline{)14}$

1.7 Long Division

Many of you may already have guessed what our next topic in division will be. It's **long division,** and we use it whenever the divisor is made up of two or more digits. Let's consider such a problem. Suppose we want to divide 3,842 by 14. What we're really trying to find out is how many 14s there are in 3,842. I'll give you some information. $200 \times 14 = 2{,}800$ and $300 \times 14 = 4{,}200$, so the number we are looking for lies between 200 and 300.

$$200 \times 14 = 2{,}800$$
$$? \times 14 = 3{,}842$$
$$300 \times 14 = 4{,}200$$

Well, since $200 \times 14 = 2{,}800$ and since 3,842 minus $2{,}800 = 1{,}042$, we know that

$$3{,}842 = 200 \times 14 + 1{,}042$$

How many 14s are there in 1,042? Let's try 50:

$$\begin{array}{r} 14 \\ \times 50 \\ \hline 700 \end{array}$$

1,042 minus 700 is 342: We do the same thing to the 342. Guess how many 14s there are in it? Let's try 20:

$$\begin{array}{r} 14 \\ \times 20 \\ \hline 280 \end{array}$$

Now, 342 minus 280 is 62, and how many 14s are there in 62? It looks like about 4.

$$\begin{array}{r} 14 \\ \times 4 \\ \hline 56 \end{array}$$

and 62 minus 56 is 6. We can't fit any 14s into 6, so it appears that we're finished. Now all that is left to do is to add up all our 14s.

$$\begin{aligned} 3842 &= (200 \times 14) + (50 \times 14) + (20 \times 14) + (4 \times 14) + 6 \\ &= \quad 2800 \quad + \quad 700 \quad + \quad 280 \quad + \quad 56 \quad + 6 \end{aligned}$$

So we have (200 + 50 + 20 + 4 = 274) fourteens plus 6 left over. In other words, 3842 = 274 × 14 + 6. So, we've found the answer to our original division problem:

$$\begin{array}{r} 274 \quad \text{R } 6 \\ 14\overline{)3{,}842} \end{array}$$

Finding the answer was certainly a long and involved process. As is often the case in mathematics, someone has developed a method to make our work easier. Let's use the example that we just did to illustrate this method. First, we write the problem in this manner:

$$14\overline{)3{,}842}$$

Next, we divide 14 into the first digit, 3. Since it goes zero times, we divide 14 into the first *two* digits. It goes twice with some remainder left over. Forget about the remainder for just a second.

$$\begin{array}{r} 2 \\ 14\overline{)3{,}842} \end{array}$$

We record the 2 above the 8 and then multiply the 14 by 2, writing the product 28, under the 38. Next, we subtract the 28 from the 38. This gives us 10, which is the remainder you were told to forget about a few seconds ago.

$$\begin{array}{r} 2 \\ 14\overline{)3{,}842} \\ \underline{2\,8} \\ 1\,0 \end{array}$$

What we do next is bring down the next digit, the 4, and write it after the 10, giving us 104:

$$\begin{array}{r} 2 \\ 14\overline{)3{,}842} \\ \underline{2\,8} \\ 1\,04 \end{array}$$

Now we repeat the process. How many times does 14 divide into 104? At this point we may have some guessing to do. Let's try 10. Well, 14 × 10 is 140, but that's too big; let's try 7:

$$14 \times 7 = 98$$

which looks okay.

$$\begin{array}{r} 27 \\ 14\overline{)3{,}842} \\ \underline{2\,8} \\ 1\,04 \\ \underline{98} \\ 6 \end{array}$$

We put the 7 above the 4, and multiply the 14 times 7. Now we subtract, which gives us a 6 remainder. Incidentally, if our remainder had been 14 or larger, we would have known that our quotient was not large enough; that is, 14 would have divided 104 more than 7 times. We don't have to worry about that now, because our remainder, 6, is less than 14. Now we bring down the next digit, 2, giving us 62, and repeat the process. How many times does 14 go into 62? How about 4? We enter the 4 in our quotient, multiply 14 × 4, subtract, and we're left with a 6 remainder.

$$\begin{array}{rl} 274 & \text{R } 6 \\ 14\overline{)3{,}842} & \\ \underline{2\,8} & \\ 1\,04 & \\ \underline{98} & \\ 62 & \\ \underline{56} & \\ 6 & \end{array}$$

There are no more digits to bring down, which tells us that we're finished. Our answer is 274, with a remainder of 6. Compare this answer with the one that we obtained previously.

It appears, then, that our method works. I'm sure you've noticed that the method requires you to do quite a bit of guessing. This can be a little tricky, but if you practice, you'll find that you'll become quite good at it.

EXAMPLE 3 Divide 1,436 by 24.

SOLUTION We set up like this:

```
24)1,436
```

We try to divide 1 by 24 but it won't go, so we try the first two digits. It still won't go, so we keep going. We divide 143 by 24. It appears to go about 4 times, so we put the 4 in the quotient over the 3 and multiply, which gives us 96. Now we subtract:

```
       4
24)1,435
     96
     47
```

Wait a minute! Our remainder is larger than our divisor; 47 is greater than 24. This means that 24 will divide 143 more than 4 times, so let's reconsider. We try 5.

```
       5
24)1,436
    1 20
     236
```

We calculate that 5 times 24 is 120, and subtracting gives us a remainder of 23. Since 23 is less than 24, we conclude that 5 is the correct quotient. Now we bring down the next digit, 6 and divide 236 by 24. This time, try and guess how many times 24 goes into 236. You should take a guess and then multiply 24 by whatever number you guessed on scrap paper. If you're off on the first guess, try again. What did you get? The correct answer is 9. If you guessed a number larger than 9, it would be too big, and if you tried a number smaller than 9, the remainder will be 24 or larger. So we enter the 9 in the quotient and multiply. Subtracting gives us a 20 remainder and we're finished.

```
      59  R 20
24)1,436
    1 20
     236
     216
      20
```

As was previously mentioned, you can always check a division problem by using multiplication to see if you did it correctly. Multiply the quotient by the divisor and add in the remainder.

$$\begin{array}{r} 59 \\ \times 24 \\ \hline 236 \\ 1\,18 \\ \hline 1{,}416 \\ +\quad 20 \\ \hline 1{,}436 \end{array}$$

Our answer is correct.

Now it's time to try a few on your own. Don't get discouraged, particularly if you guess wrong quite a bit at first. Practice makes perfect, so keep on trying.

1.7 Exercises

▼ Use long division in the following problems.

1. $16\overline{)583}$
2. $17\overline{)651}$
3. $27\overline{)3{,}804}$
4. $15\overline{)30{,}572}$
5. $102\overline{)38{,}567}$
6. $115\overline{)4{,}683}$
7. $562\overline{)53{,}819}$
8. $714\overline{)728{,}364}$
9. $16\overline{)24{,}186}$
10. $27\overline{)32{,}118}$
11. $317\overline{)43{,}281}$
12. $591\overline{)60{,}000}$
13. $231\overline{)6{,}271}$
14. $547\overline{)2{,}368}$
15. $714\overline{)56{,}791}$
16. $775 \div 25$

17. 4,464 ÷ 36

18. 9,856 ÷ 28

19. 4,416 ÷ 53

20. 6,712 ÷ 28

21. 9196 ÷ 85

22. 2,345 ÷ 42

23. 7,621 ÷ 89

24. 23,221 ÷ 407

25. 36,821 ÷ 214

26. 98,521 ÷ 762

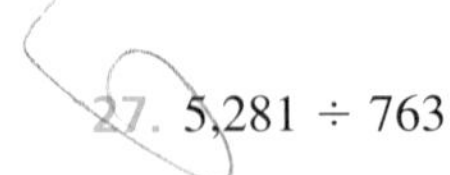

27. 5,281 ÷ 763

LIVING IN THE REAL WORLD

28. If your car traveled 1,312 miles on 41 gallons of gasoline, how many miles did it get to the gallon?

29. You earn $8 per hour. How many hurs did you work if you were paid $304?

30. An estate worth $41,185 was divided equally among 5 persons. What was the share of each?

31. If an airplane climbs at the rate of 680 feet per minute, how long does it take to reach an altitude of 11,560?

32. How many 16-inch-long boards can be cut from a board that is 180 inches long?

33. If a woman's annual salary is $20,696, how much does she earn in a week?

34. Twenty-six times some number is 1,248. What is the number?

35. The average weight of a bale of hay is 70 pounds. Approximately how many bales of hay are on a truck if the cargo weighs 25,700 pounds?

36. How many 8-oz glasses can you fill from a quart bottle (32 oz)?

37. How many 6-oz glasses can you fill from three 32-oz bottles?

38. How do you check a division problem?

39. When you are dividing, why do you sometimes get a remainder?

40. Why is division sometimes called "repeated subtraction"?

1.8 Powers of Whole Numbers

When the same factor appears many times in a multiplication, a shorthand notation can be used to save us work. For example:

$$2^4 \quad \text{means} \quad 2 \cdot 2 \cdot 2 \cdot 2 = 16$$

In the expression 2^4, the number 2 is written as a factor 4 times. Here 2 is called the **base,** the small elevated 4 is called the **exponent,** and 2^4 is read "two to the fourth power."

exponent

$$2^4 = 16$$

base

EXAMPLE 1

(a) $2^2 = 2 \cdot 2 = 4$ — 2 to the 2nd power, or "2 squared," is 4.

(b) $2^3 = 2 \cdot 2 \cdot 2 = 8$ — 2 to the 3rd power, or "2 cubed," is 8.

(c) $1^4 = 1 \cdot 1 \cdot 1 \cdot 1 = 1$ — 1 raised to any power equals 1.

(d) $5^2 = 5 \cdot 5 = 25$

(e) $7^1 = 7$ — Any number raised to the 1st power equals that number.

Zero As an Exponent

When any whole number (except zero) is raised to the zero power, the result is defined to be 1. This rule is not obvious, but it is consistent with other rules for exponents that we will investigate later.

The expression 0^0 is not defined for our purposes and we will not use it.

Definition

$a^0 = 1$ where a is any whole number except zero

EXAMPLE 2
(a) $6^0 = 1$
(b) $0^3 = 0 \cdot 0 \cdot 0 = 0$
(c) $19^0 = 1$
(d) 0^0 is undefined. (Technically it is called *indeterminant*.)
(e) $10^0 = 1$

Powers of 10

When the number 10 is used as a base, some interesting results occur, as illustrated in the following example.

EXAMPLE 3
(a) $10^0 = 1$ 1 followed by no zeros
(b) $10^1 = 10$ 1 followed by 1 zero
(c) $10^2 = 10 \cdot 10 = 100$ 1 followed by 2 zeros
(d) $10^3 = 10 \cdot 10 \cdot 10 = 1000$ 1 followed by 3 zeros
(e) What would 10^8 equal? If we follow the pattern that seems to be emerging in parts (a)–(d), we would expect, correctly, that $10^8 = 100{,}000{,}000$, or 1 followed by 8 zeros.

Remember, x^0 is **not** equal to zero; $x^0 = 1$.

1.8 Exercises

▼ Evaluate using the definition of exponents.

1. 3^2
2. 6^2
3. 3^4
4. 2^5
5. 6^3
6. 4^3
7. 2^6
8. 3^5
9. 36^2
10. 28^2
11. 19^3
12. 14^3
13. 0^4
14. 0^6
15. 9^0
16. 5^0
17. 152^0
18. 75^0

19. 0^0

20. 0^{14}

21. 1^6

22. 1^5

23. 1^{51}

24. 1^{80}

25. 10^0

26. 10^1

27. 10^2

28. 10^4

29. 10^6

30. 10^7

31. 10^{10}

32. In the expression 6^4, identify the base and the exponent.

LIVING IN THE REAL WORLD

33. It is generally thought that our universe came about over 10^{10} years ago. People migrated throughout Africa, Europe, and Asia 10^6 years ago, and even with brains about half current size, they learned to make stone tools. About 10^5 years ago, the human brain reached modern size. Evaluate each of these numbers written in exponent form.

34. There are 2^6 squares on a checker board. How many squares is this?

35. At the top of the second column is a family tree with you at the top. You are a direct descendent of everyone else in the tree. Notice that the number of people in each level of the tree can be written as a power of 2. Each level of the tree represents one generation. (a) The Declaration of Independence was signed on July 4, 1776, about 8 generations ago. There are 2^8 of your ancestors at this level of your family tree. How many ancestors is this? (b) The Pilgrims came to America on the Mayflower in 1620, about 14 generations ago. How many ancestors are in the 14th level of your family tree?

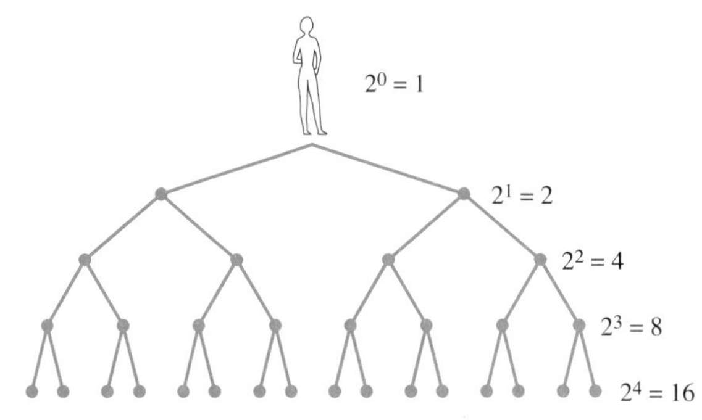

MENTAL MATHEMATICS

▼ Perform the following mentally, without using paper or pencil.

36. 7^2

37. 9^2

38. 4^2

39. 10^2

40. 7^1

41. 9^1

42. 1^3

43. 1^8

44. 6^0

45. 5^0

46. 0^5

47. 0^8

48. 0^0

49. 10^3

50. 3^3

51. 4^3

52. In the product 3^4, why don't you multiply 3 times 4 to get the result?

53. One followed by one hundred zeros is called a *googol*. How would you write this number using an exponent? Why is it to your advantage to use exponents when dealing with extremely large numbers?

54. Without doing any calculations, tell which expression represents a larger number, 10^{50} or 50^{10}. Give a reason for your answer.

1.9 Order of Operations

Consider the statement

Anna said Mary is rich.

We can punctuate it in two different ways so that it takes on two entirely different meanings:

1. "Anna," said Mary, "is rich."
2. Anna said, "Mary is rich."

Now consider the mathematical statement

$$6 + 2 \cdot 5$$

If we do the addition first, we get one answer, and if we do the multiplication first, we get another.

$6 + 2 \cdot 5 = 8 \cdot 5 = 40$ if we add first

$6 + 2 \cdot 5 = 6 + 10 = 16$ if we multiply first

Which is correct? As it turns out, the second way is the correct way, because mathematicians have agreed upon a definite order in which to perform operations. This order provides us with a kind of "mathematical punctuation" so that we cannot get two different results from the same problem. The rules for order of operations are as follows:

Order of Operations

1. Expressions inside parentheses ( ), brackets [ ], or above or below a fraction bar are done first.
2. Exponents are done next.
3. Multiplication and division are done as they occur, from left to right.
4. Addition and subtraction are done as they occur, from left to right.

EXAMPLE 1 Evaluate: $7 - 3 \cdot 2$.

SOLUTION

$7 - 3 \cdot 2 = 7 - 6$ Multiply before subtracting

$= 1$

EXAMPLE 2 Evaluate: $(7 - 3) \cdot 2$.

SOLUTION

$(7 - 3) \cdot 2 = 4 \cdot 2$ Work inside the parentheses first.

$= 8$ Then multiply

Compare this example with Example 1. The numbers are the same, but they are "punctuated" differently.

EXAMPLE 3 Evaluate: $4(6 + 9) - 3$.

SOLUTION The 4 written next to the parenthesis with no sign in between means "multiply." Remember to treat the $6 + 9$ inside the parentheses as a single unit, or a single number.

$$\begin{aligned} 4(6+9) - 3 &= 4(15) - 3) && \text{Add inside parentheses.} \\ &= 60 - 3 && \text{4(15) means "multiply."} \\ &= 57 && \text{Subtract.} \end{aligned}$$

EXAMPLE 4 Evaluate: $6 + 3(6 - 2)$.

SOLUTION

$$\begin{aligned} 6 + 3(6-2) &= 6 + 3(4) && \text{Work inside the parentheses first.} \\ &= 6 + 12 && \text{Next multiply.} \\ &= 18 && \text{Finally add.} \end{aligned}$$

EXAMPLE 5 Evaluate: $15 - 7 - 6$.

SOLUTION

$$\begin{aligned} 15 - 7 - 6 &= 8 - 6 && \text{Work from left to right.} \\ &= 2 \end{aligned}$$

EXAMPLE 6 Evaluate: $2 \cdot 3^2 - 3 \cdot 5 + 4$.

SOLUTION

$$\begin{aligned} 2 \cdot 3^2 - 3 \cdot 5 + 4 &= 2 \cdot 9 - 3 \cdot 5 + 4 && \text{Do exponents first.} \\ &= 18 - 15 + 4 && \text{Do both multiplications.} \\ &= 3 + 4 && \text{Next subtract.} \\ &= 7 && \text{Then add.} \end{aligned}$$

EXAMPLE 7 Evaluate: $\dfrac{9 + 7 - 6}{2^2 + 1}$.

SOLUTION The fraction bar separates this problem into two parts, above and below, each of which must be done separately.

$$\begin{aligned} \frac{9+7-6}{2^2+1} &= \frac{16-6}{4+1} && \text{Left to right above the bar. Powers first below the bar.} \\ &= \frac{10}{5} \\ &= 2 && \text{Perform the division last.} \end{aligned}$$

EXAMPLE 8 Evaluate: $7 - 2^3 \div 2 + 5$.

SOLUTION

$$\begin{aligned} 7 - 2^3 \div 2 + 5 &= 7 - 8 \div 2 + 5 && \text{Do exponents first.} \\ &= 7 - 4 + 5 && \text{Divide next.} \\ &= 3 + 5 && \text{Subtract and add from left to right.} \\ &= 8 \end{aligned}$$

1.9 Exercises

Evaluate each expression using the rules for order of operations.

1. $6 + 4 \cdot 7$
2. $3 + 6 \cdot 5$
3. $4 \cdot 6 - 5$
4. $3 \cdot 2 - 5$
5. $10 - (8 - 6)$
6. $15 - 7 - 8$
7. $3 \cdot 5^2$
8. $4 \cdot 3^2$
9. $16 \div 4^2$
10. $32 \div 2^3$
11. $15 - 8^0$
12. $17 - 11^0$
13. $24 \div 8 \div 4$
14. $36 \div 6 \div 2$
15. $3 + 5(6 - 1)$
16. $5 + 7(3 - 1)$
17. $4 \cdot 6 + 3 \cdot 5$
18. $8 \cdot 4 + 2 \cdot 7$
19. $8 - 2(5 - 2)$
20. $11 - 3(6 - 4)$
21. $3(2 + 6 \cdot 2)$
22. $5(8 + 2 \cdot 3)$
23. $3 \cdot 2^2 - 4 \cdot 2$
24. $6 \cdot 3^2 - 5 \cdot 9$
25. $4^2 - 2(5 + 1)$
26. $5^2 - 3(4 + 2)$
27. $17 + 50 \div 5^2$
28. $23 + 36 \div 3^2$
29. $\dfrac{7 + 1}{2 + 2}$
30. $\dfrac{9 + 11}{3 + 2}$
31. $\dfrac{3^2 + 1}{11 - 6}$
32. $\dfrac{5^2 + 1}{15 - 2}$
33. $\dfrac{3(2 + 5)}{3^2 - 2}$
34. $\dfrac{5(2 + 3)}{4^2 - 11}$
35. $\dfrac{6 + 3(6^0 + 5)}{6^2 - 3 \cdot 8}$
36. $\dfrac{10 + 3(4 + 3^0)}{7^2 - 6 \cdot 4}$

LIVING IN THE REAL WORLD

37. A car rental company advertises a rate of $22.00 per day plus a charge of $0.10 per mile driven. How much does it cost to rent a car for 4 days if you drive it a total of 438 miles?

38. To rent a motor scooter in Fort Lauderdale, Florida, costs $15.00 per day plus $0.08 per mile. What will it cost to rent a scooter for 3 days and drive it 300 miles?

MENTAL MATHEMATICS

Find the answer mentally without using pencil or paper.

39. $3 \cdot 6 - 5$

40. $6 \cdot 4 - 4$

41. $7 + 5 - 4$

42. $8 + 9 - 2$

43. $6 + 2 \cdot 5$

44. $8 + 4 \cdot 5$

45. $8 - 2^2$

46. $15 - 3^2$

47. $16 - 4^2$

48. $5 \cdot 4 + 2^2$

49. $6 \cdot 5 + 3^2$

50. $10^2 - 5^2$

51. $3^2 + 2 \cdot 4$

52. $\dfrac{6^2 - 1}{7}$

53. $\dfrac{4^2 - 1}{3}$

54. $\dfrac{7 \cdot 5 + 5}{8}$

55. $\dfrac{6 \cdot 8 + 2}{10}$

56. $\dfrac{4 \cdot 10 + 8}{6}$

57. $\dfrac{6^2 - 6}{6}$

58. $\dfrac{5^2 - 5}{5}$

59. $\dfrac{4^2 - 4}{4}$

60. Look at the answers to Exercises 57, 58, and 59. Predict the answer to other similar problems, like

$$\frac{7^2 - 7}{7} \quad \text{and} \quad \frac{8^2 - 8}{8}$$

See if you can find some pattern to these answers.

1.10 Properties of Whole Numbers

When you change the order of two numbers in an addition problem, the sum does not change.

EXAMPLE 1

(a) $6 + 3 = 9$
$3 + 6 = 9$

(b) $7 + 11 = 18$
$11 + 7 = 18$

This property is true in general and is called the **commutative property of addition.**

Commutative Property of Addition:

If a and b represent any whole numbers, then

$$a + b = b + a$$

Adding zero to any number gives us back the *identical* number we started with. It seems reasonable, then, to call **zero** the **additive identity.**

Additive Identity:

If a represents any whole number, then

$$a + 0 = a \text{ and } 0 + a = a$$

In the two preceding rules, we used the letters a and b to represent numbers. For example, what is the value of $5 + a$ if $a = 8$?

$$5 + a = 5 + 8 = 13$$

When we add three whole numbers, we have a choice of which two to add first. Does it make any difference? Let's see. We will use parentheses to indicate which two numbers we are adding first.

EXAMPLE 2 Add $3 + 2 + 6$

SOLUTION **(a)** $(3 + 2) + 6 = 5 + 6$ Add 3 + 2 first.
$= 11$

(b) $3 + (2 + 6) = 3 + 8$ Add 2 + 6 first.
$= 11$

Our example illustrates that when we add three whole numbers, it makes no difference which two we add first. This property is always true and is called the **associative property of addition.**

Associative Property of Addition:

If a, b and c represent any whole numbers, then

$$(a + b) + c = a + (b + c)$$

As in the case of addition, if you change the order of two whole numbers in multiplication, the result remains the same.

EXAMPLE 3 **(a)** $6 \cdot 7 = 42$
$7 \cdot 6 = 42$

(b) $4 \cdot 5 = 20$
$5 \cdot 4 = 20$

This property is called the **commutative property of multiplication** and is stated formally as follows:

Commutative Property of Multiplication:

If a and b represent any whole numbers, then

$$a \cdot b = b \cdot a$$

The associative property also holds for multiplication of whole numbers, as illustrated in the following example.

EXAMPLE 4 Multiply: $3 \cdot 6 \cdot 2$.

SOLUTION **(a)** $(3 \cdot 6) \cdot 2 = 18 \cdot 2$ Multiply $3 \cdot 6$ first.
$= 36$

(b) $3 \cdot (6 \cdot 2) = 3 \cdot 12$ Multiply $6 \cdot 2$ first.
$= 36$

As you can see, we obtained the same result no matter which two numbers we multiplied first.

Associative Property of Multiplication:

If a, b, and c represent any whole numbers, then

$$(a \cdot b) \cdot c = a \cdot (b \cdot c)$$

If we multiply any whole number by 1, we get back the *identical* number we started with. For this reason we call 1 the **multiplicative identity.**

Multiplicative Identity:

If a represents any whole number, then

$$a \cdot 1 = 1 \cdot a = a$$

Study the next example carefully to discover still another important property of whole numbers.

EXAMPLE 5 **(a)** Evaluate $4 \cdot (2 + 7)$.

(b) Evaluate $4 \cdot 2 + 4 \cdot 7$.

SOLUTION **(a)** $4 \cdot (2 + 7) = 4 \cdot 9$ Add inside the parentheses first.
$= 36$ Multiply.

(b) $4 \cdot 2 + 4 \cdot 7 = 8 + 28$ Multiply first.
$= 36$ Add.

Since the two results are the same, we see that

$$4 \cdot (2 + 7) = (4 \cdot 2) + (4 \cdot 7)$$

This property is true in general and is called the **distributive property of multiplication over addition.**

Distributive Property of Multiplication over Addition:

If a, b, and c are whole numbers, then

$$a \cdot (b + c) = (a \cdot b) + (a \cdot c)$$

The following application illustrates the distributive property nicely.

EXAMPLE 6 A saleswoman travels two days a week, on Monday and Tuesday. Her car gets 20 miles per gallon. If she used 13 gallons of gasoline on Monday and 17 gallons on Tuesday, how many miles did she travel? The problem can be solved two different ways.

SOLUTION **(a)** $20 \dfrac{\text{miles}}{\text{gallon}} \cdot 13 \text{ gallons} + 20 \dfrac{\text{miles}}{\text{gallon}} \cdot 17 \text{ gallons}$
$= 260 \text{ miles} + 340 \text{ miles}$
$= 600 \text{ miles}$

(b) $20 \dfrac{\text{miles}}{\text{gallon}} (13 \text{ gallons} + 17 \text{ gallons}) = 20 \dfrac{\text{miles}}{\text{gallon}} \cdot 30 \text{ gallons}$
$= 600 \text{ miles}$

As you can see, both methods provide a correct solution.

1.10 Exercises

Replace the question mark with the correct number, and tell which property of whole numbers you used to get your answer.

1. $5 + 8 = 8 + ?$
2. $6 + 9 = ? + 6$
3. $4 \cdot ? = 3 \cdot 4$
4. $7 \cdot ? = 3 \cdot 7$
5. $6 \cdot 1 = ?$
6. $12 \cdot 1 = ?$

7. $0 + ? = 8$

8. $0 + ? = 11$

9. $m + 6 = 6 + ?$

10. $9 + n = n + ?$

11. $x + y = y + ?$

12. $s + t = t + ?$

▼ Identify which property of whole numbers that each of the following illustrates.

13. $6 + 9 = 9 + 6$

14. $8 + 2 = 2 + 8$

15. $6 \cdot 7 = 7 \cdot 6$

16. $8 \cdot 4 = 4 \cdot 8$

17. $6 \cdot 1 = 6$

18. $1 \cdot 3 = 3$

19. $5 + 0 = 5$

20. $0 + 6 = 6$

21. $1 \cdot 8 = 8$

22. $15 \cdot 1 = 15$

23. $5 \cdot 6 = 6 \cdot 5$

24. $13 + 8 = 8 + 13$

25. $(3 + 6) + 7 = 3 + (6 + 7)$

26. $(5 + 8) + 2 = 5 + (8 + 2)$

27. $7 \cdot (2 \cdot 4) = (7 \cdot 2) \cdot 4$

28. $5 \cdot (3 \cdot 6) = (5 \cdot 3) \cdot 6$

29. $3 \cdot (4 + 5) = (3 \cdot 4) + (3 \cdot 5)$

30. $2 \cdot (7 + 3) = (2 \cdot 7) + (2 \cdot 3)$

▼ Using complete sentences, write the meaning of each of the properties in Exercises 31–33.

31. Commutative property of addition

32. Additive identity

33. Associative property of multiplication

34. Why is 1 called the multiplicative identity?

35. Why isn't subtraction commutative for whole numbers?

SUMMARY—CHAPTER 1

The numbers in brackets refer to the section in the text in which each concept is presented.

		EXAMPLE
[1.1]	The **whole numbers** are 0, 1, 2, 3, 4, and so on.	[1.1] Number line: 0 1 2 3 4 5 6 →
[1.1]	**Place value:** a digit takes on a specific value depending on what place it occupies in a number.	[1.1] 8 3 4 — 8: hundreds, 3: tens, 4: units
[1.1]	**Expanded notation** is a method of writing a number as sums of its parts according to the place value of each digit.	[1.1] $682 = 600 + 80 + 2$
[1.2]	To **round off** a whole number, the digit in the place to be rounded is (a) not changed if the digit immediately to the right is less than 5; (b) increased by 1 if the digit immediately to the right is 5 or greater. All digits to the right of the round-off place are replaced with zeros.	[1.2] See table below

Original number	Rounded to hundreds place
38,246	38,200
4,189	4,200
5,953	6,000

[1.3] Two whole numbers are **added** by totalling the digits in each place.

[1.3]
$$\begin{array}{r} 3{,}841 \\ +1{,}895 \\ \hline 5{,}736 \end{array}$$

[1.4] **Subtraction** of two whole numbers involves finding the difference between the digits in each place.

[1.4]
$$\begin{array}{r} 3{,}821 \\ -1{,}465 \\ \hline 2{,}356 \end{array}$$

[1.5] **Multiplication** of whole numbers is a method of repeated addition. The **factors** are the numbers that are multiplied together. The **product** is the answer to a multiplication problem.

[1.5]
$$\begin{array}{rl} 329 & \longleftarrow \text{factor} \\ \times\ 14 & \longleftarrow \text{factor} \\ \hline 1316 & \\ 329 & \\ \hline 4{,}606 & \longleftarrow \text{product} \end{array}$$

[1.6] **Division** of a whole number is the process of splitting it into a specific number of parts. The **dividend** is the number being divided up into parts. The **divisor** is the number that we divide by. The answer to a division problem is the **quotient.** The **reminder** is the number that is left over when the divisor does not divide the dividend an exact number of times.

[1.6] divisor: 23; quotient: 21; dividend: 492; remainder: 9

$$\begin{array}{r} 21 \\ 23\overline{)492} \\ \underline{46} \\ 32 \\ \underline{23} \\ 9 \end{array}$$

Division by zero is undefined.

[1.6] $\frac{0}{14} = 0$

$\frac{14}{0}$ is undefined.

$\frac{0}{0}$ is indeterminant.

EXAMPLE

[1.8] **Exponents** are a shorthand notation for repeated multiplication.

$3 \times 3 \times 3 \times 3 = 3^4$ (exponent: 4; base: 3)

Any nonzero number raised to the **zero power** equals 1.

$8^0 = 1$

0^0 is undefined.

[1.9] **Order of operations:**

1. Expressions inside parentheses ( ) or brackets [ ] or above and below a fraction bar are done first.
2. Powers are done next.
3. Multiplication and division are done from left to right.
4. Addition and subtraction are done from left to right.

$$18 - 3(3 + 2) = 18 - 3(5) = 18 - 15 = 3$$

[1.10] **Properties of whole numbers** If *a, b,* and *c* represent any whole numbers, the following are true:

1. **Commutative property of addition:**

$$a + b = b + a$$

$6 + 9 = 9 + 6$

2. **Associative property of addition:**

$$a + (b + c) = (a + b) + c$$

$3 + (4 + 2) = (3 + 4) + 2$

3. **Commutative property of multiplication:**

$$a \cdot b = b \cdot a$$

$5 \cdot 8 = 8 \cdot 5$

4. **Associative property of multiplication:**

$$a \cdot (b \cdot c) = (a \cdot b) \cdot c$$

$5 \cdot (3 \cdot 4) = (5 \cdot 3) \cdot 4$

5. **Distributive property of multiplication over addition:**

$$a \cdot (b + c) = (a \cdot b) + (a \cdot c)$$

$3 \cdot (5 + 2) = (3 \cdot 5) + (3 \cdot 2)$

6. **Zero** is the **additive identity:**

$$a + 0 = 0 + a = a$$

$4 + 0 = 4$

7. **One** is the **multiplicative identity:**

$$a \cdot 1 = 1 \cdot a = a$$

$6 \times 1 = 6$

REVIEW EXERCISES—CHAPTER 1

The numbers in brackets refer to the section in the text in which similar problems are presented.

[1.1] 1. Name the place value of the 5 in the number 15,024.

[1.1] 2. Write in words: 31,006.

[1.1] 3. Write as a number: fifty-two thousand, six hundred nine.

[1.1] 4. Write 18,026 in expanded notation.

[1.2] 5. Round 24,687 to the nearest hundred.

[1.2] 6. Round 15,962 to the nearest hundred.

[1.3] 7. Add:

$$\begin{array}{r} 5{,}814 \\ 3{,}822 \\ +\ 746 \\ \hline \end{array}$$

[1.4] 8. Subtract:

$$\begin{array}{r} 50{,}024 \\ -\ 3{,}792 \\ \hline \end{array}$$

[1.5] 9. Multiply:

$$\begin{array}{r} 432 \\ \times\ 160 \\ \hline \end{array}$$

[1.5] 10. Multiply:

$$\begin{array}{r} 481 \\ \times\ 205 \\ \hline \end{array}$$

[1.6] 11. Divide using short division:

$$8\overline{)30{,}214}$$

[1.6] 12. How many seconds are there in 24 hours?

[1.6] In Exercises 13–15, divide:

13. $\frac{0}{5}$ 14. $\frac{0}{0}$ 15. $\frac{14}{0}$

[1.7] 16. Divide using long division:

$$26\overline{)3{,}356}$$

[1.7] 17. If you traveled 756 miles on 28 gallons of gasoline, how many miles per gallon did you get?

[1.8] In Exercises 18–20, evaluate the powers.

18. 10^5 19. 3^3 20. 7^0

[1.9] 21. Evaluate using the correct order of operations:

$$9 - 36 \div 3^2 - (5 - 2)$$

[1.10] In Exercises 22–24 identify the property of whole numbers that is illustrated.

22. $6 \cdot (3 + 4) = 6 \cdot 3 + 6 \cdot 4$

23. $8 + (3 + 5) = (8 + 3) + 5$

24. $1 \cdot 8 = 8$

[1.1] 25. Write a six-digit number with an 8 in the ten thousands place, a 4 in the tens place, a 5 in the units place, and a 7 in each of the other positions.

[1.3–1.4] 26. Mr. Robinson drove from Los Angeles to Chicago. If he drove 420 miles the first day, 365 miles the second day, and 382 miles the third day, how far does he still have to go if the total distance from Los Angeles to Chicago is 1850 miles?

▼ [1.5] 27. How many hours are there in 1 year?

▼ [1.7] 28. If a man earns $20,640 per year, how much does he earn per month?

▼ [1.7] 29. Two hundred forty-six times some number is 19,188. What is the number?

▼ [1.7] 30. If you can fit 47 books on a shelf, how many shelves are necessary if you have 826 books?

▼ [1.5] 31. What number multiplied by 86 will give the same product as 163 multiplied by 430?

▼ [1.3–1.7] 32. Write down 4,617, multiply it by 12, divide the product by 9, add 365 to the quotient, and subtract 5521 from the sum. What is the result?

CRITICAL THINKING EXERCISES—CHAPTER 1

For Exercises 1–5, tell whether each of the following statements is true or false, and give reasons why.

1. If you add two four-digit numbers together, you will always get a four-digit number for an answer.

2. Every whole number divided by itself equals 1.

3. Division of whole numbers is commutative.

4. Any whole number raised to the zero power is equal to zero.

5. Any whole number raised to the zero power is equal to one.

Exercises 6–10 each contains an error. Can you find it?

6.

	$346 \approx 350$	rounded to the nearest ten
and	$350 \approx 400$	rounded to the nearest hundred
Therefore	$346 \approx 400$	rounded to the nearest hundred.

7. $4 + 6 \div 2 = 10 \div 2 = 5$

8. $\dfrac{9}{10} = \dfrac{6+3}{7+3} = \dfrac{6+\cancel{3}^{1}}{7+\cancel{3}_{1}} = \dfrac{7}{8}$

9. $$\begin{aligned} 3 \cdot 3^2 - 5 + 2 &= 3 \cdot 9 - 5 + 2 \\ &= 3 \cdot 4 + 2 \\ &= 12 + 2 \\ &= 14 \end{aligned}$$

10. There are 52 weeks in one year, so there are 52 weeks $\times$ 7 days/week or $52 \times 7 = 364$ days in one year.

11. When rounding to the nearest ten, list all of the two-digit numbers that round to 50.

12. When rounding to the nearest ten, list all of the three-digit numbers that round to 150.

NAME ▮▮▮▮▮ CLASS

ACHIEVEMENT TEST—CHAPTER 1

1. Write a seven-digit number with a 5 in the hundreds place, a 6 in the thousands place, a 4 in the hundred-thousands place, and a 3 in all of the other positions. 1. ____

2. Write 5,084 in words. 2. ____

3. Write sixty-two thousand, eighty-four as a numeral. 3. ____

4. Write 5,280 in expanded notation. 4. ____

▼ Round 61,952 to the nearest:

5. ten 5. ____

6. hundred 6. ____

7. thousand 7. ____

8. Add: 3,684 + 1,881 + 2,609 + 86. 8. ____

9. If a United Fund campaign has raised \$40,307, how much more does it need to reach a goal of \$50,000? 9. ____

10. Multiply: $\begin{array}{r} 639 \\ \times 208 \\ \hline \end{array}$ 10. ____

11. Divide using short division: 11. ____

$$7\overline{)93{,}821}$$

12. Divide: $\frac{0}{0}$. 12. ____

13. Divide: $6\overline{)0}$. 13. ____

14. Divide: $\frac{9}{0}$.

15. How much money does each of 126 employees receive as a bonus, if each receives an equal share of \$22,050?

16. A plane leaves New York bound for Los Angeles with 236 passengers on board. It stops in Chicago, where 47 passengers leave and 58 new passengers board. How many passengers are on board when the plane lands in Los Angeles?

17. Evaluate 49^0.

18. Evaluate 4^4.

19. Simplify using the correct order of operations:

$$\frac{3 \cdot 2^2 - 4}{9 - 7}$$

20. $4 \cdot (3 \cdot 9) = (4 \cdot 3) \cdot 9$ illustrates what property of whole numbers?

14. ____________

15. ____________

16. ____________

17. ____________

18. ____________

19. ____________

20. ____________

____

CHAPTER 2

Fractions

INTRODUCTION

This chapter is all about fractions. We will define them, simplify or reduce them, and then learn to add, subtract, multiply, and divide them. We will also investigate prime numbers and learn how we can put them to work for us. Finally you will see how fractions are used in everyday real-world situations.

CHAPTER 2—NUMBER KNOWLEDGE

Tool sizes are but one of many uses for fractions. The following list is taken from a catalogue. As you can see, there are fractions on it everywhere!

COMBINATION WRENCHES STANDARD SIZES

Size, inches	Length, inches	Wt. lbs.	Each
$\frac{1}{4}$	$3\frac{15}{16}$	0.10	$3.89
$\frac{5}{16}$	$4\frac{3}{8}$	0.10	3.99
$\frac{11}{32}$	$4\frac{13}{16}$	0.10	4.69
$\frac{3}{8}$	$5\frac{1}{4}$	0.10	4.89
$\frac{7}{16}$	$5\frac{13}{16}$	0.10	5.29
$\frac{1}{2}$	$6\frac{1}{2}$	0.20	5.79
$\frac{9}{16}$	$7\frac{3}{16}$	0.20	6.29
$\frac{5}{8}$	8	0.30	6.79
$\frac{11}{16}$	$8\frac{25}{32}$	0.40	7.29
$\frac{3}{4}$	$9\frac{5}{8}$	0.50	7.79
$\frac{13}{16}$	$10\frac{9}{16}$	0.60	8.29
$\frac{7}{8}$	$11\frac{1}{2}$	0.69	8.79
$\frac{15}{16}$	$12\frac{1}{2}$	0.69	10.99
1	$13\frac{1}{2}$	1.00	11.49
$1\frac{1}{16}$	$14\frac{5}{8}$	1.13	13.99
$1\frac{1}{8}$	$15\frac{9}{16}$	1.13	17.99
$1\frac{1}{4}$	$16\frac{7}{8}$	1.13	20.99
$1\frac{5}{16}$	$18\frac{1}{4}$	2.00	23.99

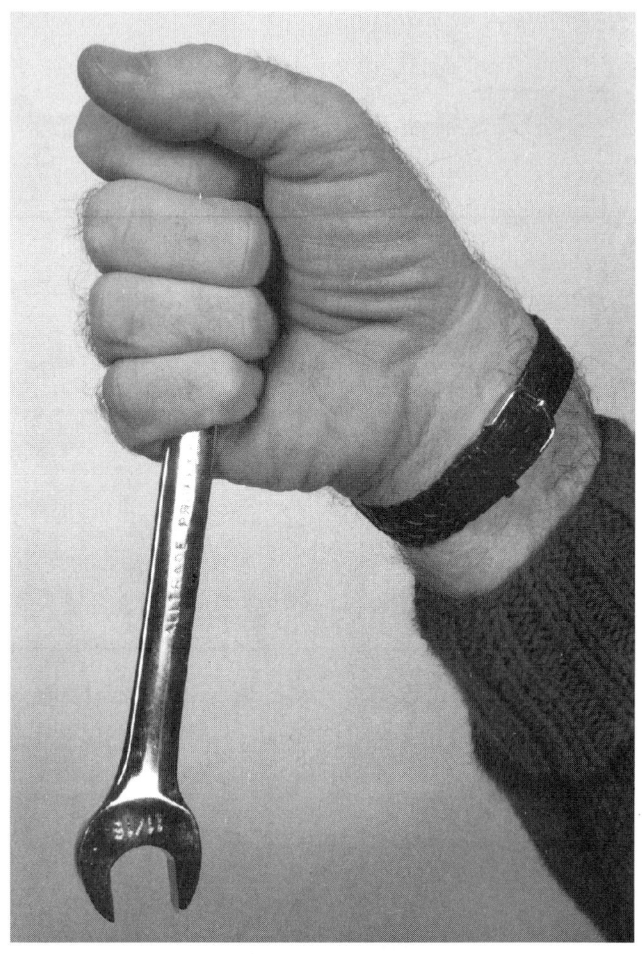

BOX-END WRENCHES STANDARD SIZES

Size, inches	Length, inches	Wt. lbs.	Each
$\frac{3}{8}\times\frac{7}{16}$	$4\frac{3}{4}$	0.10	$7.19
$\frac{1}{2}\times\frac{9}{16}$	$5\frac{1}{2}$	0.20	7.19
$\frac{5}{8}\times\frac{3}{4}$	$6\frac{3}{8}$	0.30	7.19
$\frac{1}{4}\times\frac{5}{16}$	$7\frac{1}{16}$	0.13	5.39
$\frac{3}{8}\times\frac{7}{16}$	$7\frac{3}{4}$	0.20	5.59
$\frac{7}{16}\times\frac{1}{2}$	$8\frac{7}{16}$	0.25	6.19
$\frac{1}{2}\times\frac{9}{16}$	$9\frac{1}{8}$	0.31	6.89
$\frac{9}{16}\times\frac{5}{8}$	$9\frac{3}{4}$	0.40	7.09
$\frac{19}{32}\times\frac{25}{32}$	10	0.50	8.49
$\frac{5}{8}\times\frac{3}{4}$	$10\frac{1}{2}$	0.44	8.49
$\frac{11}{16}\times\frac{13}{16}$	$11\frac{7}{8}$	0.60	9.59
$\frac{3}{4}\times\frac{7}{8}$	$12\frac{5}{8}$	0.70	11.99
$\frac{15}{16}\times1$	$15\frac{3}{8}$	1.00	12.49
$1\frac{1}{16}\times1\frac{1}{4}$	$17\frac{1}{16}$	1.40	20.99
$1\frac{1}{8}\times1\frac{5}{16}$	$18\frac{11}{16}$	2.13	21.99
$1\frac{7}{16}\times1\frac{1}{2}$	$22\frac{3}{32}$	2.50	27.99

OPEN-END WRENCHES STANDARD SIZES

Size, inches	Length, inches	Wt. lbs.	Each
$\frac{1}{4}\times\frac{5}{16}$	$4\frac{1}{2}$	0.05	$4.59
$\frac{3}{8}\times\frac{7}{16}$	$5\frac{3}{16}$	0.10	4.59
$\frac{1}{2}\times\frac{9}{16}$	$6\frac{1}{8}$	0.20	5.99
$\frac{9}{16}\times\frac{5}{8}$	7	0.44	6.29
$\frac{19}{32}\times\frac{11}{16}$	7	0.31	6.49
$\frac{5}{8}\times\frac{3}{4}$	$7\frac{13}{16}$	0.40	6.99
$\frac{11}{16}\times\frac{13}{16}$	$8\frac{5}{8}$	0.50	7.29
$\frac{3}{4}\times\frac{7}{8}$	$9\frac{5}{16}$	0.50	10.49
$\frac{25}{32}\times\frac{13}{16}$	10	0.50	10.99
$\frac{15}{16}\times1$	$10\frac{21}{32}$	0.80	11.49
$1\frac{1}{16}\times1\frac{1}{8}$	12	1.13	13.99
$1\frac{1}{4}\times1\frac{5}{16}$	$14\frac{11}{16}$	1.80	23.49
$1\frac{3}{8}\times1\frac{7}{16}$	$16\frac{1}{4}$	2.10	28.99
$1\frac{1}{2}\times1\frac{5}{8}$	$17\frac{9}{16}$	2.70	29.99

2.1 Prime Numbers

Tests for Divisibility

We know that $3 \times 7 = 21$. Recall that in this expression, 21 is called the product and 3 and 7 are called factors of 21. There are certain tests for divisibility that will be most useful to us in finding the factors of a number.

A number is divisible by:

- 2 if it is even (ends in 0, 2, 4, 6, or 8);
- 3 if the sum of its digits is divisible by 3;
- 5 if it ends in a 5 or a 0.

EXAMPLE 1

(a) 386 is divisible by 2 because it ends in an even number, 6.

(b) 288 is divisible by 3 because the sum of its digits, $2 + 8 + 8$, is 18, which is divisible by 3.

(c) 965 is divisible by 5 because it ends in a 5.

(d) 1162 is not divisible by 3, since $1 + 1 + 6 + 2 = 10$ is not divisible by 3.

Prime Numbers

The number 12 can be written as the product of two factors in several ways:

$$12 = 3 \cdot 4$$

$$12 = 2 \cdot 6$$

$$12 = 1 \cdot 12$$

The number 7 can be written as the product of two factors in exactly one way:

$$7 = 1 \cdot 7$$

Numbers like seven are called *prime numbers* and are defined in the following way:

Prime Numbers

Whole numbers greater than 1 that cannot be written as a product of factors other than 1 and the number itself are called **prime numbers.**

Our list of **prime numbers** begins like this:

2, 3, 5, 7, 11, 13, 17, 19, 23, 29, 31, . . .

Every number in this list can be written as the product of itself and 1 as its *only* factors. There is no largest prime number, and so the list continues on forever.

Composite Numbers

Any whole number greater than 1 that is not a prime number is called a **composite number.** Every composite number can be written as the product of two factors other than 1 and itself.

NOTE The number 1 is considered to be neither prime nor composite.

EXAMPLE 2 Is the number 32 prime or composite?

SOLUTION 32 is a composite number because it can be written $32 = 4 \cdot 8$ or $32 = 2 \cdot 16$, as well as $32 = 1 \cdot 32$.

EXAMPLE 3 Is the number 23 prime or composite?

SOLUTION 23 is a prime number because it cannot be written as the product of factors other than 1 and 23. In other words, $23 = 1 \cdot 23$ and that's it—there are no other factors.

Writing Composite Numbers as a Product of Prime Numbers

Consider the number 18. We know that it is a composite number because it can be written as the product of factors other than 18 and 1. For example, $18 = 2 \cdot 9$ and $18 = 3 \cdot 6$. Notice that some of these factors are prime whereas others are composite. The red factors are prime:

$$18 = 2 \cdot 9$$

$$18 = 3 \cdot 6$$

The other factors are composite and therefore can be written as the product of two factors other than 1 and the number itself. Now let's replace the 9 by its factors $(3 \cdot 3)$, and replace the 6 by its factors $(2 \cdot 3)$.

$$18 = 2 \cdot 9 = 2 \cdot 3 \cdot 3$$

$$18 = 3 \cdot 6 = 3 \cdot 2 \cdot 3$$

In both cases, we find that all the factors are prime numbers. In other words, 18 has been written as the product of prime factors. You should also notice that in both cases the prime factors are the same, two 3s and a 2. The order is different, but that makes no difference to us, since multiplication is commutative. The fact that we obtained the same prime numbers as factors of 18, even though we started in two different ways, $18 = 2 \cdot 9$ and $18 = 3 \cdot 6$, is more than a coincidence and is, in fact, a very important discovery in mathematics. It is true for all composite numbers. **Every composite number can be written as the product of exactly one set of prime numbers.** Remember that we do not care about the order of these prime factors. We can also write the factors using exponents:

$$18 = 2 \cdot 3 \cdot 3 = 2 \cdot 3^2$$

Now that we know that any composite number can be written as the product of prime factors, the next step is to learn how to find them. We shall consider two methods for writing a number as the product of prime factors.

EXAMPLE 4 Write the number 60 as the product of prime factors.

SOLUTION **Method 1**

- **Step 1.** Test to see if the smallest prime number, 2, is a factor of 60. This is easy to see. If the number is even, 2 is a factor. If the number is odd, 2 is not a factor.

$$60 = 2 \cdot 30$$

- **Step 2.** One of the factors, 2, is prime and the other factor, 30, is composite. Test to see if 2 is a factor of 30. Since $30 = 2 \cdot 15$, we replace 30 by $2 \cdot 15$.

$$60 = 2 \cdot 30 = 2 \cdot \overbrace{(2 \cdot 15)}$$

- **Step 3.** Again, test to see if 2 is a factor of 15. Since 15 is odd, 2 is not a factor of it, and we try the next larger prime number, 3. Since $15 = 3 \cdot 5$, 3 is a factor of 15. Again, replace 15 by its equal, $3 \cdot 5$,

$$60 = 2 \cdot 2 \cdot 15 = 2 \cdot 2 \cdot \overbrace{(3 \cdot 5)}$$

- **Step 4.** You can tell when you have completed the job by noticing when all of the factors are prime numbers. Since 5 is a prime number, we are finished. Now 60 has been written as a product of prime factors:

$$\begin{aligned} 60 &= 2 \cdot 2 \cdot 3 \cdot 5 \\ &= 2^2 \cdot 3 \cdot 5 \end{aligned}$$

Method 2

In some cases you may see two factors of the given numbers immediately. For example, suppose that you recognized that $60 = 6 \cdot 10$. You should then proceed as follows:

$$60 = 6 \cdot 10$$

Neither factor is a prime, so we will consider the factors one at a time and write each factor as the product of prime numbers. Notice that $6 = 2 \cdot 3$, so we replace 6 by $2 \cdot 3$, giving

$$60 = 6 \cdot 10 = \overbrace{(2 \cdot 3)} \cdot 10$$

Next, $10 = 2 \cdot 5$, so we replace it, which gives us

$$60 = 2 \cdot 3 \cdot 10 = 2 \cdot 3 \cdot \overbrace{(2 \cdot 5)}$$

Since all our factors are now prime, we're finished.

Instead of recognizing that $60 = 6 \cdot 10$, suppose you saw that $60 = 4 \cdot 15$. Proceed in exactly the same way. Since $4 = 2 \cdot 2$, we have

$$60 = 4 \cdot 15 = \overbrace{(2 \cdot 2)} \cdot 15$$

and replacing 15 by $3 \cdot 5$ gives us

$$60 = 2 \cdot 2 \cdot 15 = 2 \cdot 2 \cdot \overbrace{(3 \cdot 5)}$$

It is important to notice that all the methods give us exactly the same set of prime factors:

$$60 = 2 \cdot 2 \cdot 3 \cdot 5 = 2^2 \cdot 3 \cdot 5$$

EXAMPLE 5 Write 340 as the product of prime numbers.

SOLUTION We test to see if 2 is a factor. Since $340 \div 2$ is 170,

$$340 = 2 \cdot 170$$

We test with 2 again. Since $170 = 2 \cdot 85$, we replace 170 by $2 \cdot 85$.

$$\begin{aligned} 340 &= 2 \cdot 170 \\ &= 2 \cdot 2 \cdot 85 \end{aligned}$$

Tests with 2 and 3 show that neither 2 nor 3 is a factor of 85, but 5 is. (Recall that 5 will always be a factor of a number ending in 0 or 5.) Thus $85 = 5 \cdot 17$, which gives us

$$340 = 2 \cdot 2 \cdot 5 \cdot 17 = 2^2 \cdot 5 \cdot 17$$

Since all the factors are prime, we are finished.

EXAMPLE 6 Write the prime factorization of 100.

SOLUTION Using Method 2, we recognize that

$$100 = 10 \cdot 10$$

Since we know that $10 = 2 \cdot 5$, we replace each 10 by $2 \cdot 5$:

$$100 = 10 \cdot 10 = (2 \cdot 5) \cdot (2 \cdot 5)$$

and our answer is

$$100 = 2 \cdot 5 \cdot 2 \cdot 5$$

Although there is nothing incorrect about our answer in its present form, it is usually accepted practice to arrange the order of prime factors from smallest to largest as we go from left to right. This would give us $100 = 2 \cdot 2 \cdot 5 \cdot 5$ or $2^2 \cdot 5^2$. ❚❚

When we try to find the prime factors of a number, we need not try any prime factors with a square greater than the given number. This is illustrated in the next example.

EXAMPLE 7 Write 97 as a product of prime factors.

SOLUTION

- 2 is not a factor of 97
- 3 is not a factor of 97
- 5 is not a factor of 97
- 7 is not a factor of 97

We do not need to try to divide 97 by the next prime number 11, since $11^2 = 121$ is greater than 97. We conclude that 97 is a prime number, since it has no factors other than 1 and 97. ❚❚

2.1 Exercises

▼ Test each of the numbers for divisibility by 2, 3, and 5 using the appropriate test for each.

1. 18 **2.** 30 **3.** 19

4. 45 **5.** 6,001 **6.** 14,680

7. 6,000 **8.** 3,120 **9.** 724,137

10. 352 **11.** 7,811 **12.** 9,240

▼ Determine whether each of the following numbers is prime or composite.

13. 6 **14.** 1 **15.** 21

16. 7 **17.** 18 **18.** 29

19. 17 **20.** 30 **21.** 27

22. 58 **23.** 37 **24.** 87

▼ Write each of the following numbers as the product of prime factors using either of the methods that you have learned. Some of the numbers may be prime already.

25. 18 **26.** 50 **27.** 98

28. 62 **29.** 32 **30.** 36

31. 180 **32.** 56 **33.** 200

34. 59 **35.** 46 **36.** 64

37. 77 **38.** 88

LIVING IN THE REAL WORLD

39. A 12-ounce (oz.) soft drink can contains 355 milliliters (ml). Write each of these numbers as a product of prime factors.

40. A coffee cup holds about 300 ml. Write this number as a product of prime numbers.

41. At sea level, the boiling point of water is 212° F and the freezing point is 32° F. Express each of these numbers as a product of prime numbers.

42. Write 5,280 as a product of prime factors. Do you recognize this as the number of feet in one mile?

MENTAL MATHEMATICS

▼ Do the following exercises in your head without using pencil or paper.

▼ Test the following numbers for divisibility by 2, 3, and 5.

43. 26 **44.** 40 **45.** 45 **46.** 62

47. 23 **48.** 53 **49.** 72 **50.** 128

51. 621 **52.** 420 **53.** 500 **54.** 720

▼ Which of the following numbers are prime numbers?

55. 9 **56.** 26 **57.** 13

58. 31 **59.** 35 **60.** 56

61. 19 **62.** 28 **63.** 63

64. 41 **65.** 49 **66.** 81

67. 150 **68.** 260 **69.** 123

70. 321

71. Is it possible for the product of two prime numbers to be a prime number? Explain your answer.

72. Is it possible for the sum of two prime numbers to be a prime number?

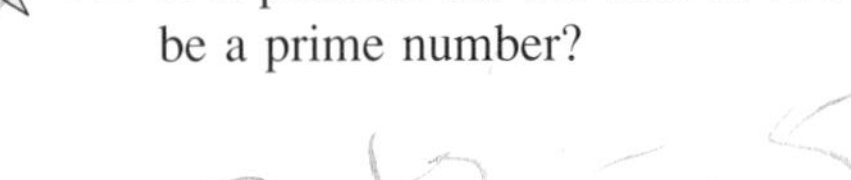

73. Is it possible for the sum of two composite numbers to be a prime number?

74. What is the only *even* prime number? Why can't any other prime number be even?

2.2 Definition of Fractions

When something is divided into equal parts, the number expressing the relation of one or more of these parts to the total number of parts is called a **fraction.** For example, if we divide a box into 8 equal parts and consider 5 of these equal parts, we are talking about the fraction five-eighths, which is written $\frac{5}{8}$.

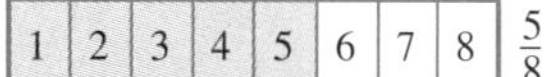

The top number of a fraction, in this case 5, is called the **numerator** and the bottom number, 8, is called the **denominator.** If you remember that *d* stands for **down** and **denominator,** you won't confuse the meaning of numerator and denominator.

The denominator of the fraction tells us how many equal parts something is divided into. The numerator tells how many of these equal parts we are interested in. For example, the fraction $\frac{1}{6}$ means that we divided something into 6 equal parts and we're considering one of these parts. If we cut a pie into 6 equal parts and ate one of them, we would have eaten $\frac{1}{6}$ of the pie. Then 5 of the 6 pieces would still remain, or $\frac{5}{6}$ of the pie would be left uneaten.

> **A proper fraction** is a fraction in which the numerator is less than the denominator. An **improper fraction** is one in which the numerator is equal to or greater than the denominator.

EXAMPLE 1 The fractions $\frac{1}{2}$, $\frac{2}{3}$, and $\frac{15}{16}$ are all proper fractions. The fractions $\frac{9}{4}$, $\frac{7}{7}$, and $\frac{5}{1}$ are all improper fractions. Every whole number can be written as a fraction by using the number 1 as the denominator.

$$0 = \frac{0}{1}, \quad 1 = \frac{1}{1}, \quad 2 = \frac{2}{1}, \quad \text{and so on.}$$

Equivalent Fractions

Consider the fractions $\frac{3}{4}$, $\frac{6}{8}$, and $\frac{12}{16}$. Each of these is represented in the following diagram:

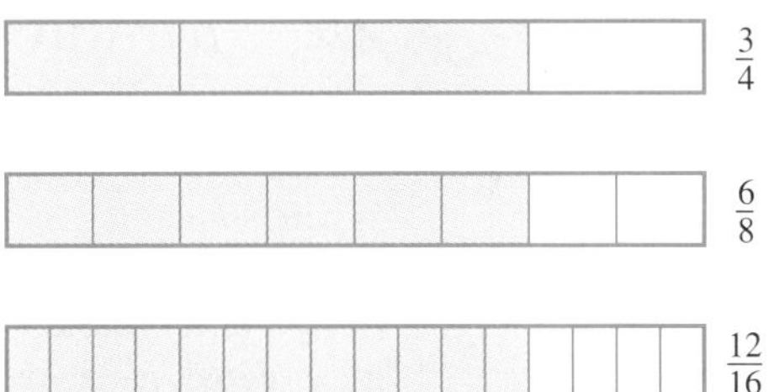

According to the diagram, each fraction represents the same amount. We say that $\frac{3}{4}$, $\frac{6}{8}$, and $\frac{12}{16}$ are all different names for the same number or that they are **equivalent** to each other. If we examine these equivalent fractions more closely, we shall find some interesting relationships between them. If we multiply both the numerator and denominator of the fraction $\frac{3}{4}$ by 2, we get $\frac{6}{8}$.

$$\frac{3 \cdot 2}{4 \cdot 2} = \frac{6}{8}$$

What we really did was to multiply the fraction $\frac{3}{4}$ by the number 1, because we multiplied $\frac{3}{4}$ by $\frac{2}{2}$, and $\frac{2}{2}$ is another way of writing 1. Since any number multiplied by 1 gives you back the number that you started with, we should expect that $\frac{3}{4} \cdot 1$, or $\frac{3}{4} \cdot \frac{2}{2} = \frac{6}{8}$, will be equivalent to $\frac{3}{4}$.

Suppose that we multiply $\frac{3}{4}$ by $\frac{4}{4}$. Again, we are multiplying by a special form of the number 1, so we should expect our answer to be a fraction that is equivalent to $\frac{3}{4}$. Our diagram shows that, indeed,

$$\frac{3}{4} = \frac{3 \cdot 4}{4 \cdot 4} = \frac{12}{16}$$

So $\frac{3}{4}$ and $\frac{12}{16}$ are equivalent fractions. All of these examples should suggest a rule for finding fractions that are equivalent to a given fraction.

EXAMPLE 2 Name three other fractions equivalent to the fraction $\frac{5}{8}$.

The Fundamental Principal of Equivalent Fractions

If both the numerator and the denominator of a fraction are multiplied by the same nonzero number, the result will be an equivalent fraction.

SOLUTION Multiplying numerator and denominator of the fraction $\frac{5}{8}$ by the same number will yield an equivalent fraction. For example, $\frac{5}{8} \cdot \frac{2}{2} = \frac{10}{16}$, $\frac{5}{8} \cdot \frac{5}{5} = \frac{25}{40}$, and $\frac{5}{8} \cdot \frac{3}{3} = \frac{15}{24}$. Therefore $\frac{5}{8}$, $\frac{10}{16}$, $\frac{25}{40}$, and $\frac{15}{24}$ are equivalent fractions. Of course, there are many more fractions that are equivalent to the fraction $\frac{5}{8}$. ❚❚

2.2 Exercises

▼ Name the fraction given by each of the following diagrams.

1.

2.

3. 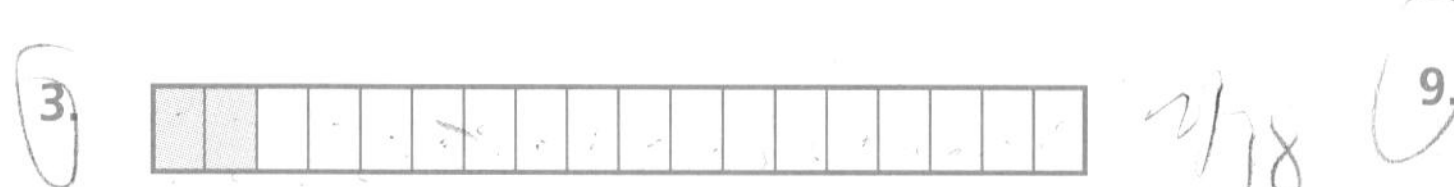

4.

5. 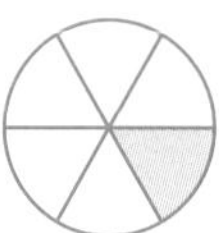

6. 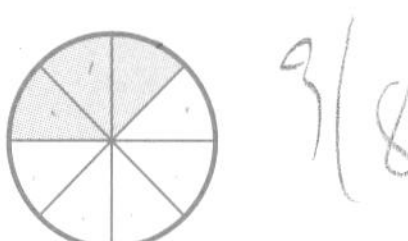

7. 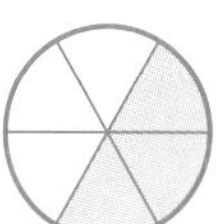

8. 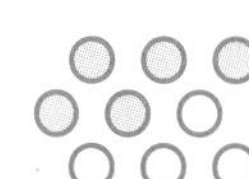

9. 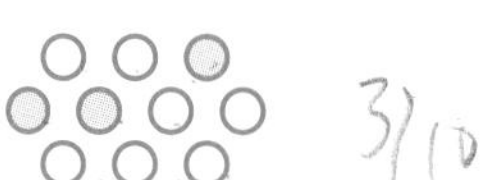

10. 

▼ Shade the second figure to make it represent a fraction that is equivalent to the fraction given by the first figure. Name each of the equivalent fractions.

11. 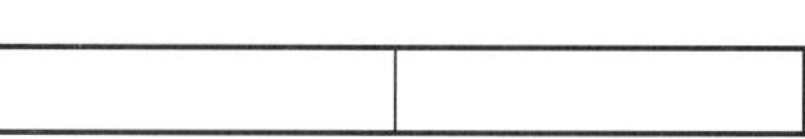

12.

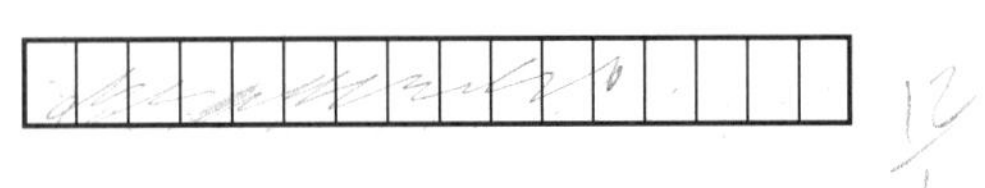

13.

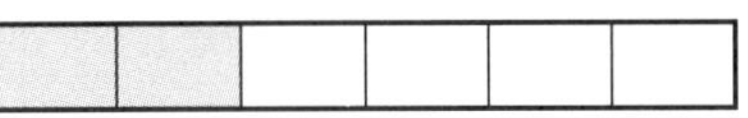

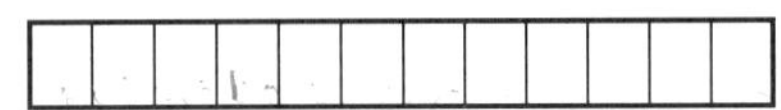

14.

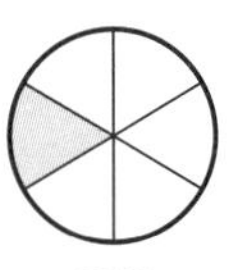

15.

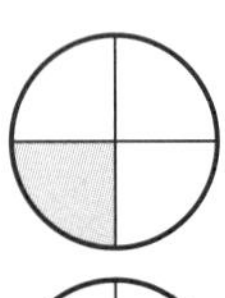

16.

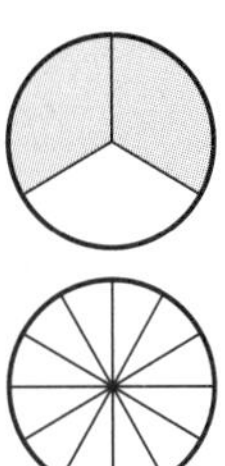

▼ Give any three other fractions equivalent to each of the given fractions.

17. $\frac{1}{2}$

18. $\frac{5}{6}$

19. $\frac{3}{8}$

20. $\frac{9}{10}$

21. $\frac{1}{4}$

22. $\frac{7}{8}$

23. $\frac{6}{11}$

24. $\frac{4}{3}$

25. $\frac{8}{7}$

26. $\frac{7}{4}$

27. $\frac{3}{10}$

28. $\frac{5}{12}$

29. $\frac{16}{5}$

30. $\frac{5}{16}$

MENTAL MATHEMATICS

▼ Without using paper or pencil, give *any* fraction that is equivalent to the given fraction.

31. $\frac{2}{3}$

32. $\frac{4}{5}$

33. $\frac{1}{8}$

34. $\frac{3}{5}$

35. $\frac{2}{7}$

36. $\frac{5}{8}$

37. $\frac{10}{7}$

38. $\frac{5}{4}$

39. $\frac{8}{9}$

40. $\frac{1}{15}$

41. $\frac{4}{15}$

42. $\frac{1}{25}$

2.3 Simplifying Fractions

The fractions $\frac{1}{2}$ and $\frac{4}{8}$ are equivalent fractions, because multiplying numerator and denominator of $\frac{1}{2}$ by the number 4 gives us $\frac{4}{8}$. It seems reasonable that we can go back the other way by reversing the process. That is, if we divide both numerator and denominator of the fraction $\frac{4}{8}$ by 4, we get the equivalent fraction $\frac{1}{2}$:

$$\frac{4 \div 4}{8 \div 4} = \frac{1}{2}$$

This leads us to another rule about equivalent fractions.

> If both numerator and denominator of a fraction are divided by the same nonzero number, **the result will be an equivalent fraction.**

EXAMPLE 1 $\frac{14}{21}$ is equivalent to $\frac{2}{3}$, because $\frac{14 \div 7}{21 \div 7} = \frac{2}{3}$.

If we are given a fraction such as $\frac{5}{8}$, we find that there is no number other than 1 that will divide both 5 and 8 evenly. Such a fraction is said to be **reduced to lowest terms** or **in simplest form.**

Sometimes it is not so easy at first glance to decide whether a fraction is in simplest form. If we use our previous knowledge of prime numbers, our task becomes simpler.

Suppose that we are asked to determine whether the fraction $\frac{42}{66}$ is in simplest form and, if not, to reduce it to lowest terms. First, we write both the numerator and the denominator of the given fraction as a product of prime factors:

$$\frac{42}{66} = \frac{2 \cdot 3 \cdot 7}{2 \cdot 3 \cdot 11}$$

Next, we look to see if any prime factors are included in **both** the numerator and the denominator. In our example we find a 2 and a 3 in both numerator and denominator, so our fraction is not yet in simplest form. To reduce it to lowest terms, we must divide numerator and denominator by each of these prime factors. Then the resulting fraction will be in lowest terms.

$$\frac{42}{66} = \frac{\overset{1}{\cancel{2}} \cdot \overset{1}{\cancel{3}} \cdot 7}{\underset{1}{\cancel{2}} \cdot \underset{1}{\cancel{3}} \cdot 11} = \frac{7}{11}$$

So the simplest form of $\frac{42}{66}$ is $\frac{7}{11}$.

This process is called **simplifying the fraction** or **reducing it to lowest terms.**

To reduce a fraction to lowest terms:

1. divide both numerator and denominator by obvious whole number divisors;
2. if there are no obvious common divisors, write both numerator and denominator as a product of prime factors;
3. divide numerator and denominator by any common factors.

EXAMPLE 2 Reduce the fraction $\frac{6}{16}$ to lowest terms.

SOLUTION We write the numerator and the denominator as products of prime factors, and then we divide both the numerator and the denominator by any prime factors found in **both** the numerator and the denominator.

$$\frac{6}{16} = \frac{\overset{1}{\cancel{2}} \cdot 3}{\underset{1}{\cancel{2}} \cdot 2 \cdot 2 \cdot 2} = \frac{3}{2 \cdot 2 \cdot 2} = \frac{3}{8}$$

▮▮

STOP The slashes in Example 2 indicate that both numerator and denominator have been divided by 2. Frequently the 1s are not written when this is done. A common error is to leave the numerator out of the answer or to call it zero.

$$\frac{3}{12} = \frac{\cancel{3}}{2 \cdot 2 \cdot \cancel{3}} = \frac{1}{4}$$

This 1 is important. Do not leave it out or call it zero.

EXAMPLE 3 Simplify: $\frac{18}{24}$.

SOLUTION $\frac{18}{24} = \frac{\overset{1}{\cancel{2}} \cdot 3 \cdot \overset{1}{\cancel{3}}}{\underset{1}{\cancel{2}} \cdot 2 \cdot 2 \cdot \underset{1}{\cancel{3}}} = \frac{3}{2 \cdot 2} = \frac{3}{4}$

You can save yourself some time if you realize immediately that 6 divides both 18 and 24:

$$\frac{\overset{3}{\cancel{18}}}{\underset{4}{\cancel{24}}} = \frac{3}{4}$$

▮▮

EXAMPLE 4 Simplify: $\frac{8}{105}$

SOLUTION $\frac{8}{105} = \frac{2 \cdot 2 \cdot 2}{3 \cdot 5 \cdot 7}$

Since there are no prime factors common to both numerator and denominator, we conclude that the fraction $\frac{8}{105}$ is already in simplest form.

2.3 Exercises

In Exercise 1–12, reduce each fraction to lowest terms without writing out the prime factors for the numerator and denominator. Some fractions may already be in simplest form.

1. $\frac{9}{45}$
2. $\frac{3}{24}$
3. $\frac{25}{35}$
4. $\frac{6}{10}$
5. $\frac{5}{12}$
6. $\frac{27}{28}$
7. $\frac{11}{77}$
8. $\frac{18}{36}$
9. $\frac{42}{49}$
10. $\frac{25}{125}$
11. $\frac{16}{48}$
12. $\frac{18}{24}$

In Exercises 13–24, reduce each fraction to lowest terms using the prime factorization method. Some fractions may already be in simplest form.

13. $\frac{75}{125}$
14. $\frac{88}{33}$
15. $\frac{56}{14}$
16. $\frac{34}{17}$
17. $\frac{32}{18}$
18. $\frac{35}{12}$
19. $\frac{27}{81}$
20. $\frac{21}{36}$
21. $\frac{12}{16}$
22. $\frac{23}{32}$
23. $\frac{24}{25}$
24. $\frac{28}{35}$
25. $\frac{36}{16}$
26. $\frac{20}{18}$
27. $\frac{28}{87}$
28. $\frac{8}{27}$
29. $\frac{28}{98}$
30. $\frac{49}{70}$
31. $\frac{64}{48}$
32. $\frac{19}{36}$
33. $\frac{28}{15}$
34. $\frac{66}{55}$
35. $\frac{18}{12}$
36. $\frac{100}{75}$

LIVING IN THE REAL WORLD

37. If a recipe calls for $\frac{12}{18}$ of a cup of sugar, you would be suspicious that an error had occurred. Why? How should the fraction $\frac{12}{18}$ be written?

38. Which of the fractions $\frac{8}{12}, \frac{10}{18}, \frac{6}{9}$, and $\frac{14}{21}$ does not reduce to $\frac{2}{3}$?

39. Which of the fractions $\frac{10}{16}, \frac{25}{32}, \frac{40}{64}$, and $\frac{15}{24}$ does not reduce to $\frac{5}{8}$?

40. What fraction of an hour is 15 minutes? Reduce your answer to lowest terms.

41. What fraction of an hour is 20 minutes? Reduce your answer to lowest terms.

42. What fraction of a 24-hour day is 4 hours? Reduce your answer to lowest terms.

43. If you sleep for 8 hours, what fraction of a 24-hour day have you slept? Reduce your answer to lowest terms.

44. Mandy bought a used car for \$800. In the first year her insurance cost \$1200, and she spent \$400 for gas and repairs. What fraction of her total expense, in lowest terms, was the cost of the car?

MENTAL MATHEMATICS

▼ Reduce each fraction to lowest terms mentally, without using pencil or paper.

45. $\frac{6}{8}$ 46. $\frac{3}{9}$ 47. $\frac{5}{10}$

48. $\frac{7}{14}$ 49. $\frac{8}{24}$ 50. $\frac{9}{36}$

51. $\frac{5}{35}$ 52. $\frac{9}{24}$ 53. $\frac{15}{10}$

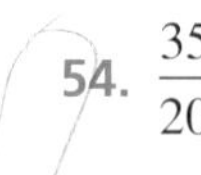

54. $\frac{35}{20}$ 55. $\frac{16}{24}$ 56. $\frac{18}{45}$

57. $\frac{9}{16}$ 58. $\frac{14}{45}$ 59. $\frac{26}{13}$

60. $\frac{45}{90}$

61. What's wrong with the following?

$$\frac{5}{6} = \frac{\cancel{3} + 2}{\cancel{3} + 3} = \frac{2}{3}$$

What is the correct answer?

62. What's wrong with the following?

$$\frac{4}{12} = \frac{\cancel{2} \cdot \cancel{2}}{\cancel{2} \cdot \cancel{2} \cdot 3} = \frac{0}{3} = 0$$

What is the correct answer?

63. Explain how to reduce a fraction to lowest terms.

2.4 Multiplication of Fractions

The procedure for multiplying two fractions is quite simple: **multiply the two numerators and multiply the two denominators.** For example, to multiply $\frac{3}{8} \cdot \frac{2}{3}$, multiply the numerators, 3 and 2, and write the result over the product of the denominators, 8 and 3. The answer must then be reduced to lowest terms:

$$\frac{3}{8} \cdot \frac{2}{3} = \frac{3 \cdot 2}{8 \cdot 3} = \frac{6}{24} = \frac{1}{4}$$

Since both 6 and 24 contain the factor 6, our final answer is $\frac{1}{4}$.

We can eliminate the need to reduce our answer if we divide out **all** the factors that appear in **any** numerator and **any** denominator before we multiply. Let's illustrate using our previous example. The first denominator, 8, and the second numerator, 2, each contain a factor of 2, so we divide each of them by 2:

$$\frac{3}{\cancel{8}_4} \cdot \frac{\cancel{2}^1}{3} = \frac{\cancel{3}^1}{4} \cdot \frac{1}{\cancel{3}_1} = \frac{1 \cdot 1}{4 \cdot 1} = \frac{1}{4}$$

The first numerator and the second denominator both contain a factor of 3, since they are both equal to 3. We divide each of the 3s by 3. Since no more factors appear in both a numerator and a denominator (except 1), we multiply the numerators and we multiply the denominators to get the final answer, $\frac{1}{4}$. You should always remove the common factors before you multiply.

Stated formally; we have the following rule:

To multiply two fractions:

Divide out all like factors of any numerator with any denominator. Then multiply the resulting numerators together and multiply the resulting denominators together. The answer will automatically be in lowest terms.

EXAMPLE 1 Multiply: $\frac{3}{20} \cdot \frac{4}{9}$.

SOLUTION We divide the first numerator and the second denominator by 3. We usually cross out in the original problem rather than rewriting the entire problem, like this:

$$\frac{\cancel{3}^1}{20} \cdot \frac{4}{\cancel{9}_3}$$

Now we divide the 20 and the 4 by 4 and we have

$$\frac{\cancel{3}^1}{\cancel{20}_5} \cdot \frac{\cancel{4}^1}{\cancel{9}_3} = \frac{1 \cdot 1}{5 \cdot 3} = \frac{1}{15}$$

EXAMPLE 2 Multiply: $\frac{7}{16} \cdot \frac{12}{15}$.

SOLUTION We divide the 16 and the 12 by 4, giving

$$\frac{7}{\underset{4}{\cancel{16}}} \cdot \frac{\overset{3}{\cancel{12}}}{15}$$

The new second numerator, 3, and the second denominator, 15, each contain a factor of 3. Dividing each by 3 gives us

$$\frac{7}{\underset{4}{\cancel{16}}} \cdot \frac{\overset{\overset{1}{\cancel{3}}}{\cancel{12}}}{\underset{5}{\cancel{15}}}$$

Finally, multiplying yields

$$\frac{7 \cdot 1}{4 \cdot 5} = \frac{7}{20}$$

EXAMPLE 3 Multiply: $\frac{6}{7} \cdot \frac{3}{5}$.

SOLUTION Examination of the problem reveals that neither numerator has any factors in common with either denominator. Therefore we multiply immediately:

$$\frac{6}{7} \cdot \frac{3}{5} = \frac{6 \cdot 3}{7 \cdot 5} = \frac{18}{35}$$

EXAMPLE 4 Multiply: $\frac{6}{8} \cdot \frac{3}{7}$.

SOLUTION $\frac{\overset{3}{\cancel{6}}}{\underset{4}{\cancel{8}}} \cdot \frac{3}{7} = \frac{9}{28}$

EXAMPLE 5 Multiply: $\frac{12}{32} \cdot \frac{18}{24}$.

SOLUTION $\frac{\overset{1}{\cancel{12}}}{\underset{16}{\cancel{32}}} \cdot \frac{\overset{9}{\cancel{18}}}{\underset{2}{\cancel{24}}} = \frac{9}{32}$

There are many different ways in which the numerators and denominators in this problem can be divided by a common divisor. However, the same answer, $\frac{9}{32}$, will result, regardless of the order you choose.

We can also choose to use prime factors, as illustrated in the following example.

EXAMPLE 6 Multiply: $\frac{36}{14} \cdot \frac{21}{18}$.

SOLUTION $\frac{36}{14} \cdot \frac{21}{18} = \frac{\overset{1}{\cancel{2}} \cdot \overset{1}{\cancel{2}} \cdot \overset{1}{\cancel{3}} \cdot \overset{1}{\cancel{3}}}{\underset{1}{\cancel{2}} \cdot \underset{1}{\cancel{7}}} \cdot \frac{3 \cdot \overset{1}{\cancel{7}}}{\underset{1}{\cancel{2}} \cdot \underset{1}{\cancel{3}} \cdot \underset{1}{\cancel{3}}} = \frac{3}{1} = 3$

EXAMPLE 7 Multiply: $\frac{3}{4} \cdot \frac{18}{7} \cdot \frac{28}{36} \cdot \frac{12}{16}$.

SOLUTION Any numerator and any denominator can be divided by a common divisor.

$$\frac{\overset{1}{\cancel{3}}}{\underset{1}{\cancel{4}}} \cdot \frac{\overset{9}{\cancel{18}}}{\underset{1}{\cancel{7}}} \cdot \frac{\overset{\overset{1}{\cancel{7}}}{\cancel{28}}}{\underset{\underset{1}{\cancel{3}}}{\cancel{36}}} \cdot \frac{\overset{1}{\cancel{12}}}{\underset{8}{\cancel{16}}} = \frac{9}{8}$$

2.4 Exercises

Multiply. Be sure to divide out all common factors before multiplying.

1. $\frac{3}{8} \cdot \frac{4}{6}$

2. $\frac{5}{27} \cdot \frac{9}{10}$

3. $\frac{2}{7} \cdot \frac{3}{9}$

4. $\frac{14}{16} \cdot \frac{4}{35}$

5. $\frac{3}{26} \cdot \frac{13}{18}$

6. $\frac{15}{8} \cdot \frac{3}{5}$

7. $\frac{24}{16} \cdot \frac{4}{36}$

8. $\frac{5}{8} \cdot \frac{1}{6}$

9. $\frac{14}{15} \cdot \frac{10}{21}$

10. $\frac{16}{30} \cdot \frac{9}{12}$

11. $\frac{6}{8} \cdot \frac{30}{48}$

12. $\frac{5}{7} \cdot \frac{14}{15}$

13. $\frac{1}{2} \cdot \frac{1}{2}$

14. $\frac{45}{49} \cdot \frac{14}{63}$

15. $\frac{35}{36} \cdot \frac{4}{7}$

16. $\frac{32}{9} \cdot \frac{45}{8}$

17. $8 \cdot \frac{5}{6}$ $\left(\text{write 8 as } \frac{8}{1}\right)$

18. $6 \cdot \frac{5}{9}$

19. $\frac{5}{8} \cdot 16$

20. $\frac{2}{3} \cdot 24$

21. $\frac{2}{7} \cdot \frac{7}{16} \cdot \frac{4}{5}$

22. $\frac{18}{16} \cdot \frac{24}{9} \cdot \frac{1}{32}$

23. $\frac{3}{4} \cdot \frac{12}{32} \cdot \frac{24}{18}$

24. $\frac{12}{8} \cdot \frac{7}{15} \cdot \frac{8}{21}$

25. $\frac{7}{8} \cdot \frac{25}{35} \cdot \frac{16}{5}$

26. $\frac{15}{14} \cdot \frac{55}{56} \cdot \frac{8}{45}$

27. $\frac{26}{14} \cdot \frac{35}{39} \cdot \frac{3}{2}$

28. $\frac{12}{30} \cdot \frac{6}{15} \cdot \frac{75}{9}$

29. $\frac{6}{8} \cdot \frac{2}{3} \cdot \frac{1}{2} \cdot \frac{1}{3}$

30. $\frac{5}{8} \cdot \frac{7}{6} \cdot \frac{2}{5} \cdot \frac{3}{10}$

31. $\frac{3}{4} \cdot \frac{6}{9} \cdot \frac{5}{8} \cdot \frac{3}{12}$

32. $\frac{6}{8} \cdot \frac{4}{9} \cdot 8 \cdot \frac{3}{12}$

33. $\frac{5}{8} \cdot 9 \cdot \frac{2}{3} \cdot \frac{8}{9} \cdot \frac{3}{5} \cdot \frac{1}{2}$

34. $\frac{12}{16} \cdot \frac{32}{8} \cdot \frac{2}{3} \cdot \frac{14}{15} \cdot \frac{20}{24} \cdot 6$

35. Multiply $\frac{15}{16}$ by $\frac{4}{5}$

36. Find $\frac{3}{8}$ of $\frac{2}{3}$

37. Find the product of $\frac{9}{16}$ and $\frac{4}{3}$.

38. What is the product of $\frac{24}{25}$ and $\frac{5}{8}$?

LIVING IN THE REAL WORLD

39. There are 270 students in the graduation class of Middletown High School, and $\frac{3}{5}$ of the students are boys. How many boys are in the class?

40. If $\frac{2}{7}$ of a shipment of 1792 light bulbs arrived damaged, how many light bulbs were damaged?

41. If a recipe calls for $\frac{3}{4}$ of a cup of sugar and you are making only $\frac{1}{2}$ of the recipe, how much sugar should you use?

42. If you plan to triple a recipe that calls for $\frac{2}{3}$ of a cup of oil, how much oil will you need?

43. Max and Lucy order a pizza, half of it plain for Max and half pepperoni for Lucy. If Max eats half of his portion, what part of the pizza did he eat? If the pizza was cut into 8 slices, how many slices did he eat?

44. Three out of every five students attending North County Community College own their own car. If there are 2500 students at this college, how many of them own their own car?

45. If 8 out of every 10 students at Orange County Community College own a television set and the total number of students at the college is 4550, how many students own a television?

46. There are 220 students in the graduating class at Monroe High School. If 6 out of every 10 students are female, how many females are graduating?

2.5 Division of Fractions

The Reciprocal of a Fraction

The **reciprocal** of a fraction is found by turning it upside down, or **inverting** it. For example, the reciprocal of $\frac{2}{3}$ is $\frac{3}{2}$. The reciprocal of a whole number such as 9 can be found by first writing it as a fraction, $\frac{9}{1}$. Then the reciprocal of $\frac{9}{1}$ is $\frac{1}{9}$. In symbols, the reciprocal of $\frac{a}{b}$ is $\frac{b}{a}$. Remember that 0 has no reciprocal, since $\frac{1}{0}$ is undefined. Division of fractions makes use of reciprocals, as you will see.

Division of Fractions

If you have learned multiplication of fractions well, then you will have no problem whatsoever with division of fractions. It involves only one additional step.

To divide one fraction by another, multiply the first fraction (the dividend) by the reciprocal of the second fraction (the divisor).

EXAMPLE 1 Divide: $\frac{3}{4} \div \frac{3}{2}$.

SOLUTION We multiply $\frac{3}{4}$ by $\frac{2}{3}$, which is the reciprocal of the second fraction:

$$\frac{3}{4} \div \frac{3}{2} = \frac{3}{4} \cdot \frac{2}{3}$$

Now using the rules that we learned for multiplication of fractions gives us

$$\frac{3}{4} \div \frac{3}{2} = \frac{\overset{1}{\cancel{3}}}{\underset{2}{\cancel{4}}} \cdot \frac{\overset{1}{\cancel{2}}}{\underset{1}{\cancel{3}}} = \frac{1}{2}$$

That's all there is to it! ▮▮

EXAMPLE 2 Divide: $\frac{1}{2} \div \frac{5}{8}$.

SOLUTION We invert the second fraction, $\frac{5}{8}$, and multiply:

$$\frac{1}{2} \div \frac{5}{8} = \frac{1}{\underset{1}{\cancel{2}}} \cdot \frac{\overset{4}{\cancel{8}}}{5} = \frac{4}{5}$$ ▮▮

EXAMPLE 3 Divide: $\frac{3}{5} \div \frac{7}{5}$.

SOLUTION $\frac{3}{5} \div \frac{7}{5} = \frac{3}{\underset{1}{\cancel{5}}} \cdot \frac{\overset{1}{\cancel{5}}}{7} = \frac{3}{7}$ ▮▮

EXAMPLE 4 Divide: $\frac{6}{7} \div \frac{3}{5}$.

SOLUTION $\frac{6}{7} \div \frac{3}{5} = \frac{\overset{2}{\cancel{6}}}{7} \cdot \frac{5}{\underset{1}{\cancel{3}}} = \frac{10}{7}$ ▮▮

Now that we've learned the method, let's take a minute to convince ourselves that it should work. Suppose that we consider the problem $\frac{3}{4} \div \frac{3}{2}$ again. Another way that we could write this would be $\frac{\frac{3}{4}}{\frac{3}{2}}$. Now, multiply both the numerator, $\frac{3}{4}$, and the denominator, $\frac{3}{2}$, by the reciprocal of $\frac{3}{2}$, which is $\frac{2}{3}$. Remember that we're multiplying our fraction by $\frac{\frac{2}{3}}{\frac{2}{3}}$, which is a special way of writing the number 1, so our answer should be equivalent to our original fraction.

$$\frac{\frac{3}{4}\cdot\frac{2}{3}}{\frac{3}{2}\cdot\frac{2}{3}} = \frac{\frac{\overset{1}{\cancel{3}}}{\underset{2}{\cancel{4}}}\cdot\frac{\overset{1}{\cancel{2}}}{\underset{1}{\cancel{3}}}}{\frac{\overset{1}{\cancel{3}}}{\underset{1}{\cancel{2}}}\cdot\frac{\overset{1}{\cancel{2}}}{\underset{1}{\cancel{3}}}} = \frac{\frac{1}{2}}{1} = \frac{1}{2}$$

Observe what happens in the denominator: we get 1. Since any number divided by 1 gives us back the number, our answer is equal to the numerator. But what is the numerator? It is equal to $\frac{3}{5}\cdot\frac{2}{3}$, which is what we get when we are given $\frac{3}{4} \div \frac{3}{2}$ and invert the divisor.

This argument tells us that our method works. Now that we know it, all we need to do is to remember the rule. Let's state it formally:

To divide one fraction by another:

Multiply the first fraction by the **reciprocal** of the second fraction.

2.5 Exercises

▼ Divide.

1. $\frac{3}{2} \div \frac{4}{8}$
2. $\frac{1}{6} \div \frac{2}{3}$
3. $\frac{5}{8} \div \frac{3}{4}$
4. $\frac{3}{8} \div \frac{3}{8}$
5. $\frac{9}{16} \div \frac{4}{3}$
6. $\frac{7}{8} \div \frac{4}{32}$
7. $\frac{2}{5} \div \frac{5}{2}$
8. $\frac{1}{1} \div \frac{7}{8}$
9. $\frac{7}{8} \div \frac{1}{1}$
10. $\frac{4}{8} \div \frac{3}{4}$
11. $\frac{20}{3} \div \frac{15}{9}$
12. $\frac{24}{5} \div \frac{3}{10}$
13. $18 \div \frac{2}{3}$
14. $16 \div \frac{8}{9}$
15. $\frac{5}{8} \div 10$
16. $\frac{3}{16} \div 12$

17. $\dfrac{7}{16} \div \dfrac{14}{4}$

18. $\dfrac{12}{18} \div \dfrac{24}{3}$

19. $\dfrac{26}{3} \div \dfrac{13}{9}$

20. $\dfrac{36}{25} \div \dfrac{18}{5}$

21. What do you get when you divide $\dfrac{3}{8}$ by $\dfrac{5}{8}$?

22. Find the quotient of $\dfrac{3}{16}$ and $\dfrac{9}{4}$.

23. How much is 8 divided by $\dfrac{1}{2}$?

24. How much is 12 divided by $\dfrac{2}{3}$?

LIVING IN THE REAL WORLD

25. How many pieces of wire, each $\dfrac{1}{2}$ inch long, can be cut from a 10-inch piece?

26. A blouse requires $\dfrac{2}{3}$ of a yard of material. How many blouses can a clothing manufacturer make from 36 yards of material?

27. A piece of wire $\dfrac{5}{8}$ of a yard long is to be cut into 16 pieces of equal length. How long will each piece be?

28. How many scarves can be made from 16 yards of material if each scarf requires $\dfrac{4}{5}$ of a yard?

29. A recipe calls for $\dfrac{3}{4}$ cup of flour, but you can only find a measuring cup that holds $\dfrac{1}{8}$ cup. How many times must you fill it in order to equal $\dfrac{3}{4}$ cup?

30. Patties weighing $\dfrac{1}{4}$ pound are to be made from 25 pounds of hamburger. How many patties can be made?

2.6 Addition and Subtraction of Like Fractions

Fractions are called **like fractions** if they have the same denominator. To add or subtract them, we simply add or subtract the numerators and keep the same denominator, reducing the answer when possible.

For instance, to add the like fractions $\frac{5}{9}$ and $\frac{2}{9}$, we add the numerators, 5 and 2, and place the result, 7, over the common denominator:

$$\frac{5}{9} + \frac{2}{9} = \frac{5+2}{9} = \frac{7}{9}$$

Many students know the procedure for adding fractions, but not too many know why it works. It is actually an application of the distributive property. If you think about the following discussion carefully, you will learn more about it.

To begin with, the fraction $\frac{5}{9}$ can be written as $5 \cdot \frac{1}{9}$ since $5 \cdot \frac{1}{9} = \frac{5}{1} \cdot \frac{1}{9} = \frac{5}{9}$. Similarly, $\frac{2}{9}$ is the same as $2 \cdot \frac{1}{9}$. To add the fractions $\frac{5}{9} + \frac{2}{9}$, we rewrite them this way and apply the distributive property:

$\frac{5}{9} + \frac{2}{9} = 5 \cdot \frac{1}{9} + 2 \cdot \frac{1}{9}$ Rewrite each factor.

$= (5 + 2)\left(\frac{1}{9}\right)$ Use the distributive property.

$= 7 \cdot \frac{1}{9}$ Add 5 + 2.

$= \frac{7}{1} \cdot \frac{1}{9} = \frac{7}{9}$

This example shows why we add the numerators but leave the denominators alone. Our rule is as follows:

To add like fractions:

Add the numerators and place the result over the like denominator:

$$\frac{a}{b} + \frac{c}{b} = \frac{a+c}{b}$$

To subtract like fractions:

Subtract the numerators and place the result over the like denominator.

$$\frac{a}{b} - \frac{c}{b} = \frac{a-c}{b}$$

Always reduce your answer to lowest terms when necessary.

EXAMPLE 1 $\frac{7}{11} + \frac{3}{11}$

SOLUTION $7 + 3 = 10$ will be the numerator. We keep the same denominator, 11.

$$\frac{7}{11} + \frac{3}{11} = \frac{7+3}{11} = \frac{10}{11}$$

EXAMPLE 2 $\frac{7}{8} - \frac{4}{8}$

SOLUTION This time we're asked to subtract $\frac{4}{8}$ from $\frac{7}{8}$. We handle it the same way we did addition: We subtract the numerators, $7 - 4 = 3$, and place the result over the common denominator, 8.

$$\frac{7}{8} - \frac{4}{8} = \frac{7-4}{8} = \frac{3}{8}$$

STOP

When adding or subtracting fractions, you must always check to see if your result can be reduced to lower terms. For example, neither of the fractions $\frac{2}{9}$ or $\frac{4}{9}$ can be reduced to lower terms, but their sum can.

$$\frac{2}{9} + \frac{4}{9} = \frac{6}{9} = \frac{2}{3}$$

The sum can be reduced to lower terms.

EXAMPLE 3 $\frac{5}{12} + \frac{1}{12} = \frac{6}{12}$

SOLUTION This time we can reduce the answer, since $\frac{6}{12} = \frac{1}{2}$, so

$$\frac{5}{12} + \frac{1}{12} = \frac{6}{12} = \frac{1}{2}$$

EXAMPLE 4 $\frac{5}{32} + \frac{7}{32} = \frac{12}{32} = \frac{3}{8}$

EXAMPLE 5 $\frac{7}{18} - \frac{5}{18} = \frac{2}{18} = \frac{1}{9}$

EXAMPLE 6 $\frac{5}{37} + \frac{4}{37} + \frac{2}{37} = \frac{11}{37}$

EXAMPLE 7 $\frac{9}{13} - \frac{7}{13} + \frac{6}{13} = \frac{8}{13}$

EXAMPLE 8 $\frac{5}{12} + \frac{4}{12} + \frac{7}{12} = \frac{16}{12} = \frac{4}{3}$

EXAMPLE 9 $\frac{18}{5} - \frac{7}{5} = \frac{11}{5}$

Notice that Examples 8 and 9 involve answers in which the numerator is larger than the denominator. Remember that fractions of this type are called *improper fractions*. We will learn how to deal further with them soon.

2.6 Exercises

Do all

▼ Combine each of the following. Reduce the answer whenever possible.

1. $\frac{5}{8} + \frac{1}{8}$
2. $\frac{5}{8} + \frac{2}{8}$
3. $\frac{1}{9} + \frac{5}{9}$
4. $\frac{5}{9} + \frac{2}{9}$
5. $\frac{10}{3} - \frac{8}{3}$
6. $\frac{5}{7} - \frac{3}{7}$
7. $\frac{11}{4} - \frac{7}{4}$
8. $\frac{13}{5} - \frac{8}{5}$
9. $\frac{7}{28} - \frac{4}{28}$
10. $\frac{15}{16} - \frac{5}{16}$
11. $\frac{7}{12} + \frac{5}{12}$
12. $\frac{11}{15} + \frac{4}{15}$
13. $\frac{3}{11} + \frac{2}{11} + \frac{5}{11}$
14. $\frac{21}{32} - \frac{8}{32} + \frac{2}{32}$
15. $\frac{9}{16} - \frac{5}{16} + \frac{7}{16}$
16. $\frac{7}{8} - \frac{6}{8} + \frac{2}{8}$

17. $\frac{6}{8} - \frac{4}{8} - \frac{2}{8}$

18. $\frac{15}{17} - \frac{8}{17} - \frac{2}{17}$

23. $\frac{27}{45} + \frac{18}{45} - \frac{3}{45}$

24. $\frac{13}{80} + \frac{32}{80} - \frac{15}{80}$

19. $\frac{5}{8} + \frac{7}{8} + \frac{2}{8} + \frac{1}{8}$

20. $\frac{5}{12} + \frac{4}{12} + \frac{7}{12} + \frac{1}{12}$

25. $\frac{3}{4} + \frac{5}{4} - \frac{1}{4} + \frac{5}{4}$

26. $\frac{2}{9} + \frac{4}{9} + \frac{3}{9} - \frac{5}{9}$

21. $\frac{2}{3} + \frac{1}{3} + \frac{5}{3} + \frac{7}{3}$

22. $\frac{7}{16} + \frac{5}{16} + \frac{3}{16} + \frac{1}{16}$

27. $\frac{5}{8} - \frac{3}{8} + \frac{4}{8} - \frac{1}{8}$

2.7 Addition and Subtraction of Fractions with Different Denominators

Finding the Least Common Denominator (LCD)

In the preceding section we added fractions that had the same denominators. Now we must consider the question of adding fractions that have different denominators, such as $\frac{1}{6} + \frac{3}{4}$. We cannot add fractions with different denominators unless we first change each of them to an equivalent fraction so that they each have the same denominator. The **least common denominator,** abbreviated **LCD,** is the smallest number that is exactly divisible (has zero remainder) by each of the given denominators. When we consider the fractions $\frac{1}{6}$ and $\frac{3}{4}$, the smallest number that is exactly divisible by both 6 and 4 is 12, so 12 is the LCD. Sometimes, you can see the LCD right away. However, many times the LCD cannot be seen immediately, and we use of the following method for finding it.

This method makes use of the work that we did previously with prime numbers. For example, suppose we want to find the LCD of the fractions $\frac{5}{24}$ and $\frac{3}{18}$. First we write each denominator as the product of prime numbers:

$$24 = 2 \cdot 2 \cdot 2 \cdot 3$$

$$18 = 2 \cdot 3 \cdot 3$$

Now we pick out the greatest number of times that each prime number appears in any of the prime factorizations. For example, the factor 2 appears three times in the first prime factorization, so we pick 2 three times. The greatest number of times that the factor 3 occurs is twice (in the second factorization), so we pick the factor 3 two times. To find the LCD, we multiply together all the factors that we have chosen:

$$\text{LCD} = 2 \cdot 2 \cdot 2 \cdot 3 \cdot 3 = 72$$

So the LCD of the fractions $\frac{5}{24}$ and $\frac{3}{18}$ is 72.

There is another way of looking at this process, which some students find easier. It works as follows. First, write down one of the prime factorizations. Then look at the other prime factorization and add to the first prime factorization anything that was not already written down. Study this example carefully to grasp this process fully. Again we will find the LCD for 24 and 18.

$$24 = 2 \cdot 2 \cdot 2 \cdot 3$$
$$18 = 2 \cdot 3 \cdot 3$$

We write down the first prime factorization.

$$2 \cdot 2 \cdot 2 \cdot 3$$

Now, looking at the second prime factorization, we find one 2. Since we already have three 2s written down, we are okay. We also find two 3s. But we have only one 3 written down, so we write down another one, giving us

$$2 \cdot 2 \cdot 2 \cdot 3 \cdot 3$$

This equals 72, which is the LCD.

EXAMPLE 1 Find the LCD of the fractions $\frac{11}{18}, \frac{2}{9}, \frac{3}{4}$, and $\frac{1}{8}$.

SOLUTION We write each denominator as the product of prime factors.

$$18 = 2 \cdot 3 \cdot 3$$
$$9 = 3 \cdot 3$$
$$4 = 2 \cdot 2$$
$$8 = 2 \cdot 2 \cdot 2$$

Now we pick out the greatest number of times each factor occurs. The number 2 appears once in the first factorization, twice in the third, and three times in the fourth, so we choose 2 three times. The number 3 occurs twice in both the first and second factorizations, so our LCD is the product of three 2s and two 3s:

$$\text{LCD} = 2 \cdot 2 \cdot 2 \cdot 3 \cdot 3 = 72$$

By the second method we described, we write down the first prime factorization:

$$2 \cdot 3 \cdot 3$$

The next prime factorization is $3 \cdot 3$, but we already have two 3s written down, so we go on to the next one, $2 \cdot 2$. We only have one 2 written down, so we put down another one:

$$2 \cdot 2 \cdot 3 \cdot 3$$

The last prime factorization is $2 \cdot 2 \cdot 2$. We only have two 2s written down and we need three, so putting another 2 in the product gives us the LCD:

$$2 \cdot 2 \cdot 2 \cdot 3 \cdot 3 = 72$$

EXAMPLE 2 Find the LCD of $\frac{3}{25}$, $\frac{7}{30}$, and $\frac{8}{75}$.

SOLUTION

$25 = 5 \cdot 5$

$30 = 2 \cdot 3 \cdot 5$

$75 = 3 \cdot 5 \cdot 5$

We choose one 2, one 3, and two 5s, and multiply together:

$$\text{LCD} = 2 \cdot 3 \cdot 5 \cdot 5 = 150$$

EXAMPLE 3 Find the LCD of $\frac{7}{12}$, $\frac{4}{15}$, and $\frac{9}{20}$

SOLUTION

$12 = 2 \cdot 2 \cdot 3$

$15 = 3 \cdot 5$

$20 = 2 \cdot 2 \cdot 5$

$\text{LCD} = 2 \cdot 2 \cdot 3 \cdot 5 = 60$

EXAMPLE 4 Find the LCD of $\frac{3}{32}$, $\frac{5}{8}$, and $\frac{1}{4}$.

SOLUTION You may be able to see by inspection that the smallest number that is exactly divisible by 32, 8, and 4 is 32. If so, you have found the LCD to be 32 and you are done. If not, we can use one of our methods.

$$32 = 2 \cdot 2 \cdot 2 \cdot 2 \cdot 2$$
$$8 = 2 \cdot 2 \cdot 2$$
$$4 = 2 \cdot 2$$
$$\text{LCD} = 2 \cdot 2 \cdot 2 \cdot 2 \cdot 2 = 32$$

Adding and Subtracting Unlike Fractions

Now we will return to our original problem: adding fractions with different denominators. Now that we know how to find the LCD of the given fractions, all we need to do is to write each of the given fractions as an equivalent fraction with the LCD as the denominator.

To illustrate, consider $\frac{1}{6} + \frac{3}{4}$. We determined previously that the LCD is 12. Our problem now is to write each of the fractions $\frac{1}{6}$ and $\frac{3}{4}$ as equivalent fractions with 12 as the denominator. We have learned that if we multiply both numerator and denominator of a given fraction by the same number, the result will be an equivalent fraction. What number must we multiply numerator and denominator of the fraction $\frac{1}{6}$ by to get an equivalent fraction with denominator 12? Looking at the denominator,

$$\frac{1}{6} = \frac{?}{12}$$

We see that if we multiply $6 \cdot 2$, we get 12. Therefore, we must also multiply the numerator by 2, and we get

$$\frac{1}{6} = \frac{1 \cdot 2}{6 \cdot 2} = \frac{2}{12}$$

Looking at our second fraction, $\frac{3}{4}$, we must multiply the denominator by 3 to get 12. Accordingly, we must also multiply the numerator by 3. Our result is $\frac{3 \cdot 3}{4 \cdot 3} = \frac{9}{12}$.

We now have $\frac{1}{6} = \frac{2}{12}$ and $\frac{3}{4} = \frac{9}{12}$, so our problem $\frac{1}{6} + \frac{3}{4}$ is equivalent to $\frac{2}{12} + \frac{9}{12}$, which we know how to do. The answer is

$$\frac{1}{6} + \frac{3}{4} = \frac{2}{12} + \frac{9}{12} = \frac{11}{12}$$

EXAMPLE 5 Add: $\frac{1}{3} + \frac{3}{8}$.

SOLUTION Our first step is to find the LCD.

$$3 = 3 \qquad \text{(since 3 is already prime)}$$

$$8 = 2 \cdot 2 \cdot 2$$

$$\text{LCD} = 2 \cdot 2 \cdot 2 \cdot 3 = 24$$

Next we must write $\frac{1}{3}$ and $\frac{3}{8}$ as equivalent fractions with 24 as the denominator.

$$\frac{1}{3} = \frac{?}{24} \quad \text{and} \quad \frac{3}{8} = \frac{?}{24}$$

What must we multiply the first denominator, 3, by to get 24? The answer is 8, since $3 \cdot 8 = 24$. If we multiply the denominator by 8, we must also multiply the numerator by 8, giving us $\frac{1}{3} = \frac{1 \cdot 8}{3 \cdot 8} = \frac{8}{24}$. In the second fraction, $\frac{3}{8}$, since we must multiply $8 \cdot 3$ to get 24 we must also multiply the 3 in the numerator by 3.

$$\frac{3}{8} = \frac{3 \cdot 3}{8 \cdot 3} = \frac{9}{24}$$

Since our new equivalent fractions have the same denominator, we can add them by adding their numerators. The answer is

$$\frac{1}{3} + \frac{3}{8} = \frac{8}{24} + \frac{9}{24} = \frac{17}{24}$$

▮▮

EXAMPLE 6 Subtract: $\frac{7}{16} - \frac{5}{12}$.

SOLUTION First we find the LCD.

$$16 = 2 \cdot 2 \cdot 2 \cdot 2$$
$$12 = 2 \cdot 2 \cdot 3$$
$$\text{LCD} = 2 \cdot 2 \cdot 2 \cdot 2 \cdot 3 = 48$$

Next we write each fraction as an equivalent fraction with 48 as the denominator.

$$\frac{7}{16} = \frac{7 \cdot 3}{16 \cdot 3} = \frac{21}{48}$$
$$\frac{5}{12} = \frac{5 \cdot 4}{12 \cdot 4} = \frac{20}{48}$$

Now we subtract:

$$\frac{21}{48} - \frac{20}{48} = \frac{1}{48}$$

▮▮

EXAMPLE 7 Add: $\frac{1}{3} + \frac{1}{4} + \frac{1}{8} + \frac{1}{6}$.

SOLUTION We begin by finding the LCD.

$$3 = 3$$
$$4 = 2 \cdot 2$$
$$8 = 2 \cdot 2 \cdot 2$$
$$6 = 2 \cdot 3$$
$$\text{LCD} = 2 \cdot 2 \cdot 2 \cdot 3 = 24$$

We then write each fraction as an equivalent fraction with 24 as the denominator:

$$\frac{1}{3} = \frac{1 \cdot 8}{3 \cdot 8} = \frac{8}{24}$$

$$\frac{1}{4} = \frac{1 \cdot 6}{4 \cdot 6} = \frac{6}{24}$$

$$\frac{1}{8} = \frac{1 \cdot 3}{8 \cdot 3} = \frac{3}{24}$$

$$\frac{1}{6} = \frac{1 \cdot 4}{6 \cdot 4} = \frac{4}{24}$$

Now we add all the numerators of our equivalent fractions to get the final answer:

$$\frac{8}{24} + \frac{6}{24} + \frac{3}{24} + \frac{4}{24} = \frac{21}{24} = \frac{7}{8}$$

Notice that our answer could be simplified, since both the numerator and the denominator of $\frac{21}{24}$ are divisible by 3.

EXAMPLE 8 Subtract: $\frac{11}{16} - \frac{5}{8}$.

SOLUTION You may be able to see that the LCD is 16 just by looking at the problem. If not, we can use our prime factorization method as follows:

$$16 = 2 \cdot 2 \cdot 2 \cdot 2$$

$$8 = 2 \cdot 2 \cdot 2$$

$$\text{LCD} = 2 \cdot 2 \cdot 2 \cdot 2 = 16$$

We write each fraction as an equivalent fraction with 16 as the denominator. The first fraction, $\frac{11}{16}$, already has 16 as its denominator, so we leave it as it is. For the second fraction,

$$\frac{5}{8} = \frac{5 \cdot 2}{8 \cdot 2} = \frac{10}{16}$$

Subtracting gives us the final answer:

$$\frac{11}{16} - \frac{5}{8} = \frac{11}{16} - \frac{10}{16} = \frac{1}{16}$$

EXAMPLE 9 Frank saves $\frac{1}{2}$ of his part-time summer earnings for college and spends $\frac{1}{3}$ of his earnings for clothes for the fall. What portion of his summer salary does this amount to?

SOLUTION A graph or a picture sometimes helps us to see better what is happening mathematically.

Savings $= \frac{1}{2}$ or $\frac{3}{6}$

$\frac{1}{2} = \frac{3}{6}$

Total $= \frac{3}{6} + \frac{2}{6} = \frac{5}{6}$

Clothes $= \frac{1}{3}$ or $\frac{2}{6}$

$\frac{1}{3} = \frac{2}{6}$

Algebraically, our problem is done as before. By inspection, the LCD is 6. We change $\frac{1}{2}$ and $\frac{1}{3}$ to sixths:

$$\frac{1}{2} = \frac{1 \cdot 3}{2 \cdot 3} = \frac{3}{6} \quad \text{and} \quad \frac{1}{3} = \frac{1 \cdot 2}{3 \cdot 2} = \frac{2}{6}$$

Adding gives us $\frac{3}{6} + \frac{2}{6} = \frac{5}{6}$.

2.7 Exercises

In Exercises 1–16, assume that the numbers in each group are denominators of fractions. Find the LCD using the method of prime factorization.

1. 14, 6
2. 10, 12
3. 18, 24
4. 4, 9
5. 16, 24
6. 10, 5, 15
7. 4, 7, 9

8. 3, 11, 6

9. 9, 6, 15, 10

10. 21, 6, 9, 14

11. 24, 40, 36

12. 28, 14, 21

13. 33, 15, 22

14. 16, 18, 24

15. 16, 24, 18, 36

16. 32, 24, 45, 18

MENTAL MATHEMATICS

▼ For each group of fractions, find the LCD mentally, by inspection.

17. $\frac{3}{4}, \frac{1}{5}$ 18. $\frac{2}{3}, \frac{1}{7}$ 19. $\frac{3}{5}, \frac{1}{10}$

20. $\frac{5}{6}, \frac{1}{3}$ 21. $\frac{1}{3}, \frac{2}{3}, \frac{5}{6}$ 22. $\frac{1}{9}, \frac{5}{18}, \frac{4}{9}$

23. $\frac{1}{2}, \frac{3}{4}, \frac{1}{3}$ 24. $\frac{3}{8}, \frac{2}{3}, \frac{1}{4}$ 25. $\frac{7}{8}, \frac{1}{2}, \frac{3}{5}$

▼ In Exercises 17–32, add or subtract as indicated. Simplify your answer whenever possible.

26. $\frac{1}{4} + \frac{3}{8}$ 27. $\frac{1}{2} + \frac{1}{3}$

28. $\frac{1}{2} + \frac{3}{16} + \frac{2}{9}$ 29. $\frac{1}{5} + \frac{2}{9} + \frac{2}{15}$

30. $\frac{17}{32} - \frac{3}{8}$ 31. $\frac{2}{3} + \frac{1}{8} + \frac{3}{16}$

32. $\frac{11}{12} - \frac{2}{9}$ 33. $\frac{3}{16} + \frac{1}{18} + \frac{7}{24}$

34. $\frac{2}{9} + \frac{3}{9} + \frac{1}{6}$ 35. $\frac{1}{56} + \frac{5}{112} + \frac{7}{8}$

36. $\frac{9}{16} + \frac{3}{9} + \frac{1}{5}$ 37. $\frac{1}{8} + \frac{5}{16} - \frac{1}{4}$

38. $\frac{1}{2} + \frac{1}{4} + \frac{1}{8} + \frac{1}{16}$

39. $\frac{3}{4} + \frac{1}{5} - \frac{3}{10}$

40. $\frac{7}{8} - \frac{2}{3} + \frac{1}{6}$

41. $\frac{13}{16} - \frac{3}{8} + \frac{2}{3}$

42. Find the sum of $\frac{7}{8}$ and $\frac{1}{12}$.

43. Find the sum of $\frac{1}{3}$, $\frac{5}{12}$, and $\frac{2}{7}$.

44. Find the difference between $\frac{9}{16}$ and $\frac{1}{2}$.

45. Find the difference between $\frac{9}{14}$ and $\frac{2}{21}$.

LIVING IN THE REAL WORLD

46. A recipe calls for $\frac{1}{2}$ cup of sugar and $\frac{2}{3}$ cup of flour. What is the total volume of these ingredients?

47. A family spends $\frac{1}{5}$ of their income on food, $\frac{1}{12}$ on clothing, and $\frac{1}{2}$ on rent. What is the total fraction of their income spent on these three items?

48. On an English exam, $\frac{1}{6}$ of the students received an A, $\frac{1}{5}$ received a B, $\frac{1}{4}$ received a C, and $\frac{1}{5}$ received a D. Answer the following:

a. What fraction of the class received a B or better?

b. What fraction of the class received a C or D?

c. If the remainder of the class received an F, what fraction is this?

49. If Karen ate $\frac{1}{12}$ of a pizza and Frank ate $\frac{1}{3}$ of the same pizza, what fraction of the pizza remains?

50. An average person spends $\frac{1}{3}$ of the day sleeping, $\frac{1}{3}$ of the day working, and $\frac{1}{12}$ of the day eating. What total portion of the day is spent on these three activities?

51. Christine needs $\frac{3}{4}$ of a cup of sugar, but she only has $\frac{3}{8}$ of a cup. How much does she still need?

52. A board $\frac{7}{8}$ inch thick has $\frac{3}{16}$ inch removed from it. How thick is the part that remains?

53. Frank has $\frac{1}{8}$ of a teaspoon of vanilla ready to put into a batch of fudge he is making; then he notices that the recipe calls for $\frac{3}{4}$ teaspoon. How much more must he add?

54. What is wrong with the following?

$$\frac{3}{4} + \frac{2}{3} = \frac{5}{7}$$

55. When is the LCD equal to the product of all the denominators in the fractions being added? Consider the two problems $\frac{3}{4} + \frac{2}{3}$ and $\frac{5}{6} + \frac{1}{2}$ in your explanation.

SUMMARY—CHAPTER 2

The numbers in brackets refer to the section in the text in which each concept is presented.

	EXAMPLES
[2.1] A number is **divisible** by • **2** if it is an even number; • **3** if the sum of its digits is divisible by 3; • **5** if it ends in a 5 or 0.	[2.1] 136 is divisible by 2. 522 is divisible by 3, since $5 + 2 + 2 = 9$. 365 is divisible by 5.
[2.1] A **prime number** is divisible only by itself and 1.	[2.1] 17 is a prime number, since only 1 and 17 divide it.
[2.1] A **composite number** is divisible by some number other than itself or 1.	[2.1] 26 is a composite number, since $26 = 2 \cdot 13$.
[2.1] **1** is neither prime nor composite.	
[2.1] Every composite number can be written as a product of prime factors.	[2.1] $24 = 2 \cdot 2 \cdot 2 \cdot 3 = 2^3 \cdot 3$
[2.2] A **fraction** is a number that expresses the relation to the whole of one or more equal parts of a whole.	[2.2] ▧□□ $= \frac{1}{3}$
[2.2] The **numerator** of a fraction is the top number of the fraction.	[2.2] $\frac{5}{16}$ (5 ← numerator; 16 ← denominator)
[2.2] The **denominator** of a fraction is the number on the bottom of the fraction.	
[2.2, 2.3] An **equivalent fraction** is the fraction obtained by multiplying or dividing both the numerator and the denominator of a given fraction by the same nonzero number.	[2.2, 2.3] $\frac{2}{3} = \frac{2 \cdot 5}{3 \cdot 5} = \frac{10}{15}$
[2.3] A fraction is in **simplest form** or **reduced to lowest terms** if the numerator and the denominator do not contain any identical factors.	[2.3] $\frac{6}{10} = \frac{\overset{1}{\cancel{2}} \cdot 3}{\underset{1}{\cancel{2}} \cdot 5} = \frac{3}{5}$
[2.4] **To multiply two fractions,** divide out all like factors of any numerator with any denominator. Then multiply the resulting numerators together and multiply the denominators together to form a single fraction.	[2.4] $\frac{7}{\underset{4}{\cancel{8}}} \cdot \frac{\overset{1}{\cancel{2}}}{3} = \frac{7 \cdot 1}{4 \cdot 3} = \frac{7}{12}$
[2.5] **To divide one fraction by another,** multiply the first fraction by the reciprocal of the second.	[2.5] $\frac{5}{8} \div \frac{10}{3} = \frac{\overset{1}{\cancel{5}}}{8} \cdot \frac{3}{\underset{2}{\cancel{10}}} = \frac{3}{16}$
[2.6] **To add (or subtract) fractions with the same denominator,** add (or subtract) the numerators and keep the same denominator. Reduce the result to lowest terms.	[2.6] $\frac{9}{16} - \frac{5}{16} = \frac{4}{16} = \frac{1}{4}$

[2.7] The **least common denominator** of two or more fractions is the smallest nonzero number that is evenly divisible by all the denominators of the given fractions.

[2.7] **To add (or subtract) two fractions with different denominators,** find the LCD, change each fraction to an equivalent fraction with the LCD as the new denominator, and then add (or subtract) as you would for like fractions. Reduce the answer to lowest terms.

EXAMPLES

[2.7] The LCD for the fractions $\frac{2}{3}, \frac{1}{2}$, and $\frac{5}{8}$ is 24.

[2.7] $\frac{1}{6} + \frac{5}{8} = \frac{4}{24} + \frac{15}{24} = \frac{19}{24}$

REVIEW EXERCISES—CHAPTER 2

The numbers in brackets refer to the section in the text in which similar problems are presented.

[2.1]

1. Test 1,782 for divisibility by 2, 3, and 5.

[2.1]

2. Write 76 as a product of prime factors.

[2.2]

3. In the fraction $\frac{7}{16}$, the number 7 is called the ______ and 16 is called the ________________.

[2.2]

4. The shaded portion of the accompanying figure represents what fraction?

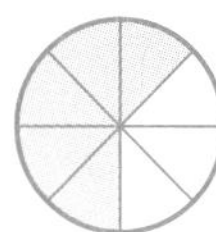

[2.2-2.3]

5. Write three other fractions that are equivalent to the fraction $\frac{3}{21}$.

[2.3]

6. Reduce each fraction to lowest terms.

a. $\frac{21}{36}$ **b.** $\frac{14}{35}$ **c.** $\frac{9}{216}$ **d.** $\frac{18}{35}$

[2.4-2.7] In Exercises 7–16, perform the calculations indicated. Reduce your answer if possible.

7. $\frac{2}{5} \cdot \frac{3}{8}$

8. $\frac{2}{3} \cdot \frac{9}{16} \cdot \frac{2}{3}$

9. $\frac{2}{5} + \frac{3}{8}$

10. $\frac{3}{4} \div \frac{4}{9}$

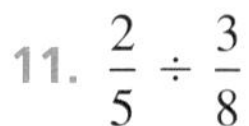

11. $\frac{2}{5} \div \frac{3}{8}$

12. $\frac{3}{8} + \frac{1}{3} - \frac{3}{16}$

13. $\frac{2}{3} + \frac{5}{16} + \frac{1}{4}$

14. $\frac{1}{5} \cdot \frac{7}{8} \cdot \frac{5}{14}$

15. $\frac{4}{9} - \frac{1}{8}$

16. $\frac{2}{3} \div \frac{6}{12}$

[2.4-2.7] In Exercises 17–22, work the part of the problem inside the parentheses first.

17. $\left(\frac{3}{4} - \frac{1}{3}\right) \cdot \frac{3}{8}$

18. $\left(\frac{3}{16} + \frac{2}{9}\right) \div \frac{1}{2}$

19. $\left(\frac{1}{3} \cdot \frac{3}{8}\right) \div \frac{5}{8}$

20. $\left(\frac{5}{8} - \frac{3}{10}\right) \div \frac{3}{4}$

21. $\frac{5}{9} \div \left(\frac{5}{24} \div \frac{1}{3}\right)$

22. $\left(\frac{5}{9} \div \frac{5}{24}\right) \div \frac{1}{3}$

▼ [2.4]

23. Find the product of $\frac{7}{18}$ and $\frac{9}{35}$.

▼ [2.5]

24. What is the quotient of $\frac{9}{25}$ and $\frac{3}{15}$?

LIVING IN THE REAL WORLD

▼ [2.5]

25. If a recipe calls for $\frac{3}{4}$ of a cup of flour and you are making only half the recipe, how much flour do you need?

▼ [2.5]

26. A piece of wire $\frac{2}{3}$ of a yard long is to be cut into 15 pieces of equal length. How long will each piece be?

▼ [2.5]

27. Kim can find only a $\frac{1}{8}$-teaspoon measure. Her recipe calls for $\frac{3}{4}$ of a teaspoon of vanilla. How many times must she fill the $\frac{1}{8}$ teaspoon?

▼ [2.7]

28. What is the sum of $\frac{5}{12}$ and $\frac{2}{3}$?

▼ [2.7]

29. Find the difference between $\frac{11}{12}$ and $\frac{3}{4}$.

▼ [2.7]

30. A farmer orders $\frac{1}{2}$ of a ton of oats, $\frac{3}{4}$ of a ton of hay and $\frac{3}{8}$ of a ton of corn. How many tons of animal food did he order altogether?

▼ [2.7]

31. James spends $\frac{1}{12}$ of his day eating, $\frac{1}{4}$ sleeping, and $\frac{1}{2}$ socializing. What portion of his day is spent on these three activities? If he spends the remainder of his day studying, what fraction is this?

▼ [2.7]

32. A board $\frac{15}{16}$ of an inch thick is put through a sanding process that results in a board that is $\frac{7}{8}$ of an inch thick. How much wood was removed in the sanding process?

CRITICAL THINKING EXERCISES—CHAPTER 2

Tell whether each of the following statements is true or false, and explain why.

1. $\frac{3}{9}$ and $\frac{4}{12}$ are equivalent fractions.

2. Adding the same number to both the numerator and the denominator of any fraction yields an equivalent fraction.

3. Multiplying the numerator and the denominator of a fraction by zero will always give zero as an answer.

4. Every whole number can be written as a fraction.

5. $\frac{3}{5} + \frac{4}{5} = \frac{7}{10}$

Find the error in each of the following:

6. $\frac{2}{3} \div \frac{3}{4} = \frac{3}{2} \cdot \frac{3}{4} = \frac{9}{8}$

7. $\frac{3}{4} \cdot \frac{4}{9} = \frac{12}{36}$

8. $\frac{6}{18} = \frac{\overset{1}{\cancel{6}}}{12 + \underset{1}{\cancel{6}}} = \frac{1}{13}$

9. $\frac{8}{24} = \frac{\cancel{2} \cdot \cancel{2} \cdot \cancel{2}}{\cancel{2} \cdot \cancel{2} \cdot \cancel{2} \cdot 3} = \frac{0}{3} = 0$

10. $\frac{3}{4} + \frac{5}{8} = \frac{8}{12} = \frac{2}{3}$

NAME ▮▮▮▮▮ CLASS

ACHIEVEMENT TEST—CHAPTER 2

1. Test 2,055 for divisibility by 2, 3, and 5. 1. ______

2. Write 180 as a product of prime factors. 2. ______

3. In the fraction $\frac{5}{11}$, the number 11 is called the ______ and 5 is called the ______. 3. ______

4. The shaded portions of the accompanying figures represent what fractions? 4. ______ ______

a. 

b.  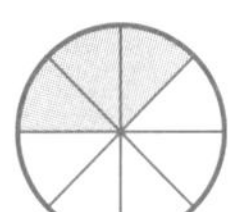

5. Write three fractions that are equivalent to the fraction $\frac{10}{15}$. 5. ______

6. Reduce to lowest terms: 6. ______ ______

a. $\frac{27}{60}$

b. $\frac{9}{35}$

7. Fill in the blank: 7. ______ ______

a. $\frac{7}{16} = \frac{\quad}{48}$

b. $\frac{2}{3} = \frac{\quad}{33}$

8. Find the LCD for the fractions $\frac{3}{16}, \frac{5}{8}$, and $\frac{7}{12}$.

9. $\frac{3}{4} \cdot \frac{12}{18} =$

10. $\frac{6}{12} \cdot \frac{2}{3} \cdot \frac{5}{8} =$

11. $\frac{7}{16} \div \frac{3}{4} =$

12. $\frac{9}{11} - \frac{5}{11} =$

13. $\frac{7}{16} + \frac{5}{12} =$

14. $\frac{1}{4} + \frac{3}{8} + \frac{5}{32} =$

15. Find the difference between $\frac{5}{8}$ and $\frac{5}{12}$.

8. ______

9. ______

10. ______

11. ______

12. ______

13. ______

14. ______

15. ______

16. A family spends $\frac{1}{6}$ of its income on food, $\frac{1}{4}$ on housing, $\frac{1}{12}$ on clothing, and $\frac{1}{6}$ on travel. What fraction of the family income is spent on these four items?

16. ______________________

17. Tim has $\frac{3}{8}$ of a cup of milk ready to put into a bowl of pancake batter when he realizes that the recipe calls for $\frac{3}{4}$ cup of milk. How much additional milk does he need?

17. ______________________

CHAPTER 3

Mixed Numbers

INTRODUCTION

In the first two chapters we learned about whole numbers and fractions. In this chapter we will combine whole numbers and fractions to form mixed numbers and we will use the operations of multiplication, division, addition, and subtraction to combine mixed numbers. We will also consider the situation in which there are fractions within fractions.

Practical applications of mixed numbers will be found throughout the chapter.

NUMBER KNOWLEDGE—CHAPTER 3

Recipes are but one of many places where you will find mixed numbers.

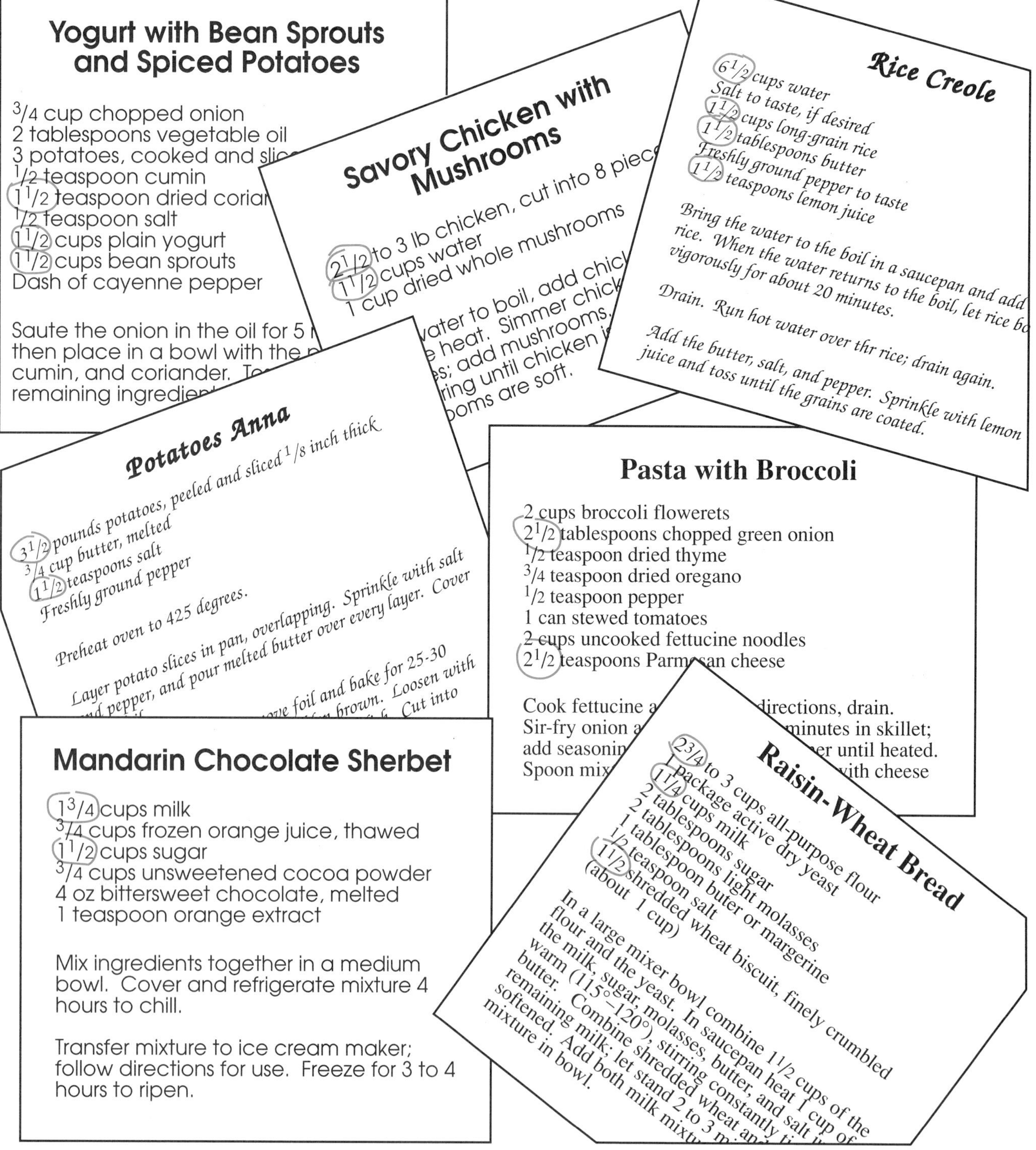

3.1 Mixed Numbers and Improper Fractions

Changing a Mixed Number to an Improper Fraction

A **mixed number** is a combination of a whole number and a fraction; for example, $2\frac{1}{3}$ is a mixed number. Each mixed number is actually a whole number and a fraction added together, that is, $2\frac{1}{3}$ actually means $2 + \frac{1}{3}$. However, we read $2\frac{1}{3}$ as "two and one third" and agree to leave out the plus sign. Every mixed number can be written as a fraction. For example, to change $2\frac{1}{3}$ to a fraction, we write it as $2 + \frac{1}{3}$.

$$2 + \frac{1}{3} = \frac{2}{1} + \frac{1}{3} \qquad \text{The LCD is 3.}$$

$$= \frac{6}{3} + \frac{1}{3} \qquad \frac{2}{1} = \frac{2}{1} \cdot \frac{3}{3} = \frac{6}{3}$$

$$= \frac{7}{3}$$

We can see, then, that $2\frac{1}{3}$ can be written as $\frac{7}{3}$. The following diagram demonstrates that $2\frac{1}{3}$ and $\frac{7}{3}$ represent the same amount.

$$1 + 1 + \frac{1}{3} = 2\frac{1}{3}$$

These are the same.

$$\frac{3}{3} + \frac{3}{3} + \frac{1}{3} = \frac{7}{3}$$

The fraction $\frac{7}{3}$ is called an **improper fraction** because the numerator is larger than the denominator. Actually, there is nothing really improper about an improper fraction at all! It's a perfectly good number, so don't let the name influence you. There is a short way to write a mixed number as an improper fraction, and you should learn it well.

To Write a Mixed Number as an Improper Fraction:

1. Multiply the whole number by the denominator of the fraction;
2. Add this result to the numerator of the fraction.
3. This number becomes the new numerator in the improper fraction. The denominator stays the same.

Using our previous example, let's use this method to change $2\frac{1}{3}$ to an improper fraction.

$$2\frac{1}{3} = \frac{(2 \cdot 3) + 1}{3} = \frac{6 + 1}{3} = \frac{7}{3}$$

Multiply the 2 by the 3 and add the 1 to give 7 for the numerator. Keep the same denominator, 3.

EXAMPLE 1 Change $5\frac{3}{4}$ to an improper fraction.

SOLUTION We multiply the 5 times the 4 and add the 3 to obtain the new numerator. We keep the same denominator, 4.

$$5\frac{3}{4} = \frac{(5 \cdot 4) + 3}{4} = \frac{23}{4}$$

EXAMPLE 2 Write $3\frac{5}{8}$ as an improper fraction.

SOLUTION

$$3\frac{5}{8} = \frac{(3 \cdot 8) + 5}{8} = \frac{29}{8}$$

EXAMPLE 3 Write $1\frac{3}{4}$ as an improper fraction.

SOLUTION

$$1\frac{3}{4} = \frac{(4 \cdot 1) + 3}{4} = \frac{7}{4}$$

It is common to confuse mixed-number notation with notation for multiplication. The mixed number $2\frac{1}{4}$ means 2 *plus* $\frac{1}{4}$, not 2 *times* $\frac{1}{4}$.

Changing an Improper Fraction to a Mixed Number

There are times when we would like to change an improper fraction to a mixed number. If we keep in mind that the fraction bar (the line between the numerator and denominator of a fraction) also means to divide, the procedure for changing an improper fraction to a mixed number seems quite reasonable. For example, the fraction $\frac{7}{3}$ can also be thought of as 7 divided by 3. We can divide 3 into 7 twice with a remainder of 1, or

$$\frac{7}{3} = 2\frac{1}{3}$$

EXAMPLE 4 Convert the improper fraction $\frac{23}{4}$ to a mixed number.

SOLUTION If we divide 23 by 4, we see that 4 divides into 23 five times with a remainder of 3, so

$$\frac{23}{4} = 5\frac{3}{4}$$

EXAMPLE 5 Write $\frac{13}{5}$ as a mixed number.

SOLUTION

$$\frac{13}{5} = 2\frac{3}{5}$$

EXAMPLE 6 Write $\frac{86}{9}$ as a mixed number.

SOLUTION

$$\frac{86}{9} = 9\frac{5}{9}$$

3.1 Exercises

▼ Give both the mixed number and the improper fraction represented by the shaded portions in each of the following diagrams.

1.

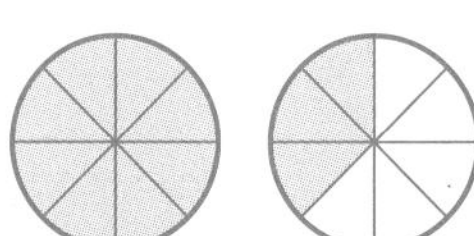

2.

3.

4.

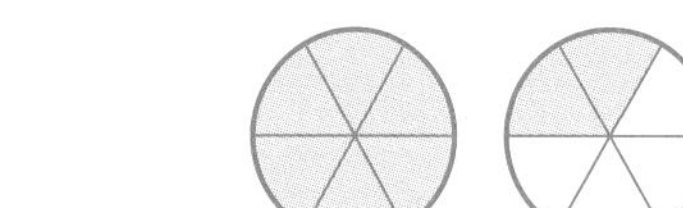

5.

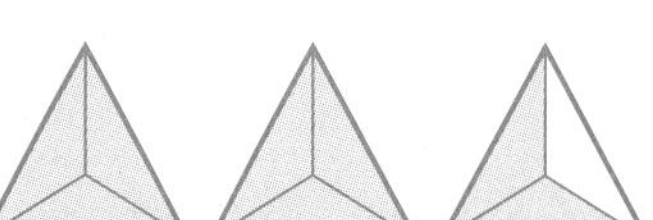

6.

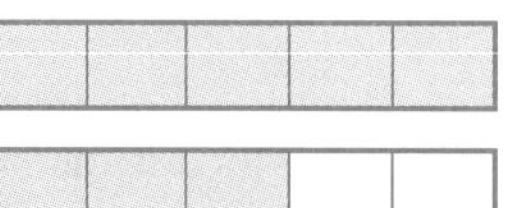

7.

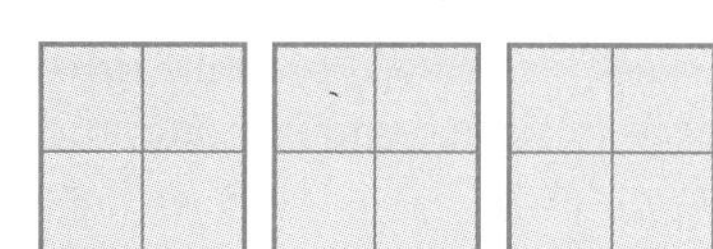

▼ Write each mixed number as an improper fraction.

8. $3\frac{3}{4}$ 9. $1\frac{5}{6}$ 10. $2\frac{5}{8}$

11. $1\frac{7}{16}$ 12. $1\frac{5}{8}$ 13. $6\frac{2}{7}$

14. $9\frac{2}{5}$ 15. $11\frac{3}{5}$ 16. $12\frac{3}{8}$

17. $16\frac{2}{3}$ 18. $12\frac{5}{18}$ 19. $12\frac{5}{22}$

▼ Write each improper fraction as a mixed number.

20. $\frac{23}{6}$ 21. $\frac{5}{5}$ 22. $\frac{7}{5}$

23. $\frac{58}{3}$ 24. $\frac{57}{8}$ 25. $\frac{28}{9}$

26. $\frac{120}{23}$ 27. $\frac{41}{15}$ 28. $\frac{164}{13}$

29. $\frac{181}{5}$ 30. $\frac{112}{19}$ 31. $\frac{132}{17}$

LIVING IN THE REAL WORLD

32. Janine is 5 ft 5 in. tall. Her height can also be written as $5\frac{5}{12}$ ft. Represent this mixed number as an improper fraction.

33. Bill is 6 ft 1 in. tall, which is the same as $6\frac{1}{12}$ ft. Write this number as an improper fraction.

34. A recipe calls for $1\frac{1}{3}$ cups of flour. How many times do you have to fill a $\frac{1}{3}$-cup measure to obtain the $1\frac{1}{3}$ cups that you need? Write $1\frac{1}{3}$ as an improper fraction.

35. How many times do you need to fill a $\frac{1}{4}$-teaspoon measure in order to obtain $1\frac{3}{4}$ teaspoons? Write $1\frac{3}{4}$ as a mixed number.

36. A stock on the American Stock Exchange is listed at $23\frac{5}{8}$. Express this number as an improper fraction.

37. A stock on the New York Stock Exchange is listed at $31\frac{1}{8}$. Write this number as an improper fraction.

38. How many times do you need to fill a $\frac{1}{4}$-cup measure in order to obtain $2\frac{1}{4}$ cups? Write $2\frac{1}{4}$ as an improper fraction.

MENTAL MATHEMATICS

▼ Write each mixed number as an improper fraction and each improper fraction as a mixed number. Do them mentally, using a pencil or pen only to record your results.

39. $3\frac{1}{5}$ 40. $4\frac{1}{8}$ 41. $3\frac{1}{4}$

42. $2\frac{2}{3}$ 43. $1\frac{4}{5}$ 44. $7\frac{1}{2}$

45. $6\frac{3}{4}$ 46. $\frac{11}{3}$ 47. $\frac{13}{4}$

48. $\frac{15}{4}$ 49. $\frac{9}{2}$ 50. $2\frac{1}{8}$

51. $7\frac{2}{3}$ 52. $\frac{21}{5}$ 53. $\frac{19}{4}$

54. $\frac{13}{3}$ 55. $9\frac{3}{5}$ 56. $\frac{47}{7}$

57. What is a mixed number?

58. Why is $5\frac{6}{6}$ the same as 6?

3.2 Multiplication and Division of Mixed Numbers

Multiplying Mixed Numbers

To multiply one mixed number by another, we convert each mixed number to an improper fraction and then multiply the two fractions together. For example, to multiply $2\frac{1}{3} \cdot 3\frac{3}{4}$, we change each mixed number to an improper fraction:

$$2\frac{1}{3} = \frac{7}{3} \quad \text{and} \quad 3\frac{3}{4} = \frac{15}{4}$$

Now we multiply the two improper fractions together:

$$\frac{7}{\cancel{3}_{1}} \cdot \frac{\cancel{15}^{5}}{4} = \frac{35}{4} \quad \text{or} \quad 8\frac{3}{4}$$

Our answer is $\frac{35}{4}$, or $8\frac{3}{4}$. Either form is usually acceptable, although one form might be more useful than the other in a particular situation.

EXAMPLE 1 Multiply: $2\frac{2}{3} \cdot 5\frac{1}{2}$.

SOLUTION We convert each mixed number to an improper fraction:

$$2\frac{2}{3} = \frac{8}{3} \quad \text{and} \quad 5\frac{1}{2} = \frac{11}{2}$$

Now we multiply the two fractions:

$$\frac{\cancel{8}^{4}}{3} \cdot \frac{11}{\cancel{2}_{1}} = \frac{44}{3} \quad \text{or} \quad 14\frac{2}{3}$$

In summary, our rule is as follows:

To multiply mixed numbers:

Change each mixed number to an improper fraction and then multiply the fractions.

EXAMPLE 2 Multiply: $5\frac{2}{5} \cdot 2\frac{2}{9}$.

SOLUTION

$$5\frac{2}{5} \cdot 2\frac{2}{9} = \frac{\overset{3}{\cancel{27}}}{\underset{1}{\cancel{5}}} \cdot \frac{\overset{4}{\cancel{20}}}{\underset{1}{\cancel{9}}} = \frac{12}{1} = 12$$

EXAMPLE 3 Multiply: $4\frac{1}{6} \cdot 2\frac{4}{5}$.

SOLUTION

$$4\frac{1}{6} \cdot 2\frac{4}{5} = \frac{\overset{5}{\cancel{25}}}{\underset{3}{\cancel{6}}} \cdot \frac{\overset{7}{\cancel{14}}}{\underset{1}{\cancel{5}}} = \frac{35}{3} \quad \text{or} \quad 11\frac{2}{3}$$

Dividing Mixed Numbers

Dividing one mixed number by another is done in a similar manner. For example, to divide $4\frac{1}{8}$ by $2\frac{3}{4}$, we first change each mixed number to an improper fraction:

$$4\frac{1}{8} = \frac{33}{8} \quad \text{and} \quad 2\frac{3}{4} = \frac{11}{4}$$

Now we divide the two improper fractions:

$$4\frac{1}{8} \div 2\frac{3}{4} = \frac{33}{8} \cdot \frac{11}{4}$$ Change each mixed number to an improper fraction.

$$= \frac{33}{8} \cdot \frac{4}{11}$$ To divide, multiply by the reciprocal of the divisor.

$$= \frac{\overset{3}{\cancel{33}}}{\underset{2}{\cancel{8}}} \cdot \frac{\overset{1}{\cancel{4}}}{\underset{1}{\cancel{11}}}$$ Cancel and multiply.

$$= \frac{3}{2} \quad \text{or} \quad 1\frac{1}{2}$$

In summary:

To divide mixed numbers:

Change each mixed number to an improper fraction and divide the fractions (by multiplying by the reciprocal of the divisor).

EXAMPLE 4 Divide: $3\frac{3}{5} \div 5\frac{1}{4}$.

SOLUTION We change each mixed number to an improper fraction and then we follow the rules for division of fractions.

$$3\frac{3}{5} \div 5\frac{1}{4} = \frac{18}{5} \div \frac{21}{4} = \frac{\overset{6}{\cancel{18}}}{5} \cdot \frac{4}{\underset{7}{\cancel{21}}} = \frac{24}{35}$$

EXAMPLE 5 Divide: $8\frac{1}{3} \div 2\frac{1}{3}$.

SOLUTION

$$8\frac{1}{3} \div 2\frac{1}{3} = \frac{25}{3} \div \frac{7}{3} = \frac{25}{\underset{1}{\cancel{3}}} \div \frac{\overset{1}{\cancel{3}}}{7} = \frac{25}{7} \quad \text{or} \quad 3\frac{4}{7}$$

EXAMPLE 6 Divide: $3\frac{1}{4} \div \frac{2}{3}$

SOLUTION

$$3\frac{1}{4} \div \frac{2}{3} = \frac{13}{4} \div \frac{2}{3} = \frac{13}{4} \cdot \frac{3}{2} = \frac{39}{8} \quad \text{or} \quad 4\frac{7}{8}$$

EXAMPLE 7 Divide: $6 \div 2\frac{2}{3}$

SOLUTION We write 6 as $\frac{6}{1}$ and $2\frac{2}{3}$ as $\frac{8}{3}$. Then

$$6 \div 2\frac{2}{3} = \frac{6}{1} \div \frac{8}{3} = \frac{\overset{3}{\cancel{6}}}{1} \cdot \frac{3}{\underset{4}{\cancel{8}}} = \frac{9}{4} \quad \text{or} \quad 2\frac{1}{4}$$

STOP When multiplying or dividing mixed numbers, you do *not* need to find the LCD. The correct method is to change all mixed numbers to improper fractions and then multiply or divide these improper fractions. Multiplying or dividing fractions does not require finding an LCD.

3.2 Exercises

▼ Multiply or divide as indicated.

1. $2\frac{3}{4} \cdot 3\frac{1}{2}$

2. $4\frac{3}{4} \cdot 1\frac{1}{3}$

3. $5\frac{1}{8} \div 9\frac{1}{3}$

4. $2\frac{5}{8} \div 3\frac{1}{9}$

5. $7\frac{1}{8} \cdot 3\frac{3}{7}$

6. $5\frac{2}{3} \cdot 7\frac{1}{7}$

7. $6\frac{4}{5} \div 4\frac{1}{6}$

8. $3\frac{2}{3} \div 2\frac{1}{4}$

9. $2\frac{1}{4} \cdot 3\frac{5}{9}$

10. $7\frac{6}{7} \cdot 4\frac{5}{8}$

11. $\frac{9}{16} \div 5\frac{1}{4}$

12. $\frac{4}{13} \div 3\frac{1}{4}$

13. $16 \cdot 4\frac{1}{8}$

14. $24 \cdot 5\frac{5}{12}$

15. $1\frac{2}{3} \div 9$

16. $3\frac{1}{5} \div 8$

17. $14 \div 2\frac{1}{3}$

18. $16 \div 5\frac{1}{6}$

19. $4\frac{2}{3} \div 7$

20. $5\frac{5}{8} \div 9$

21. $2\frac{1}{3} \cdot 3\frac{1}{8} \cdot 1\frac{1}{2}$

22. $2\frac{2}{3} \cdot 1\frac{1}{4} \cdot 1\frac{1}{5}$

23. $4\frac{1}{8} \cdot 9 \cdot 4\frac{4}{11}$

24. $5\frac{2}{5} \cdot 6 \cdot \frac{5}{9}$

25. $6 \cdot \frac{7}{8} \cdot 1\frac{5}{6}$

26. $9 \cdot \frac{2}{3} \cdot 3\frac{1}{3}$

27. $3\frac{1}{5} \cdot \frac{3}{8} \div \frac{3}{10}$

28. $4\frac{2}{5} \cdot 3\frac{3}{4} \div 5\frac{1}{2}$

29. $3\frac{1}{4} \cdot 2\frac{2}{3} \div 1\frac{2}{3}$

30. $7\frac{1}{2} \cdot 1\frac{3}{5} \div 1\frac{1}{5}$

31. $2\frac{1}{3} \div 2\frac{2}{3} \cdot 2\frac{2}{7}$

32. $1\frac{1}{9} \div 1\frac{2}{3} \cdot 1\frac{1}{2}$

33. Find the product of $5\frac{1}{4}$ and $2\frac{2}{7}$.

34. What is the product of $4\frac{2}{3}$ and $3\frac{3}{7}$?

35. What is the quotient of $\frac{7}{8}$ and $2\frac{1}{4}$?

36. How much is $6\frac{2}{3}$ divided by $\frac{3}{10}$?

37. Divide $5\frac{1}{4}$ by $2\frac{1}{3}$, and then multiply this result by $\frac{4}{9}$.

38. Find the product of $3\frac{3}{5}$ and $\frac{1}{3}$, and then divide this result by 6.

39. What is the result if you divide 9 by $6\frac{3}{5}$?

40. What is the result if you multiply $7\frac{4}{5}$ by 20?

LIVING IN THE REAL WORLD

41. If $2\frac{2}{3}$ yards of material is required to make a blouse, how much material will be needed to make 9 blouses?

42. Darla needs $3\frac{1}{3}$ yards of fabric to make a skirt. How much fabric is required to make 3 skirts?

43. Carol is making shelves that are $4\frac{1}{4}$ feet long. How many shelves can she cut from a 17-foot board?

44. Jonathan needs pieces of cord $3\frac{7}{16}$ in. long. How many such pieces can be cut from 110 in. of cord?

45. Kerry is using a recipe that calls for $3\frac{3}{4}$ cups of flour. How much flour is needed to make $\frac{1}{3}$ of the recipe?

46. How much sugar will be needed if a recipe is being tripled and the original recipe calls for $1\frac{3}{4}$ cups?

47. When squaring a mixed number, can you square the whole-number part and add the result to the square of the fractional part? Does $(3\frac{1}{2})^2 = 3^2 + \left(\frac{1}{2}\right)^2$?

48. What is the proper procedure for squaring a mixed number? Illustrate your explanation with an example.

3.3 Addition of Mixed Numbers

When we are asked to add two mixed numbers, such as $3\frac{1}{4} + 5\frac{2}{3}$, it is generally easiest to add the fractional parts together and then add the whole numbers together separately. Adding the fractions $\frac{1}{4} + \frac{2}{3}$ first requires that we find the LCD, which is 12. Then writing the given fractions as equivalent fractions with 12 as the new denominator gives us

$$\frac{1}{4} + \frac{2}{3} = \frac{3}{12} + \frac{8}{12} = \frac{11}{12}$$

Now we add the whole-number part, $3 + 5 = 8$, and our answer is $8\frac{11}{12}$.

The mixed numbers being added are usually written in a column. The work then looks like this:

$$\begin{array}{rcccl} 3\frac{1}{4} & = & 3\frac{1}{4}\cdot\frac{3}{3} & = & 3\frac{3}{12} \\ +5\frac{2}{3} & = & 5\frac{2}{3}\cdot\frac{4}{4} & = & 5\frac{8}{12} \\ \hline & & & & 8\frac{11}{12} \end{array}$$

EXAMPLE 1 Add: $6\frac{1}{3} + 7\frac{2}{5}$.

SOLUTION We first look at the fractional parts and find that the LCD is 15. We rewrite the fractions as equivalent fractions with 15 as the new denominator:

$$\frac{1}{3} = \frac{1}{3}\cdot\frac{5}{5} = \frac{5}{15} \quad \text{and} \quad \frac{2}{5} = \frac{2}{5}\cdot\frac{3}{3} = \frac{6}{15}$$

We can now do the addition. We add the fractional parts and the whole-number parts separately.

$$\begin{array}{rcl} 6\frac{1}{3} & = & 6\frac{5}{15} \\ +7\frac{2}{5} & = & 7\frac{6}{15} \\ \hline & & 13\frac{11}{15} \end{array}$$

▮▮

EXAMPLE 2 Add: $9\frac{3}{4} + 6\frac{2}{3}$.

SOLUTION Finding the LCD for the fractional parts gives us

$$\frac{3}{4} = \frac{3}{4}\cdot\frac{3}{3} = \frac{9}{12} \quad \text{and} \quad \frac{2}{3} = \frac{2}{3}\cdot\frac{4}{4} = \frac{8}{12}$$

Then adding gives us

$$\begin{array}{rcl} 9\frac{3}{4} & = & 9\frac{9}{12} \\ +6\frac{2}{3} & = & 6\frac{8}{12} \\ \hline & & 15\frac{17}{12} = 15 + \frac{17}{12} = 15 + 1\frac{5}{12} = 16\frac{5}{12} \end{array}$$

When the sum of the fractional parts is an improper fraction, we rewrite it as a mixed number, $\frac{17}{12} = 1\frac{5}{12}$, and add the whole-number part, 1, to the whole numbers in the original problem.

▮▮

EXAMPLE 3 Add: $3\frac{7}{8} + 4\frac{1}{2} + 7\frac{3}{4}$.

SOLUTION After noting that

$$\frac{7}{8} = \frac{7}{8}, \quad \frac{1}{2} = \frac{1}{2} \cdot \frac{4}{4} = \frac{4}{8} \quad \text{and} \quad \frac{3}{4} = \frac{3}{4} \cdot \frac{2}{2} = \frac{6}{8}$$

we write

$$\begin{array}{rcl} 3\frac{7}{8} & = & 3\frac{7}{8} \\ 4\frac{1}{2} & = & 4\frac{4}{8} \\ +7\frac{3}{4} & = & 7\frac{6}{8} \\ \hline & & 14\frac{17}{8} = 14 + \frac{17}{8} = 14 + 2\frac{1}{8} = 16\frac{1}{8} \end{array}$$

EXAMPLE 4 Add: $4\frac{2}{3} + 2\frac{7}{8}$.

SOLUTION

$$\frac{2}{3} = \frac{2}{3} \cdot \frac{8}{8} = \frac{16}{24} \quad \text{and} \quad \frac{7}{8} = \frac{7}{8} \cdot \frac{3}{3} = \frac{21}{24}$$

$$\begin{array}{rcl} 4\frac{2}{3} & = & 4\frac{16}{24} \\ +2\frac{7}{8} & = & 2\frac{21}{24} \\ \hline & & 6\frac{37}{24} = 6 + \frac{37}{24} = 6 + 1\frac{13}{24} = 7\frac{13}{24} \end{array}$$

Another Method of Adding Mixed Numbers

If the whole-number parts of the mixed numbers to be added are small, it may be convenient to convert the mixed numbers to improper fractions and then follow the rules that we have already learned for adding fractions.

EXAMPLE 5 Add: $2\frac{1}{5} + 1\frac{2}{3}$.

SOLUTION We convert each mixed number to an improper fraction and add the fractions:

$$2\frac{1}{5} + 1\frac{2}{3} = \frac{11}{5} + \frac{5}{3} = \frac{33}{15} + \frac{25}{15} = \frac{58}{15} = 3\frac{13}{15}$$

EXAMPLE 8 Add: $1\frac{3}{8} + 4\frac{1}{2} + 1\frac{3}{4}$.

SOLUTION

$$1\frac{3}{8} + 4\frac{1}{2} + 1\frac{3}{4} = \frac{11}{8} + \frac{9}{2} + \frac{7}{4} = \frac{11}{8} + \frac{36}{8} + \frac{14}{8} = \frac{61}{8} = 7\frac{5}{8}$$

EXAMPLE 7 Add: $4\frac{1}{6} + 3\frac{1}{3}$.

SOLUTION

$$4\frac{1}{6} + 3\frac{1}{3} = \frac{25}{6} + \frac{10}{3} = \frac{25}{6} + \frac{20}{6} = \frac{45}{6} = 7\frac{3}{6} = 7\frac{1}{2}$$

Summarizing these examples gives us a rule for adding mixed numbers.

To add mixed numbers:

Method 1 Add the fractions separately from the whole numbers using the rules for adding fractions. If the result is an improper fraction, change it to a mixed number and add the whole-number part to the whole numbers from the original problem.

Method 2 Use this method only when the whole-number parts are small. Change each mixed number to an improper fraction and add the fractions using the rules for adding fractions.

Do not attempt to use Method 2 if the whole-number parts are large. For example, the following problem is easily solved using Method 1.

$$\begin{array}{rcr} 126\frac{3}{16} & = & 126\frac{3}{16} \\ +\ 518\frac{1}{8} & = & 518\frac{2}{16} \\ \hline & & 644\frac{5}{16} \end{array}$$

To use Method 2, we would have to multiply 126 by 16 and also 518 by 8. Although we know how to do these calculations, performing them increases our chances of making an error. Since we can avoid these calculations by using Method 1, we probably should do so.

3.3 Exercises

▼ Add:

1. $3\frac{3}{5} + 1\frac{1}{2}$
2. $3\frac{1}{9} + 4\frac{3}{4}$
3. $7\frac{5}{8} + 3\frac{1}{16}$
4. $4\frac{2}{5} + 5\frac{3}{10}$
5. $8\frac{7}{9} + 7\frac{5}{6}$
6. $7\frac{4}{7} + 5\frac{2}{3}$

7. $6\frac{3}{16} + 4\frac{4}{3}$

8. $9\frac{5}{8} + 5\frac{7}{16}$

9. $7\frac{7}{8} + 4\frac{7}{10}$

10. $8\frac{7}{9} + 7\frac{2}{3}$

11. $5\frac{1}{6} + 4\frac{3}{4} + 2\frac{7}{8}$

12. $10\frac{1}{2} + 9\frac{3}{14} + 7\frac{5}{7}$

13. $2\frac{1}{2} + 5\frac{1}{8} + 7\frac{3}{4} + 6\frac{3}{8}$

14. $7\frac{5}{12} + 7\frac{1}{8} + 3\frac{3}{4} + 9\frac{1}{2}$

15. $6\frac{2}{3} + 5\frac{1}{4}$

16. $7\frac{1}{4} + 3\frac{5}{16}$

17. $9\frac{11}{12} + 4\frac{5}{8}$

18. $6\frac{3}{5} + 7\frac{2}{3}$

19. $4\frac{7}{9} + 11\frac{5}{6}$

20. $14\frac{11}{12} + 6\frac{3}{8}$

21. $32\frac{11}{14} + 19\frac{9}{21}$

22. $41\frac{7}{11} + 32\frac{23}{22}$

23. $16\frac{1}{20} + 14\frac{3}{10} + 6\frac{4}{5}$

24. $14\frac{7}{12} + 19\frac{2}{3} + 10\frac{3}{16}$

25. $18\frac{13}{15} + 16\frac{2}{3} + 5\frac{1}{5}$

26. $16\frac{2}{3} + 19\frac{1}{4} + 17\frac{5}{6}$

27. $121\frac{5}{8} + 76\frac{3}{4} + 108\frac{1}{8}$

28. $119\frac{7}{8} + 114\frac{5}{8} + 66\frac{1}{2}$

LIVING IN THE REAL WORLD

29. A stock listed on the New York Stock Exchange at $32\frac{3}{8}$ goes up $2\frac{7}{8}$. What is its new value?

30. A stock worth $141\frac{5}{8}$ increases by $5\frac{1}{4}$. What is its new value?

31. On Monday a stock lists by $98\frac{7}{8}$. On Tuesday it goes up $1\frac{1}{4}$, and on Wednesday its value increases by $3\frac{1}{4}$. What is its new value?

32. The opening price for a stock on the American Stock Exchange is $23\frac{1}{2}$. In the next two days, it goes up $\frac{7}{8}$ and $1\frac{3}{4}$, respectively. What is its new price?

33. Find the sum of $7\frac{3}{4}$ in., $5\frac{1}{2}$ in., and $1\frac{1}{16}$ in.

34. What is the total length of two pieces of wood measuring $14\frac{1}{2}$ in. and $17\frac{7}{8}$ in.?

35. What is the shortest-length bolt that will pass through a piece of metal $1\frac{1}{8}$ in. thick, a second piece of metal $1\frac{7}{16}$ in. thick, a washer $\frac{3}{32}$ in. thick, and a nut $\frac{1}{4}$ in. thick?

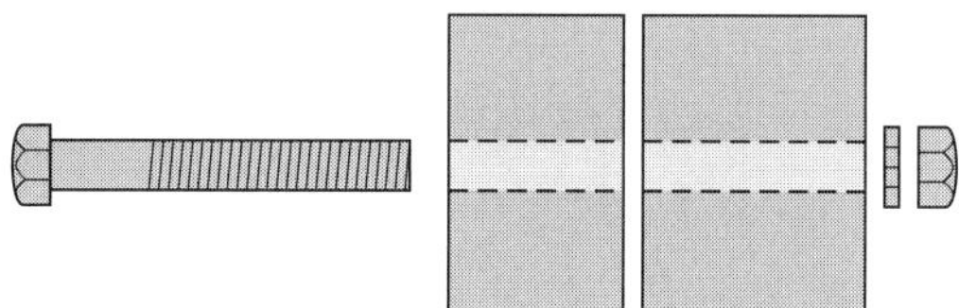

36. Max works part-time as a cashier in a supermarket. He put in $6\frac{1}{2}$ hours on Saturday, $7\frac{1}{4}$ hours on Sunday, $2\frac{3}{4}$ hours on Tuesday, and $3\frac{3}{4}$ hours on Thursday. How many hours did Max work altogether?

37. Tom has a babysitting job in which he works 5 days a week. If he worked $2\frac{1}{2}$ hours on both Monday and Tuesday and $1\frac{3}{4}$ hours each day on Wednesday, Thursday, and Friday, how many hours did he work in total?

38. Marcy is making an outfit that requires $3\frac{7}{8}$ yards of fabric for the skirt, $1\frac{7}{8}$ yards for the vest, and $\frac{3}{4}$ of a yard for a matching scarf. How many yards of fabric does she need to buy to make the outfit?

39. Jennifer studied $2\frac{1}{2}$ hours on Monday, $3\frac{3}{4}$ hours on Tuesday, and $4\frac{3}{4}$ hours on Saturday. How many hours did she study altogether?

40. What is the total distance around a room that measures $13\frac{3}{4}$ ft by $17\frac{1}{2}$ ft?

41. A standard sheet of typewriter paper measures $8\frac{1}{2}$ in. by 11 in. What is the total distance around a standard sheet of paper?

42. Which number is larger, $3\frac{3}{8}$ or $3\frac{5}{16}$? How can you tell?

3.4 Subtraction of Mixed Numbers

To subtract one mixed number from another, we use a method similar to that of addition: Subtract the fractional part first and then subtract the whole-number part. Consider the problem $5\dfrac{5}{8} - 3\dfrac{1}{3}$. First, finding the LCD of the fractions gives us

$$\frac{5}{8} = \frac{5}{8}\cdot\frac{3}{3} = \frac{15}{24} \quad \text{and} \quad \frac{1}{3} = \frac{1}{3}\cdot\frac{8}{8} = \frac{8}{24}$$

Now subtracting, we have

$$\begin{array}{rcr} 5\dfrac{5}{8} & = & 5\dfrac{15}{24} \\ -3\dfrac{1}{3} & = & 3\dfrac{8}{24} \\ \hline & & 2\dfrac{7}{24} \end{array}$$

Suppose that we are given the problem $7\frac{3}{8} - 4\frac{5}{8}$. Since both fractions have the same denominator, we can attempt to subtract immediately:

$$\begin{array}{r} 7\frac{3}{8} \\ -\ 4\frac{5}{8} \\ \hline \end{array}$$

However, we are unable to subtract $\frac{5}{8}$ from $\frac{3}{8}$. To overcome this difficulty, we must borrow 1 from the 7 and add it to the $\frac{3}{8}$. Since $1 = \frac{8}{8}$, we have $1 + \frac{3}{8} = \frac{8}{8} + \frac{3}{8} = \frac{11}{8}$. We organize our work like this:

$$\begin{array}{rcccr} 7\frac{3}{8} & = & \overset{6}{\cancel{7}}\frac{3}{8} + \frac{8}{8} & = & 6\frac{11}{8} \\ -4\frac{5}{8} & & = \frac{11}{8} & & -4\frac{5}{8} \\ \hline & & & & 2\frac{6}{8} = 2\frac{3}{4} \end{array}$$

Note that our answer, $2\frac{6}{8}$, can be reduced to $2\frac{3}{4}$.

EXAMPLE 1 Subtract: $9\frac{1}{5} - 7\frac{2}{3}$.

SOLUTION We note that $\frac{1}{5} = \frac{1}{5} \cdot \frac{3}{3} = \frac{3}{15}$ and $\frac{2}{3} = \frac{2}{3} \cdot \frac{5}{5} = \frac{10}{15}$, so

$$\begin{array}{rcr} 9\frac{1}{5} & = & 9\frac{3}{15} \\ -7\frac{2}{3} & = & -7\frac{10}{15} \\ \hline \end{array}$$

Since we are unable to subtract $\frac{10}{15}$ from $\frac{3}{15}$, we must borrow 1 from the 9. This time we write 1 as $\frac{15}{15}$, because we are going to add it to $\frac{3}{15}$. We have

$$\begin{array}{rcccr} 9\frac{1}{5} & = & \overset{8}{\cancel{9}}\frac{3}{15} + \frac{15}{15} & = & 8\frac{18}{15} \\ -7\frac{2}{3} & = & -7\frac{10}{15} & & -7\frac{10}{15} \\ \hline & & & & 1\frac{8}{15} \end{array}$$

EXAMPLE 2 Subtract: $13\frac{3}{4} - \frac{2}{3}$.

SOLUTION

$$\frac{3}{4} = \frac{3}{4} \cdot \frac{3}{3} = \frac{9}{12} \quad \text{and} \quad \frac{2}{3} = \frac{2}{3} \cdot \frac{4}{4} = \frac{8}{12}$$

$$\begin{array}{rcr} 13\frac{3}{4} & = & 13\frac{9}{12} \\ -\ \frac{2}{3} & = & -\ \frac{8}{12} \\ \hline & & 13\frac{1}{12} \end{array}$$

▮▮

EXAMPLE 3 Subtract: $13 - 3\frac{1}{3}$.

SOLUTION

$$\begin{array}{rcr} 13 & = & 12\frac{3}{3} \\ -\ 3\frac{1}{3} & = & -\ 3\frac{1}{3} \\ \hline & & 9\frac{2}{3} \end{array}$$

▮▮

EXAMPLE 4 Subtract: $9 - 5\frac{13}{20}$.

SOLUTION

$$\begin{array}{rcr} 9 & = & 8\frac{20}{20} \\ -5\frac{13}{20} & = & -5\frac{13}{20} \\ \hline & & 3\frac{7}{20} \end{array}$$

▮▮

Another Method for Subtracting Mixed Numbers

As in the case of adding mixed numbers, if the whole-number parts of the mixed numbers being subtracted are small, we can convert each mixed number to an improper fraction and follow the rules that we learned for subtracting fractions.

EXAMPLE 5 Subtract: $3\frac{1}{5} - 2\frac{1}{3}$.

SOLUTION We convert each mixed number to an improper fraction and subtract.

$$3\frac{1}{5} = \frac{16}{5} \quad \text{and} \quad 2\frac{1}{3} = \frac{7}{3}$$

$$3\frac{1}{5} - 2\frac{1}{3} = \frac{16}{5} - \frac{7}{3}$$ Change to improper fractions.

$$= \frac{48}{15} - \frac{35}{15}$$ The LCD is 15.

$$= \frac{13}{15}$$ Subtract the numerators.

EXAMPLE 6 Subtract: $2\frac{3}{8} - 1\frac{3}{16}$.

SOLUTION

$$2\frac{3}{8} - 1\frac{3}{16} = \frac{19}{8} - \frac{19}{16} = \frac{38}{16} - \frac{19}{16} = \frac{19}{16} = 1\frac{3}{16}$$

EXAMPLE 7 Subtract: $5\frac{2}{3} - 2\frac{5}{12}$.

SOLUTION

$$5\frac{2}{3} - 2\frac{5}{12} = \frac{17}{3} - \frac{29}{12} = \frac{68}{12} - \frac{29}{12} = \frac{39}{12} = 3\frac{3}{12} = 3\frac{1}{4}$$

In summary, our rule is this:

To subtract mixed numbers:

Method 1 Subtract the fractions separately, borrowing from the whole number if necessary, then subtract the whole-number parts.
Method 2 Use this method only when the whole-number parts are small. Change each mixed number to an improper fraction and subtract, using the rules for subtracting fractions.

3.4 Exercises

▼ Subtract, reducing all fractions to lowest terms.

1. $4\frac{3}{4} - 2\frac{1}{4}$

2. $7\frac{7}{9} - 5\frac{3}{9}$

3. $5\frac{7}{8} - 3\frac{1}{4}$

4. $6\frac{2}{3} - 4\frac{2}{9}$

5. $7\frac{5}{12} - 2\frac{1}{6}$

6. $11\frac{9}{16} - 8\frac{5}{12}$

7. $\begin{array}{r} 9\frac{1}{4} \\ -3\frac{1}{2} \\ \hline \end{array}$

8. $\begin{array}{r} 12\frac{1}{3} \\ -\ 4\frac{5}{6} \\ \hline \end{array}$

9. $\begin{array}{r} 26\frac{1}{8} \\ -14\frac{7}{12} \\ \hline \end{array}$

10. $\begin{array}{r} 22\frac{3}{16} \\ -15\frac{7}{8} \\ \hline \end{array}$

11. $\begin{array}{r} 8 \\ -5\frac{2}{3} \\ \hline \end{array}$

12. $\begin{array}{r} 7 \\ -4\frac{5}{8} \\ \hline \end{array}$

13. $\begin{array}{r} 16 \\ -11\frac{5}{12} \\ \hline \end{array}$

14. $\begin{array}{r} 18 \\ -14\frac{7}{9} \\ \hline \end{array}$

15. $16\frac{5}{8} - 13\frac{3}{10}$

16. $29\frac{2}{3} - 16\frac{1}{5}$

17. $8 - \dfrac{2}{3}$

18. $17 - \dfrac{4}{5}$

19. $27 - 14\frac{3}{5}$

20. $31 - 26\frac{7}{8}$

21. $123\frac{7}{16} - 118\frac{3}{4}$

22. $226\frac{2}{7} - 86\frac{1}{14}$

23. $43\frac{9}{16} - 38\frac{11}{12}$

24. $93\frac{3}{8} - 76\frac{11}{12}$

25. Find the difference between $14\frac{7}{16}$ and $9\frac{11}{12}$.

26. What is the difference between $71\frac{1}{3}$ and $58\frac{5}{6}$?

LIVING IN THE REAL WORLD

27. On Monday morning, a stock opened at $38\frac{1}{8}$. By noon it had lost $2\frac{5}{8}$. What was its value at noon?

28. A stock valued at 97 goes down $3\frac{5}{8}$. What is its new value?

29. Last week Doug Cohen worked $37\frac{1}{2}$ hours. This week he worked $19\frac{1}{2}$ hours. How many more hours did he work the previous week?

30. David wants to make pancakes for breakfast. The recipe calls for $2\frac{1}{4}$ cups of milk, but he has only $\frac{3}{4}$ cup. How much milk does he need to borrow from his neighbor?

31. Yvette Greene plans to study a total of 8 hours for a history exam that will be given on Friday. On Monday she studied $1\frac{1}{2}$ hours, and on Wednesday she put in $3\frac{3}{4}$ hours. How much more studying does she need to do?

32. Jason Cook is allowed to work 20 hours per week at his part-time job. He worked $7\frac{1}{2}$ hours on Saturday and $6\frac{3}{4}$ hours on Sunday. How many more hours must he work to make a total of 20 hours?

33. Joann is permitted to work a maximum of 12 hours per week as part of the work-study program at her college. She worked $3\frac{3}{4}$ hours on Monday, 1 hour on Tuesday, and $1\frac{3}{4}$ hours on Wednesday. How many more hours is she allowed to work this week?

34. Don Perez has $3\frac{1}{2}$ gallons of paint. It will take $1\frac{1}{2}$ gallons to paint his bedroom and $2\frac{3}{4}$ gallons to paint the living room. How much more paint does he need?

35. Two different methods for subtracting mixed numbers were described in this section. Explain when you might use one method rather than the other.

3.5 Complex Fractions

A **simple fraction** contains only whole numbers in both its numerator and denominator. Simple fractions are the only kind of fractions we have considered so far.

A **complex fraction** contains fractions in either the numerator or the denominator or both. It will be our job now to change complex fractions into simple fractions.

When more than one fraction is involved in a problem, we must be especially careful to identify the **main fraction line.** It is generally longer and thicker than the other fraction lines. Examples of complex fractions are:

$$\frac{\frac{3}{4}}{\frac{7}{3}} \qquad \frac{15}{\frac{2}{3}} \qquad \frac{\frac{3}{4}+\frac{7}{8}}{\frac{5}{12}}$$

main fraction line

EXAMPLE 1 Simplify the complex fraction $\frac{\frac{3}{4}}{\frac{5}{7}}$.

SOLUTION There is only one fraction above and one fraction below the main fraction line. Since this line means division, we will divide the numerator, $\frac{3}{4}$, by the denominator $\frac{5}{7}$:

$$\frac{\frac{3}{4}}{\frac{5}{7}} = \frac{3}{4} \div \frac{5}{7} \qquad \text{The fraction line means divide.}$$

$$= \frac{3}{4} \cdot \frac{7}{5} \qquad \text{Multiply by the reciprocal of } \frac{5}{7}.$$

$$= \frac{21}{20} \quad \text{or} \quad 1\frac{1}{20}$$

EXAMPLE 2 Simplify the complex fraction $\dfrac{\frac{1}{4}+\frac{3}{8}}{\frac{5}{12}+\frac{2}{3}}$.

SOLUTION There are two methods of simplifying this complex fraction, and we will examine each of them.

Method 1

We can add the fractions in the numerator of the complex fraction and then add the fractions in the denominator. This will give us one fraction on top, divided by a second fraction on the bottom. On top we have

$$\frac{1}{4}+\frac{3}{8}=\frac{1\cdot 2}{4\cdot 2}+\frac{3}{8} \quad \text{8 is the LCD.}$$
$$=\frac{2}{8}+\frac{3}{8} \quad \text{Add the numerators.}$$
$$=\frac{5}{8}$$

Adding the bottom fractions gives us

$$\frac{5}{12}+\frac{2}{3}=\frac{5}{12}+\frac{2\cdot 4}{3\cdot 4} \quad \text{12 is the LCD.}$$
$$=\frac{5}{12}+\frac{8}{12} \quad \text{Add the numerators.}$$
$$=\frac{13}{12}$$

Now we divide the top fraction by the bottom fraction, as we did in Example 1.

$$\frac{\frac{1}{4}+\frac{3}{8}}{\frac{5}{12}+\frac{2}{3}}=\frac{\frac{5}{8}}{\frac{13}{12}}$$
$$=\frac{5}{8}\div\frac{13}{12} \quad \text{The main fraction line means divide.}$$
$$=\frac{5}{8}\cdot\frac{12}{13} \quad \frac{12}{13} \text{ is the reciprocal of } \frac{13}{12}.$$
$$=\frac{5}{\cancel{8}_{2}}\cdot\frac{\cancel{12}^{3}}{13} \quad \text{Cancel factors from numerator and denominator and multiply.}$$
$$=\frac{15}{26}$$

Method 2

The LCD of all four fractions in the problem is 24, so if we multiply the top and bottom of the complex fraction by 24, all the denominators will divide it, leaving only whole numbers in the top and bottom.

$$\frac{\frac{1}{4}+\frac{3}{3}}{\frac{5}{12}+\frac{2}{3}} = \frac{24(\frac{1}{4}+\frac{3}{8})}{24(\frac{5}{12}+\frac{2}{3})}$$ Multiply top and bottom by 24, the LCD.

$$= \frac{\overset{6}{\cancel{24}} \cdot \dfrac{1}{\underset{1}{\cancel{4}}} + \overset{3}{\cancel{24}} \cdot \dfrac{3}{\underset{1}{\cancel{8}}}}{\overset{2}{\cancel{24}} \cdot \dfrac{5}{\underset{1}{\cancel{12}}} + \overset{8}{\cancel{24}} \cdot \dfrac{2}{\underset{1}{\cancel{3}}}}$$ Apply the distributive property.

$$= \frac{6+9}{10+16}$$ Add the whole numbers.

$$= \frac{15}{26}$$

This is the same result we obtained before. ▮▮

In summary, there are two methods for changing a complex fraction to a simple one:

To change a complex fraction to a simple fraction:

Method 1 Obtain a single fraction in both numerator and denominator, and divide the top fraction by the fraction on the bottom.
Method 2 Multiply both the top and the bottom of the complex fraction by the LCD of all the denominators in the problem and simplify the resulting fraction.

EXAMPLE 3 Simplify $\dfrac{10}{\frac{2}{3}-\frac{1}{4}}$.

SOLUTION **Method 1**

First we write 10 as $\dfrac{10}{1}$. Next we combine the fractions on the bottom.

$$\frac{2}{3}-\frac{1}{4} = \frac{2\cdot 4}{3\cdot 4} - \frac{1\cdot 3}{4\cdot 3}$$ The LCD is 12.

$$= \frac{8}{12} - \frac{3}{12} = \frac{5}{12}$$ Subtract the numerators.

Now we divide the top fraction by the bottom fraction.

$$\frac{10}{\frac{2}{3}-\frac{1}{4}} = \frac{\frac{10}{1}}{\frac{5}{12}}$$

$$= \frac{10}{1} \div \frac{5}{12}$$ The divisor is $\frac{5}{12}$.

$$= \frac{\overset{2}{\cancel{10}}}{1} \cdot \frac{12}{\underset{1}{\cancel{5}}}$$ Invert and multiply.

$$= 24$$

Method 2

We multiply the top and bottom by the LCD, which is 12:

$$\frac{10}{\frac{2}{3}-\frac{1}{4}} = \frac{12(\frac{10}{1})}{12(\frac{2}{3}-\frac{1}{4})}$$ The LCD is 12.

$$= \frac{12 \cdot 10}{\overset{4}{\cancel{12}} \cdot \frac{2}{\underset{1}{\cancel{3}}} - \overset{3}{\cancel{12}} \cdot \frac{1}{\underset{1}{\cancel{4}}}}$$ The distributive property.

$$= \frac{120}{8-3}$$ Combine the whole numbers.

$$= \frac{120}{5}$$ Simplify.

$$= 24$$ This is the same result we obtained using Method 1.

EXAMPLE 4 Simplify $\frac{3\frac{1}{3}}{2\frac{2}{9}}$.

SOLUTION Since we can easily write the mixed numbers in both the top and the bottom as improper fractions, the solution lends itself to Method 1.

$$\frac{3\frac{1}{3}}{2\frac{2}{9}} = \frac{\frac{10}{3}}{\frac{20}{9}}$$ Change to improper fractions.

$$= \frac{10}{3} \div \frac{20}{9}$$ Divide.

$$= \frac{\overset{1}{\cancel{10}}}{\underset{1}{\cancel{3}}} \cdot \frac{\overset{3}{\cancel{9}}}{\underset{2}{\cancel{20}}}$$ Multiply by the reciprocal of $\frac{20}{9}$.

$$= \frac{3}{2} \text{ or } 1\frac{1}{2}$$ The result can be written either way.

EXAMPLE 5 Simplify $\dfrac{\frac{7}{8}-\frac{1}{12}}{\frac{3}{4}+\frac{3}{8}}$.

SOLUTION In this case, it is probably more efficient to use Method 2 than Method 1, because there are so many different denominators.

$$\frac{\frac{7}{8}-\frac{1}{12}}{\frac{3}{4}+\frac{3}{8}} = \frac{24(\frac{7}{8}-\frac{1}{12})}{24(\frac{3}{4}+\frac{3}{8})} \quad \text{Multiply by 24, the LCD.}$$

$$= \frac{\overset{3}{\cancel{24}}\cdot\frac{7}{\underset{1}{\cancel{8}}} - \overset{2}{\cancel{24}}\cdot\frac{1}{\underset{1}{\cancel{12}}}}{\overset{6}{\cancel{24}}\cdot\frac{3}{\underset{1}{\cancel{4}}} + \overset{3}{\cancel{24}}\cdot\frac{3}{\underset{1}{\cancel{8}}}} \quad \text{Distribute the 24 and cancel.}$$

$$= \frac{21-2}{18+9} \quad \text{Combine.}$$

$$= \frac{19}{27}$$

3.5 Exercises

▼ Simplify the complex fractions.

1. $\dfrac{\frac{2}{3}}{\frac{1}{6}}$

2. $\dfrac{\frac{1}{2}}{\frac{3}{4}}$

3. $\dfrac{\frac{2}{5}}{\frac{2}{10}}$

4. $\dfrac{\frac{4}{7}}{\frac{1}{14}}$

5. $\dfrac{6}{\frac{12}{7}}$

6. $\dfrac{15}{\frac{9}{16}}$

7. $\dfrac{\frac{3}{4}}{12}$

8. $\dfrac{\frac{3}{8}}{9}$

9. $\dfrac{8}{\frac{5}{4}}$

10. $\dfrac{\frac{8}{5}}{16}$

11. $\dfrac{\frac{1}{4}+\frac{3}{5}}{\frac{1}{10}}$

12. $\dfrac{\frac{2}{3}+\frac{5}{6}}{\frac{1}{12}}$

13. $\dfrac{\frac{5}{6}-\frac{2}{3}}{\frac{5}{16}}$

14. $\dfrac{\frac{7}{8}-\frac{3}{4}}{\frac{3}{8}}$

15. $\dfrac{\frac{5}{9}-\frac{1}{3}}{\frac{3}{4}+\frac{3}{4}}$

16. $\dfrac{\frac{7}{10}-\frac{3}{5}}{\frac{7}{8}+\frac{1}{4}}$

17. $\dfrac{6+\frac{1}{4}}{7-\frac{3}{8}}$

18. $\dfrac{5+\frac{2}{3}}{11-\frac{1}{3}}$

19. $\dfrac{\frac{5}{18} + \frac{1}{9}}{7 - \frac{3}{8}}$

20. $\dfrac{\frac{7}{8} + \frac{5}{16}}{4 - \frac{7}{8}}$

21. $\dfrac{5\frac{2}{3}}{7\frac{3}{8}}$

22. $\dfrac{10\frac{1}{2}}{7\frac{7}{10}}$

23. $\dfrac{\frac{1}{8} + \frac{1}{4}}{3\frac{1}{4}}$

24. $\dfrac{\frac{3}{5} + \frac{4}{5}}{3\frac{3}{4}}$

25. $\dfrac{2\frac{1}{2} + 3\frac{1}{4}}{6\frac{2}{3}}$

26. $\dfrac{3\frac{1}{5} + 2\frac{2}{3}}{5\frac{1}{3}}$

27. $\dfrac{3\frac{5}{9}}{10\frac{5}{8} - 4\frac{1}{16}}$

28. $\dfrac{4\frac{1}{4}}{6\frac{3}{4} - 4\frac{1}{8}}$

29. $\dfrac{3\frac{5}{12} + 2\frac{1}{4}}{5\frac{1}{3} + 7\frac{1}{4}}$

30. $\dfrac{9\frac{5}{8} + 7\frac{1}{4}}{6\frac{3}{4} + 4\frac{1}{8}}$

31. $\dfrac{8\frac{5}{6} + 2\frac{1}{12}}{3\frac{4}{5} - 1\frac{9}{10}}$

32. $\dfrac{3\frac{4}{5} + 1\frac{3}{10}}{6\frac{5}{6} - 3\frac{3}{4}}$

33. When we multiply the top and bottom of a complex fraction by the LCD of all the fractions in the problem, we find that each of the denominators divides the LCD evenly, with no remainder. What if this doesn't happen?

SUMMARY—CHAPTER 3

The numbers in brackets refer to the section in the text in which each concept is presented.

[3.1] A **mixed number** is a combination of a whole number and a fraction.

EXAMPLES

[3.1] $7\frac{2}{3}$ is a mixed number.

[3.1] An **improper fraction** is a fraction in which the numerator is larger than the denominator.

[3.1] $\frac{41}{5}$ is an improper fraction

[3.1] **To change a mixed number to an improper fraction,** multiply the whole number by the denominator of the fraction and add this result to the numerator of the fraction. This number becomes the new numerator in the improper fraction. The denominator stays the same.

[3.1] $7\frac{2}{3} = \frac{23}{3}$

[3.1] **To change an improper fraction to a mixed number,** divide the numerator by the denominator.

[3.1] $\frac{41}{5} = 8\frac{1}{5}$

[3.2] **To multiply mixed numbers,** change each to an improper fraction and multiply the fractions.

[3.2] $6\frac{2}{3} \cdot 3\frac{1}{5} = \frac{\overset{4}{\cancel{20}}}{3} \cdot \frac{16}{\underset{1}{\cancel{5}}} = \frac{64}{3}$ or $21\frac{1}{3}$

[3.2] **To divide mixed numbers,** change each to an improper fraction and divide the fractions.

[3.2] $4\frac{1}{8} \div 1\frac{3}{4} = \frac{33}{8} \div \frac{7}{4} = \frac{33}{\underset{2}{\cancel{8}}} \cdot \frac{\overset{1}{\cancel{4}}}{7}$

$= \frac{33}{14}$ or $2\frac{5}{14}$

[3.3] **To add mixed numbers:**

Method 1 Add the fractions separately, using the rules for adding fractions. If the result is an improper fraction, change it to a mixed number and add the whole-number part in with the whole numbers from the original problem.

[3.3]

$$\begin{array}{rl} 6\frac{2}{3} = & 6\frac{8}{12} \\ +5\frac{3}{4} = & 5\frac{9}{12} \\ \hline & 11\frac{17}{12} = 11 + 1\frac{5}{12} \\ & = 12\frac{5}{12} \end{array}$$

Method 2 Use when the whole-number parts are small. Change each mixed number to an improper fraction and add the fractions, using the rules for adding fractions that we learned previously.

$2\frac{2}{5} + 1\frac{2}{3} = \frac{12}{5} + \frac{5}{3}$

$= \frac{36}{15} + \frac{25}{15}$

$= \frac{61}{15}$ or $4\frac{1}{15}$

[3.4] **To subtract mixed number:**

Method 1 Subtract the fractions separately, borrowing from the whole number if necessary. Then subtract the whole-number parts.

[3.4]

$$\begin{array}{rl} 8\frac{1}{6} = & 7\frac{7}{6} \\ -4\frac{5}{6} = & -\ 4\frac{5}{6} \\ \hline & 3\frac{2}{6} = 3\frac{1}{3} \end{array}$$

EXAMPLES

Method 2 Use when the whole-number parts are small. Change each mixed number to an improper fraction and subtract using the rules for subtraction of fractions.

$$3\frac{2}{5} - 1\frac{1}{3} = \frac{17}{5} - \frac{4}{3} = \frac{51}{15} - \frac{20}{15} = \frac{31}{15} \text{ or } 2\frac{1}{15}$$

[3.5] **To change a complex fraction to a simple fraction:**

Method 1 Obtain a single fraction in both numerator and denominator, and divide the top fraction by the bottom fraction.

[3.5]

$$\frac{\frac{2}{3}+\frac{1}{4}}{\frac{5}{6}} = \frac{\frac{8}{12}+\frac{3}{12}}{\frac{5}{6}} = \frac{\frac{11}{12}}{\frac{5}{6}} = \frac{11}{12} \div \frac{5}{6} = \frac{11}{\cancel{12}_{2}} \times \frac{\cancel{6}^{1}}{5} = \frac{11}{10} \text{ or } 1\frac{1}{10}$$

Method 2 Multiply both the top and the bottom of the complex fraction by the LCD of all of the denominators in the problem; simplify.

$$\frac{\frac{2}{3}+\frac{1}{4}}{\frac{5}{6}} = \frac{12(\frac{2}{3}+\frac{1}{4})}{12(\frac{5}{6})} = \frac{\frac{\cancel{12}^{4}}{1}\cdot\frac{2}{\cancel{3}_{1}} + \frac{\cancel{12}^{3}}{1}\cdot\frac{1}{\cancel{4}_{1}}}{\frac{\cancel{12}^{2}}{1}\cdot\frac{5}{\cancel{6}_{1}}} = \frac{8+3}{10} = \frac{11}{10} \text{ or } 1\frac{1}{10}$$

REVIEW EXERCISES—CHAPTER 3

Numbers in brackets refer to the section in the text in which similar problems are presented.

[3.1] Write each mixed number as an improper fraction.

1. $5\frac{5}{7}$ 2. $6\frac{3}{4}$ 3. $10\frac{3}{11}$

[3.1] Write each improper fraction as a mixed number.

4. $\dfrac{21}{4}$ 5. $\dfrac{49}{5}$ 6. $\dfrac{106}{9}$

[3.1]

7. Give both the mixed number and the improper fraction that is represented by the shaded portion of the diagram.

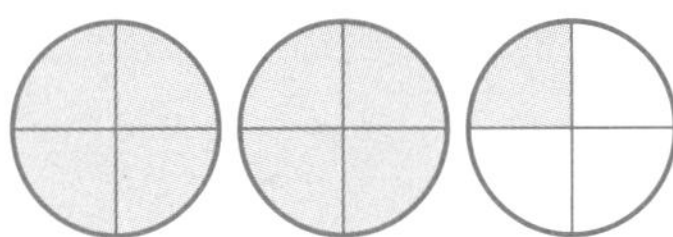

[3.2] Multiply:

8. $9\frac{3}{7} \cdot 4\frac{2}{3}$ 9. $6 \cdot 5\frac{2}{3}$

[3.2] Divide:

10. $7\frac{3}{5} \div 4\frac{3}{10}$ 11. $5\frac{5}{8} \div 4$

[3.2]

12. Multiply: $\frac{5}{16} \cdot 4\frac{3}{4} \cdot 8.$

[3.3] Add:

13. $\begin{array}{r} 6\frac{1}{3} \\ 5\frac{4}{5} \\ +12\frac{4}{15} \\ \hline \end{array}$ 14. $9\frac{2}{3} + 5\frac{5}{6}$

[3.4] Subtract:

15. $\begin{array}{r} 11\frac{1}{5} \\ -\ 6\frac{2}{3} \\ \hline \end{array}$ 16. $29\frac{5}{6} - 11\frac{3}{8}$

17. $\begin{array}{r} 62 \\ -14\frac{5}{16} \\ \hline \end{array}$ 18. $9 - \dfrac{11}{12}$

[3.5] Simplify the complex fractions:

19. $\dfrac{\frac{3}{4}}{\frac{9}{16}}$ 20. $\dfrac{9}{\frac{11}{3}}$

21. $\dfrac{7\frac{1}{3}}{8\frac{3}{4}}$ 22. $\dfrac{5\frac{1}{3}}{\frac{2}{3} + \frac{7}{12}}$

LIVING IN THE REAL WORLD

[3.2]

23. Andy Schwartz has decided to make fudge. The recipe calls for $1\frac{3}{4}$ cups of sugar. How much sugar will he need if he plans to triple the recipe?

▼ [3.3] [3.4]

24. A stock has a value of $79\frac{5}{8}$ on Monday. On Tuesday it goes up $1\frac{1}{8}$, and on Wednesday it goes down $2\frac{7}{8}$. What is its new value?

▼ [3.3] [3.4]

25. Laura works a maximum of 15 hours per week at her work-study job at the college she attends. If she works $2\frac{1}{4}$ hours one day, $3\frac{3}{4}$ hours the next day, and $2\frac{3}{4}$ hours the third day, how many more hours must she work to complete her 15 hours?

CRITICAL THINKING EXERCISES—CHAPTER 3

Tell whether each of the following is true or false and give reasons why.

1. $6\frac{2}{3} = \overset{2}{\cancel{6}} \cdot \frac{2}{\underset{1}{\cancel{3}}} = 4$

2. $3\frac{2}{5} = \frac{(3 \cdot 2) + 5}{5} = \frac{6 + 5}{5} = \frac{11}{5}$

3. The two mixed numbers $2\frac{1}{3} + 1\frac{3}{4}$ can be added either by adding the whole numbers and adding the fractions separately or by changing each mixed number to an improper fraction and then adding the improper fractions.

4. $4\frac{1}{4} \cdot 3\frac{3}{5} = (4 \cdot 3) + \left(\frac{1}{4} \cdot \frac{3}{5}\right) = 12 + \frac{3}{20} = 12\frac{3}{20}$

5. $5\frac{1}{3} \cdot \frac{3}{4} = 5 + \frac{1}{3} \cdot \frac{3}{4}$ $5\frac{1}{3}$ means $5 + \frac{1}{3}$.

$= 5 + \frac{1}{\underset{1}{\cancel{3}}} \cdot \frac{\overset{1}{\cancel{3}}}{4}$ Multiply before adding.

$= 5 + \frac{1}{4}$ Now add.

$= 5\frac{1}{4}$

▼ Find the error in each of the following.

6. $\frac{6}{2\frac{2}{3}} = \frac{\overset{3}{\cancel{6}}}{\underset{1}{\cancel{2}}\frac{2}{3}} = \frac{3}{1} \cdot \frac{3}{2} = \frac{9}{2}$

7. $3\frac{1}{4} \div \frac{2}{3} = 3\left(\frac{1}{4} \cdot \frac{3}{2}\right) = 3\frac{3}{8}$

8. Eve is using a recipe that uses $1\frac{1}{2}$ cups of sugar, but since she is making only $\frac{1}{2}$ of a recipe, she divides by $\frac{1}{2}$ to obtain

$$1\frac{1}{2} \div \frac{1}{2} = \frac{3}{\not{2}_1} \cdot \frac{\not{2}^1}{1} = 3 \text{ cups}$$

9. $2\frac{1}{3} + 1\frac{3}{4} = \frac{7}{3} + \frac{7}{4} = \frac{14}{7} = 2$

10. $\dfrac{4\frac{2}{3}}{2\frac{1}{3}} = \dfrac{4}{2} + \dfrac{\frac{2}{3}}{\frac{1}{3}} = 2 + \dfrac{2}{\not{3}_1} \cdot \dfrac{\not{3}^1}{1} = 2 + 2 = 4$

11. $\left(2\frac{1}{2}\right)^2 = \left(2 + \frac{1}{2}\right)^2 = 2^2 + \left(\frac{1}{2}\right)^2 = 4 + \frac{1}{4} = 4\frac{1}{4}$

NAME ▮▮▮▮▮ CLASS

ACHIEVEMENT TEST—CHAPTER 3

1. Write $7\frac{7}{9}$ as an improper fraction.

2. Write $\dfrac{59}{8}$ as a mixed number.

3. Multiply: $4\frac{2}{7} \cdot 3\frac{4}{5}$.

4. Divide: $12\frac{2}{3} \div 9\frac{1}{2}$.

5. Multiply: $15 \cdot 6\frac{3}{5} \cdot \frac{2}{3}$.

6. Divide: $15 \div \dfrac{5}{16}$.

7. Add: $\begin{array}{r} 7\frac{5}{14} \\ 19\frac{5}{7} \\ +11\frac{1}{2} \\ \hline \end{array}$

8. Subtract: $\begin{array}{r} 18 \\ -12\frac{7}{10} \\ \hline \end{array}$

9. Subtract: $\begin{array}{r} 138\frac{5}{9} \\ -\ 77\frac{11}{12} \\ \hline \end{array}$

1. ______

2. ______

3. ______

4. ______

5. ______

6. ______

7. ______

8. ______

9. ______

10. Simplify the following complex fraction:

$$\frac{6 + \frac{2}{3}}{\frac{1}{4} + \frac{3}{8}}$$

11. Lillian plans to make a matching skirt and blouse. She needs $1\frac{3}{4}$ yards of fabric for the skirt and $1\frac{1}{2}$ yards for the blouse. If she decides also to make a scarf that requires $\frac{5}{8}$ yards, how much fabric must she purchase altogether?

12. Marcella can type $4\frac{3}{4}$ pages per hour. How many pages will she finish if she types for $4\frac{1}{2}$ hours?

10. ______________________

11. ______________________

12. ______________________

NAME CLASS

CUMULATIVE REVIEW—CHAPTERS 1, 2, 3

1. Give the place value of the 4 in the number 2,416.

2. Write 3,046 in words.

3. Round 23,849 to the nearest thousand.

4. Add: $\begin{array}{r} 6{,}318 \\ 4{,}992 \\ +\underline{1{,}583} \end{array}$

5. Subtract: $\begin{array}{r} 60{,}135 \\ -\underline{\ \ 9{,}442} \end{array}$

6. Multiply: $\begin{array}{r} 382 \\ \times\underline{240} \end{array}$

7. Randy always gets 8 hours of sleep. How many seconds are there in 8 hours?

8. $\frac{8}{0} =$

9. Divide: $38\overline{)5{,}296}$

10. $4^3 =$

11. $6^0 =$

1. ______
2. ______
3. ______
4. ______
5. ______
6. ______
7. ______
8. ______
9. ______
10. ______
11. ______

12. Evaluate $6 + 3^2 - 4 \cdot 2$.

12. ______

13. Write a five-digit number with a 6 in the tens place, a 4 in the ten-thousands place, a 9 in the hundreds place, and 3s in all of the remaining places.

13. ______

14. Test 3640 for divisibility by 2, 3, and 5.

14. ______

15. Write 180 as a product of prime factors.

15. ______

16. Reduce $\frac{42}{56}$ to lowest terms.

16. ______

▼ Combine as indicated in Exercises 17–20. Reduce your answer if possible.

17. $\frac{6}{18} \cdot \frac{9}{5} \cdot \frac{15}{3}$

17. ______

18. $\frac{3}{16} \div \frac{9}{4}$

18. ______

19. $\frac{2}{3} + \frac{1}{5} + \frac{7}{10}$

19. ______

20. $\frac{5}{8} - \frac{1}{3}$

20. ______

21. Write $5\frac{5}{8}$ as an improper fraction.

22. Write $\dfrac{39}{5}$ as a mixed number.

23. Divide: $5\frac{1}{3} \div 2\frac{2}{3}$.

24. Add:
$$\begin{array}{r} 15\frac{1}{8} \\ 17\frac{3}{4} \\ +\ 2\frac{1}{2} \\ \hline \end{array}$$

25. Subtract:
$$\begin{array}{r} 123\frac{3}{16} \\ -\ 27\frac{5}{8} \\ \hline \end{array}$$

26. Simplify the following complex fraction:

$$\frac{\frac{2}{3} + \frac{1}{4}}{\frac{1}{2} - \frac{1}{3}}$$

27. Charles can type $4\frac{1}{2}$ pages in an hour. If he types for $2\frac{3}{4}$ hours, how many pages will he type?

28. Give both the mixed number and the improper fraction represented by the shaded portion of the accompanying diagram.

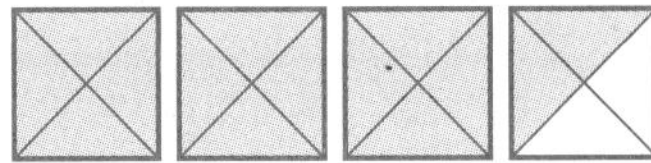

21. ______

22. ______

23. ______

24. ______

25. ______

26. ______

27. ______

28. ______

CHAPTER 4

Decimals

INTRODUCTION

In this chapter we will investigate decimals in detail. We will learn to read and write them properly and also to express them as fractions or mixed numbers, which we studied previously.

As we did with whole numbers, fractions, and mixed numbers, we will add, subtract, multiply, and divide decimals. Finally, we will round decimals to a specified place and apply all that we have learned in the solution of real-world practical problems.

CHAPTER 4—NUMBER KNOWLEDGE

Regular unleaded gas: 118.9 cents per gallon

Joan works part-time for her father and earns $4.70 per hour. This amounts to $74.03 if she works a total of 15.75 hours.

1 inch = 2.54 centimeters

A 10K race (10 kilometers) is 6.24 miles long.

Did you know that distance on land is measured differently than distance on water? Distance on land is measured in *statute miles.* Distance on water is measured in *nautical miles,* which are larger than statute miles. One nautical mile equals 1.15 statute miles.

A furlong is frequently used as a unit of distance in horseracing and is equal to 0.125 miles. In England, a fortnight is a common measure of time and is equal to 14 days. The next time you are pulled over by a policeman and he asks, "Do you know how fast you were going?" give him the answer in furlongs per fortnight and see what happens!

A single sheet of paper for a copy machine is 0.004125 inches thick. This means that a stack of 500 sheets (1 ream) is 2.0625 or $2\frac{1}{16}$ inches thick.

Ted Williams played baseball for the Boston Red Sox from 1939 to 1960. During this time he batted 7,760 times and got 2,654 hits, which gave him a batting average of 0.342 for his career:

$$\frac{2{,}654}{6{,}760} = 0.342$$

As you can see, decimals are used all the time in everyday life.

4.1 Reading and Writing Decimals

In the decimal system of numeration, we write the number 524 in expanded notation as

$$524 = 500 + 20 + 4$$

We extend this idea to include numbers that contain decimal points between various **places.** We read

- 0.6 as 6 tenths or $\frac{6}{10}$
- 0.28 as 28 hundredths or $\frac{28}{100}$
- 0.04 as 4 hundredths or $\frac{4}{100}$
- 0.146 as 146 thousandths or $\frac{146}{1000}$

Notice that we write 0.146 rather than .146, because it is easy to overlook the decimal point when no digits appear to its left. To prevent this, it is customary to put a zero at the left of the decimal point in such numbers.

All of this is consistent with the place-value system of numeration. Consider a number like 423.5674 and look at the **place** of each digit.

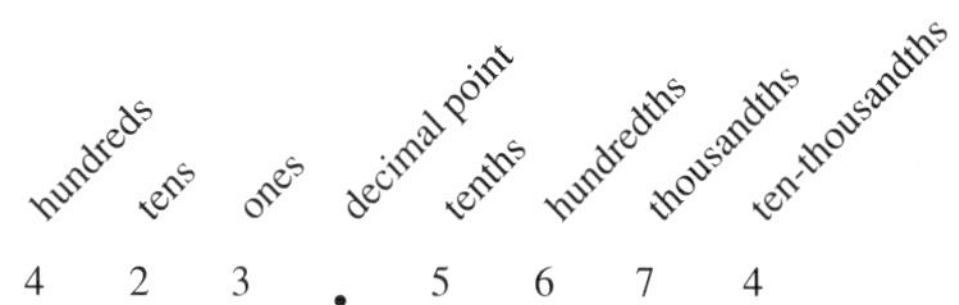

STOP Be careful to distinguish between the words that end in *s* and those that end in *ths*. As you can see, the numerals to the left of the decimal point end in *s*, while those to the right of the decimal point end in *ths*.

Now we write our number in expanded notation:

$$423.5674 = 400 + 20 + 3 + \frac{5}{10} + \frac{6}{100} + \frac{7}{1000} + \frac{1}{10{,}000}$$

EXAMPLE 1 Name the place value of the underlined digit in each of the following numbers.

SOLUTION

(a) $32.0\underline{4}2$ We find the place value of the 4 by counting from the decimal point to the right, starting with tenths, then hundredths, then thousandths, and so on. In this case the 4 is in the hundredths place.

(b) $4.26\underline{5}$ The 5 is in the thousandths place.

tenths thousandths
hundredths

(c) $\underline{9}4.261$ hundreds

(d) $43.201\underline{6}$ ten-thousandths

(e) $0.00024\underline{1}$ millionths

EXAMPLE 2 Write the number 0.031 first as a fraction then in words.

SOLUTION The numerator of the fraction is easy to determine; it's just the number without regard to the decimal point, 31. The denominator is obtained by starting at the decimal point and counting to the right starting with tenths, then hundredths, thousandths, and so on. The denominator is the name of the position on the far right.

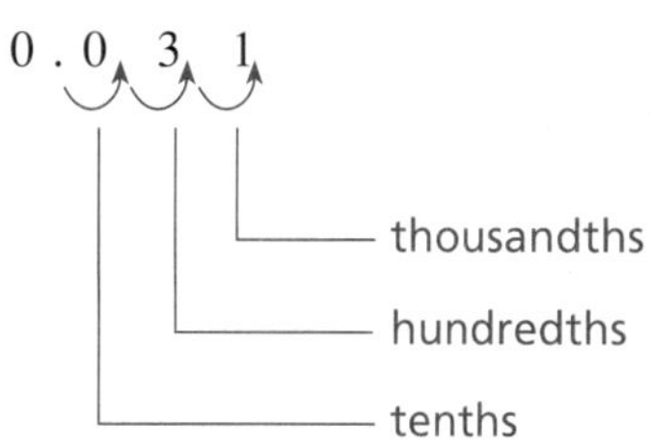

So our denominator is 1000, and we have

$$0.031 = \frac{31}{1000} = \text{thirty-one thousandths}$$

Notice that the number of zeros equals the number of digits at the right of the decimal point.

EXAMPLE 3 Write the decimal 0.57 as a fraction and then write it in words.

SOLUTION The numerator of the fraction will be 57. We count from the decimal point to the right two places: tenths, hundredths; so the denominator is 100.

$$0.57 = \frac{57}{100} = \text{fifty-seven hundredths}$$

EXAMPLE 4 Write the decimal 0.0136 as a fraction and then write it in words.

SOLUTION

$$0.0136 = \frac{136}{10{,}000} = \text{one hundred thirty-six ten-thousandths}$$

EXAMPLE 5 Write the decimal 0.001 as a fraction and also in words.

SOLUTION

$$0.001 = \frac{1}{1000} = \text{one thousandth}$$

EXAMPLE 6 Write the number 4.23 as a mixed number and also write it in words.

SOLUTION The whole number to the left of the decimal point is read as a whole number, 4; the decimal point is read as the word "and "; and the 23 to the right of the decimal point is read as 23 hundredths. Putting this all together gives us

$$4.23 = 4\,\frac{23}{100} = \text{four and twenty-three hundredths}$$

EXAMPLE 7 Write 16.041 first as a mixed number and then in words.

SOLUTION

$$16.041 = 16\,\frac{41}{1000} = \text{sixteen and forty-one thousandths}$$

EXAMPLE 8 Write 44.0061 first as a mixed number and then in words.

SOLUTION

$$44.0061 = 44\,\frac{61}{10{,}000} = \text{forth-four and sixty-one ten-thousandths}$$

EXAMPLE 9 Linda has to write a check for insurance in the amount of \$194.23. She must also write that amount in words, like this:

Linda Evans R.D. 4 - Box 641 Center City, NY 10775	1003 July 20, 19 94
Pay To The Order Of Belly Up Ins. Co.	\$ 194.23
One hundred ninety-four and 23/100	Dollars
Memo	Linda Evans

EXAMPLE 10 Write the decimal 0.34 in expanded notation.

SOLUTION As we move from the decimal point to the right, the place value of each digit goes from $\frac{1}{10}$ to $\frac{1}{100}$ to $\frac{1}{1000}$, and so on. So we have

$$0.34 = \frac{3}{10} + \frac{4}{100}$$

EXAMPLE 11 Write 0.1304 in expanded notation.

SOLUTION

$$0.1304 = \frac{1}{10} + \frac{3}{100} + \frac{0}{1000} + \frac{4}{10{,}000}$$

EXAMPLE 12 Write 0.730 in expanded notation.

SOLUTION

$$0.730 = \frac{7}{10} + \frac{3}{100} + \frac{0}{1000}$$

The zero really has no meaning in the thousandths place in this problem, except to show that it is accurate to three decimal places.

EXAMPLE 13 Write 0.00134 in expanded notation.

SOLUTION

$$0.00134 = \frac{0}{10} + \frac{0}{100} + \frac{1}{1000} + \frac{3}{10{,}000} + \frac{4}{100{,}000}$$

EXAMPLE 14 Write 37.064 in expanded notation.

SOLUTION

$$37.064 = 30 + 7 + \frac{0}{10} + \frac{6}{100} + \frac{4}{1000}$$

EXAMPLE 15 Write twelve and thirty-six hundredths as a decimal.

SOLUTION Twelve is the whole number part, "and" is written as the decimal point, and thirty-six hundredths is to the right of the decimal point: written

$$12.36$$

4.1 Exercises

▼ In Exercises 1–10, write each decimal **(a)** as a fraction, **(b)** in words, and **(c)** in expanded notation.

1. 0.358

   a.

   b.

   c.

2. 0.1234

   a.

   b.

   c.

3. 0.012

   a.

   b.

   c.

4. 0.00001

   a.

   b.

   c.

5. 0.4

   a.

   b.

   c.

6. 0.25

   a.

   b.

   c.

7. 0.9999

   a.

   b.

   c.

8. 0.3210

   a.

   b.

   c.

9. 0.364

   a.

   b.

   c.

10. 0.1001

    a.

    b.

    c.

▼ In Exercises 11–16, write each decimal (a) as a mixed number and (b) in words. Do not reduce the fraction part of the mixed number.

11. 3.7

a.

b.

12. 4.31

a.

b.

13. 17.022

a.

b.

14. 23.007

a.

b.

15. 47.234

a.

b.

16. 22.0032

a.

b.

▼ Write each written expression as a decimal.

17. Nine and twenty-one thousandths

18. Six and fifteen thousandths

19. Fourteen hundredths

20. Seventy-six hundredths

21. Eighty and twelve thousandths

22. Ninety-one and fifteen thousandths

23. Six hundred and nine hundredths

24. Four thousand seventy-seven and five hundredths

▼ Name the place value of the 3 in each of the following decimals.

25. 31.04

26. 63.56

27. 19.37

28. 24.35

29. 0.0035

30. 0.0362

31. 0.536

32. 0.439

33. 20.003

34. 16.035

35. 2304.15

36. 362.57

LIVING IN THE REAL WORLD

37. There are 2.54 centimeters in 1 inch. Write 2.54 in words.

38. One ounce is equivalent to 28.35 grams. Write 28.35 in words.

39. A gallon of gasoline is the same as 3.784 liters. Write this number in words.

40. A stick one meter in length is approximately 39.37 inches long. Express the number 39.37 in words.

41. Thrifty Savings and Loan Co. advertises interest at an annual percentage rate of 8.162. Write 8.162 in words.

42. A certificate of deposit at a local bank pays interest at a percentage rate of 7.2. Write this number in words.

In Exercises 43–50, fill in the blank checks.

43. $301.46, paid to Sullivan's Department Store.

Frank Field
201 Windy Blvd.
Chicago, IL
167
March 15, 19 94
Pay To The Order Of $
Dollars
Memo
Frank Field

44. $56.01, paid to Market Basket.

Melissa Oliver
306 Sunset Ave.
Concord, NH
2138
May 27, 19 92
Pay To The Order Of $
Dollars
Memo
Melissa Oliver

45. $684.12, paid to Oneonta Community College.

Eva Robbins
26 Thurston Ave., Apt. 4A
Oneonta, NY
341
July 6, 19 93
Pay To The Order Of $
Dollars
Memo
Eva Robbins

46. $12.63, paid to Charleston Pharmacy.

R.J. Reynolds
62 Tobacco Dr.
Charleston, SC
621
April 1, 19 94
Pay To The Order Of $
Dollars
Memo
R. J. Reynolds

47. $31.00, paid to U.S.A. Subscription Service.

Maria Rojas
114 8th St.
San Diego, CA
749
June 19, 19 92
Pay To The Order Of $
Dollars
Memo
Maria Rojas

48. $1,390.46, paid to Mohave County General Hospital.

Elizabeth J. Sutton
606 Turquoise St.
Kingman, AZ
496
May 30, 19 92
Pay To The Order Of $
Dollars
Memo
Elizabeth J. Sutton

49. $1090.09, paid to Treadwell Savings and Loan Co.

Frank E. Baca
1187 Hesperia Rd.
Hesperia, CA
1483
April 27, 19 94
Pay To The Order Of $
Dollars
Memo
Frank E. Baca

50. $2,006.60, paid to First Card.

Robert Perry
95 Vincent Dr.
Spring Valley, PA
2185
May 5, 19 92
Pay To The Order Of $
Dollars
Memo
Robert Perry

4.2 Rounding Decimals

If you were asked to cut a piece of cloth 3.5421987 feet long, what would you do? Hand the scissors to someone else! The 7 on the far right of the number represents 7 ten-millionths of a foot, which is too small a measurement to make for almost anyone's purposes. More than likely we would be asked to "round" the number to a smaller degree of accuracy, say to the nearest tenth of a foot. Since 3.5421987 is closer to 3.5 feet than it is to 3.6 feet, we say that 3.5421987 feet is 3.5 feet rounded to the nearest tenth of a foot.

Suppose that we are asked to round 7.186 to the nearest hundredth. This means that our answer will contain two digits after the decimal point. Is the number 7.186 closer to 7.19 or to 7.18? The figure

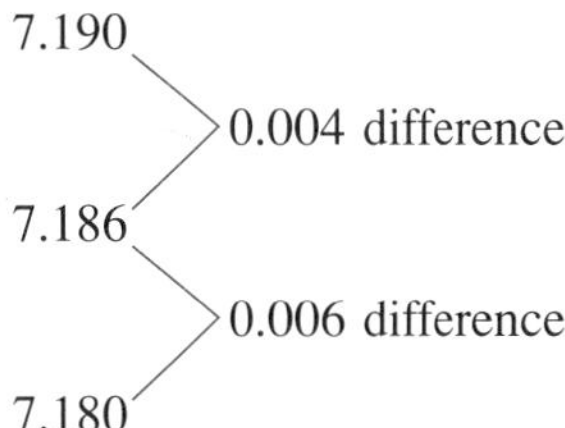

shows that the number 7.186 is only 0.004 away from 7.190, while it is 0.006 away from 7.180. As a result, 7.186 rounded to the nearest hundredth is 7.19.

It is quite easy to tell whether 7.186 should be rounded to 7.19 or 7.18 by looking at the digit 6. Since it is greater than 5, we increase the hundredths digit by 1, giving us 7.19.

The following rule is similar to the one you learned for whole numbers and shows how to round decimals.

To round a decimal:

1. Identify the place to which you want to round.
2. The digit in the desired place is
   a. not changed if the digit immediately to the right is less than 5:
   b. increased by 1 if the digit immediately to right is 5 or greater.
3. Drop all digits to the right of the rounded place.

The examples that follow will help you become more familiar with the rounding process.

EXAMPLE 1 Round 34.823 to the nearest tenth.

SOLUTION Since we are rounding to the nearest tenth, we should have one digit after the decimal point in our answer. Looking at the digit immediately following the tenths digit, we find a 2 in the hundredths position. Since 2 is less than 5, we simply drop it and all digits that follow it, and our answer is 34.8.

EXAMPLE 2 Round 0.0235 to the nearest thousandth.

SOLUTION Rounding to thousandths means that our answer will have three digits after the decimal point. The very next digit is a 5 in the ten-thousandths position, so we increase the thousandths digit by 1 and our answer is 0.024.

EXAMPLE 3 Round 23.91 to the nearest tenth.

SOLUTION

$$23.91 \approx 23.9 \qquad \text{to the nearest tenth}$$

Recall that the symbol "≈" is read "is approximately equal to," and it is appropriate to use this symbol in this case.

EXAMPLE 4 Round 5.297 to the nearest hundredth.

SOLUTION Our answer should have 2 decimal places. Since the 7 is greater than 5, we increase the 9 by 1, making it a 10. Don't let this bother you. Just put down the zero and carry the 1, giving us

$$5.297 \approx 5.30 \qquad \text{to the nearest hundredth}$$

Be careful not to drop the zero, or the number would be rounded to the nearest tenth.

STOP

If a rounded decimal ends in 0, be certain to include the 0. Without it, the rounded decimal has one fewer decimal place than it should. For example, a decimal rounded to the nearest thousandth must have three decimal places.

EXAMPLE 5 26.39848 to the nearest ten-thousandth is 26.3985

EXAMPLE 6 0.0047 to the nearest thousandth is 0.005

EXAMPLE 7 0.0078 to the nearest hundredth is 0.01

EXAMPLE 8 36.981 to the nearest whole number is 37

EXAMPLE 9 48.362 to the nearest whole number is 48

EXAMPLE 10 48.362 to the nearest tens is 50

4.2 Exercises

▼ Complete the table.

ROUND TO THE NEAREST:

	Number	Thousandth	Hundredth	Tenth
1.	0.4614			
2.	0.3729			
3.	4.58263			
4.	7.14258			
5.	62.5198			
6.	75.3195			
7.	0.0984			
8.	0.0625			
9.	3.967			
10.	17.9583			
11.	39.9926			
12.	79.9837			

13. Round 321.62 to the nearest whole number.

14. Round 675.832 to the nearest whole number.

15. Round 17.394 to the nearest whole number.

16. Round 44.275 to the nearest whole number.

17. Which of the following numbers will round to 8.6?

    8.6154 8.477 8.5999 8.6001 8.686

18. Which of the following numbers will round to 14.5?

    14.449 14.57 14.499 14.5023 14.45111

▼ For each of the following pairs of numbers, choose the number that is closest to 1.

19. 0.09 or 1.01

20. 1.05 or 1.005

21. 1.01 or 1.1

22. 0.05 or 0.06

23. 0.91 or 1.09

24. 0.94 or 1.06

25. Put the following list of numbers in order from smallest to largest:

    0.95 0.003 0.09 1.001 1.05 0.901 1.0

26. Put the following list of numbers in order from smallest to largest?

    0.3 0.23 0.32 0.03 0.023 2.3 0.003

LIVING IN THE REAL WORLD

27. A 5-lb bag of flour weighs approximately 2.273 kg. Round this number to the nearest tenth of a kilogram.

28. A used car costs $1851.85. Round this number to the nearest dollar (nearest whole number).

29. The Greek letter π (pronounced "pie") is used to represent the ratio of the circumference of any circle to its diameter. Its value is approximately 3.14159. Round π to the nearest hundredth.

30. On January 26, 1906, a Stanley Steamer automobile set a world land speed record of 127.659 miles per hour. Round this number to the nearest tenth.

31. In 1959, Lee Petty won the first Daytona 500-mile car race with an average speed of 135.521 miles per hour. Round this number to the nearet mile per hour.

4.3 Addition and Subtraction of Decimals

As in the case of whole numbers, when we add or subtract decimals, we must first make sure that the numbers are lined up in columns according to their place value. The easiest way to do this is to use the decimal points as our reference and make sure that the decimal points are all placed under one another in a vertical column.

We illustrate by finding the sum 3.064 + 0.038 + 0.5 + 26.27; we place the numbers in a vertical column and add:

$$\begin{array}{r@{.}l} 3&064 \\ 0&038 \\ 0&5 \\ \underline{26}&\underline{27} \\ 29&872 \end{array}$$

The decimal point in the sum goes right below the decimal points in the problem.

The idea of expanded notation helps us to understand why this works so simply. If we line the numbers up according to the decimal point, we automatically line them up according to place value, so that when we add, we are adding tenths to tenths, hundredths to hundredths, thousandths to thousandths, and so on.

We can illustrate this point with an example. To add 3.47 and 0.223, we line up the numbers according to the decimal points and add them.

$$\begin{array}{r@{.}l} 3&47 \\ \underline{0}&\underline{223} \\ 3&693 \end{array}$$

Now write each of the numbers in expanded notation and add the fractions with like denominators.

$$\begin{array}{rl} 3.47 = & 3 + \frac{4}{10} + \frac{7}{100} \\ 0.223 = & \frac{2}{10} + \frac{2}{100} + \frac{3}{1000} \\ \hline & 3 + \frac{6}{10} + \frac{9}{100} + \frac{3}{1000} = 3.693 \end{array}$$

This example should help you to understand why our method for adding decimals works.

Our rule is as follows:

To add or subtract decimals:

1. Write the numbers under each other so that the decimal points are placed directly under one another.
2. Add or subtract as indicated.
3. Place the decimal point in the answer directly below where it appears in the problem.

EXAMPLE 1 Find: 3.049 + 7.32 + 0.0046.

SOLUTION

Decimal points aligned
↓

$$\begin{array}{r} 3.049 \\ 7.32 \\ +\ 0.0046 \\ \hline 10.3736 \end{array}$$

EXAMPLE 2 Find: 7.348 − 4.62.

SOLUTION

$$\begin{array}{r} 7.348 \\ -4.62 \\ \hline 2.728 \end{array}$$

Any number of zeros may be placed at the far right of the decimal *after* the decimal point without changing the value of the decimal number. For example, consider the fractions $\frac{5}{10}$, $\frac{50}{100}$, and $\frac{500}{1000}$. Since they are all equivalent fractions, it follows that their equivalent decimals must be equal to each other, so we have

$$0.5 = 0.50 = 0.500 \qquad \text{and so on}$$

As you can see, placing zeros after the decimal point at the right end of the number does not change its value. We make use of this principal in the next examples.

EXAMPLE 3 Subtract: 143.7 − 81.362.

SOLUTION

$$\begin{array}{r} 143.7 \\ -\ \ 81.362 \\ \hline \end{array} = \begin{array}{r} 143.700 \\ -\ \ 81.362 \\ \hline 62.338 \end{array}$$

Notice that we placed two 0s after the seven to allow us to borrow. The next example illustrates this point again.

EXAMPLE 4 Subtract 0.00021 from 0.0036.

SOLUTION

$$\begin{array}{r} 0.00360 \\ -0.00021 \\ \hline 0.00339 \end{array}$$ ← Place one 0 on the far right.

EXAMPLE 5 Subtract 3.281 from 5.

SOLUTION

$$\begin{array}{r} 5.000 \\ -3.281 \\ \hline 1.719 \end{array}$$ ← Place three 0s after the decimal point.

Dollars and cents provide us with one of the most common applications of decimal numbers.

EXAMPLE 6 Last week Anthony wrote checks for $123.17, $58.20, $43.00, $11.58, and $176.70. If his beginning balance was $621.25, how much is in his account at the end of the week?

SOLUTION We will add all the checks that Anthony wrote and subtract that sum from his beginning balance to obtain his final balance.

$$\begin{array}{r} \$123.17 \\ 58.20 \\ 43.00 \\ 11.58 \\ +\ 176.70 \\ \hline \$412.65 \end{array} \text{ Total} \qquad \begin{array}{rl} \$621.25 & \text{Beginning balance} \\ -\ 412.65 & \text{Checks written} \\ \hline \$208.60 & \text{Final balance} \end{array}$$

4.3 Exercises

▼ Add:

1. $\begin{array}{r} 5.23 \\ +2.84 \\ \hline \end{array}$

2. $\begin{array}{r} 6.31 \\ +9.87 \\ \hline \end{array}$

3. $\begin{array}{r} 6.874 \\ +1.11 \\ \hline \end{array}$

4. $\begin{array}{r} 9.943 \\ +6.21 \\ \hline \end{array}$

5. $\begin{array}{r} 12.108 \\ 14.142 \\ +\ 8.13 \\ \hline \end{array}$

6. $\begin{array}{r} 17.381 \\ 19.426 \\ +\ 1.04 \\ \hline \end{array}$

7. $\begin{array}{r} 3.072 \\ 0.43 \\ +17.819 \\ \hline \end{array}$

8. $\begin{array}{r} 4.627 \\ 21.81 \\ +\ 0.042 \\ \hline \end{array}$

▼ Put in columns and add:

9. 7.438 + 216.09 + 0.00034

10. 3.682 + 9.81 + 0.036

11. 17.421 + 0.0094 + 86.72

12. 19.084 + 0.9211 + 21.43

13. 87 + 1.042 + 19.876 + 8.6

14. 96 + 3.275 + 86.42 + 9.1

▼ Subtract:

15. $\begin{array}{r} 72.14 \\ -\underline{18.36} \end{array}$

16. $\begin{array}{r} 24.19 \\ -\underline{17.46} \end{array}$

17. $\begin{array}{r} 176.43 \\ -\underline{\ 29.87} \end{array}$

18. $\begin{array}{r} 43.76 \\ -\underline{12.88} \end{array}$

19. $\begin{array}{r} 76.038 \\ -\underline{\ 4.91\ } \end{array}$

20. $\begin{array}{r} 96.142 \\ -\underline{\ 9.77\ } \end{array}$

21. $\begin{array}{r} 121.42 \\ -\underline{\ 91.762} \end{array}$

22. $\begin{array}{r} 38.46 \\ -\underline{18.765} \end{array}$

23. $\begin{array}{r} 46 \\ -\underline{19.4} \end{array}$

24. $\begin{array}{r} 77 \\ -\underline{34.8} \end{array}$

▼ Put in columns and subtract:

25. 36.42 − 26.19

26. 17.59 − 7.71

27. 58.12 − 36.295

28. 69.68 − 27.466

29. 39 − 18.46

30. 58 − 12.04

31. What is the sum of 3.87, 4.261, and 19.04?

32. What is the total of 7.008, 0.396, and 18.1?

33. What do you get when you subtract 14.26 from 95?

34. What number must be added to 181.42 to get 195.6?

35. What is the result when you subtract 2.864 from the sum of 8.235 and 19.42?

36. What do you get when you subtract 19.15 from the sum of 87.1 and 41.6?

37. 7.438 is how much less than 9?

38. 76.6 is how much less than 89.004?

LIVING IN THE REAL WORLD

39. If Ron earns $263.20 per week and receives a raise in pay of $31.95, how much is his new salary?

40. James earns $6.29/hr at his job. If he gets a $0.63/hr increase in pay, what is his new hourly rate?

41. Nancy buys a skirt for $41.95 and a blouse for $29.95; how much change will she receive from $75?

42. Jack purchased a sweater for \$38.88 and a shirt for \$17.99; how much change will he receive from \$60?

43. Catherine has \$321.14 in her checking account. If she writes checks for \$120.86, \$81.35, \$50.00, and \$17.15, what is her new balance?

44. Betty Jo has \$90.05 in her bank account. If she writes checks for \$19.29, \$7.00, \$9.85, and \$10.15, how much will be left in her account?

45. Lynne had \$43.91 in her checking account at the beginning of the month. During the month she made deposits of \$100 and \$312.45, and wrote checks for \$175, \$114.25, \$81.11, \$30, and \$9.45. What is her balance at the end of the month?

46. Frank had \$79.25 in his checking account at the beginning of the month. During the month he made deposits of \$78.25, \$150, and \$167.30, and wrote checks for \$38.75, \$19, \$45.18, \$16.16, and \$135.38. What is his balance at the end of the month?

47. Samantha's grocery bill is \$84.95, but she has coupons worth \$0.95, \$1.25, \$0.25, \$1.00, and \$0.15, so how much change will she get from \$85.00?

48. If George's grocery bill is \$48.91, how much change will he receive from \$45.00 if he has coupons worth \$2.00, \$1.00, \$0.35 \$0.20, and \$1.25?

49. Your lunch bill comes to \$6.31; what will your change be if you pay with a \$10 bill and a penny? What bills and what coins should you receive back in your change?

50. The tab for dinner for Mr. and Mrs. Morales comes to \$46.57, so Mr. Morales gives the cashier two \$20 bills, one \$10 bill, a nickel, and two pennies. How much change should he receive? What bills and what coins should he receive in his change?

51. If your total cost for a shirt that you bought was \$31.21, why would you consider including a penny with your payment?

4.4 Multiplication of Decimals

Since all decimals can be written as fractions, the rules that apply to multiplication of fractions also apply to multiplication of decimals. However, there are shorter ways of obtaining the same results, which we will also examine.

Suppose that we are asked to multiply 0.7×0.8. Since $0.7 = \frac{7}{10}$ and $0.8 = \frac{8}{10}$, we have

0.7×0.8	Decimal form
$= \dfrac{7}{10} \times \dfrac{8}{10}$	Fraction form
$= \dfrac{7 \times 8}{10 \times 10} = \dfrac{56}{100}$	Multiply
$= 0.56$	Decimal form

We can obtain the same result in a more efficient, shorter way. To multiply 0.7×0.8, we first multiply 7 times 8, disregarding the decimal points, giving us $7 \times 8 = 56$. Now, to place the decimal point in the correct place in our answer, we count up the total number of decimal places in the numbers being multiplied. Since 0.7 has one decimal place and 0.8 has one decimal place for a total of two we mark off two places from the right in the number 56 and obtain 0.56 as the final answer.

To multiply decimals:

1. Multiply the numbers as if the decimal points were not there.
2. Count the total number of decimal places in the factors, from the right to the decimal point.
3. Counting from right to left, mark off this many decimal places in your answer and place the decimal point there. It may be necessary to add zeros to the left of the product.

Let's look at another illustration of the rule. Suppose we want to multiply 0.31 and 0.8 together. First we multiply the two numbers together without regard to the decimal points. Then we count up the total number of decimal places in the factors and mark off this many decimal places in the answer. Always count from right to left.

$$\begin{array}{rl} 0.31 & \text{(2 decimal places)} \\ \times\ \underline{0.8} & \text{(1 decimal place)} \\ 0.248 & (2 + 1 = 3 \text{ decimal places}) \end{array}$$

EXAMPLE 1 3.64×0.82

SOLUTION

$$\begin{array}{rl} 3.64 & \text{(2 decimal places)} \\ \times\ \underline{0.82} & \text{(2 decimal places)} \\ 728 & \\ \underline{2912\ \ } & \\ 2.9848 & (2 + 2 = 4 \text{ decimal places}) \end{array}$$

Notice that it isn't necessary to line up the decimal points when multiplying decimals.

EXAMPLE 2 0.24×0.003

SOLUTION

$$\begin{array}{rl} 0.24 & \text{(2 decimal places)} \\ \times\ \underline{0.003} & \text{(3 decimal places)} \\ 0.00072 & (3 + 2 = 5 \text{ decimal places}) \end{array}$$

Notice that in order to mark off five decimal places in our answer, it was necessary to add three 0s to the left of 72. The next example also illustrates this point.

EXAMPLE 3 1.24 × 0.0008

SOLUTION

```
    1.24     (2 decimal places)
×   0.0008   (4 decimal places)
  0.000992   (6 decimal places)
```

EXAMPLE 4 Multiply 3.18 by 4.6 and round the result to the nearest hundredth.

SOLUTION

```
    3.18
×    4.6
   1 908
  12 72
  14.628
```

14.628 rounded to the nearest hundredth is \$14.63

When doing calculations involving decimals, if you are asked to round your result, be sure to do the calculations first and round the answer last. Never round the numbers before you do calculations.

4.4 Exercises

Multiply:

1. 4.5 × 0.8
2. 7.3 × 0.9
3. 0.6 × 0.8
4. 0.5 × 0.7
5. 1.45 × 2.4
6. 3.72 × 6.5
7. 4.13 × 0.07
8. 4.63 × 0.09
9. 4.171 × 0.34
10. 6.973 × 0.24
11. 0.074 × 0.81
12. 0.093 × 0.92
13. 1.13 × 0.008
14. 6.31 × 0.004
15. 0.039 × 10
16. 0.056 × 10
17. 0.055 × 100
18. 0.091 × 100
19. 0.035 × 0.1
20. 0.077 × 0.1
21. 3.027 × 0.01
22. 4.059 × 0.01
23. 21.06 × 0.001
24. 89.11 × 0.001
25. What is the product of 13.138 and 1.9?
26. What is the product of 27.344 and 2.7?
27. What is the product of 0.377 and 0.07?

28. What is the product of 0.484 and 0.36?

29. Can you draw any general conclusions about what multiplying by 10 does to the decimal point in the other factor? (See Exercises 15 and 16)

30. What does multiplying by 100 do to the decimal point in the other factor? (See Exercises 17 and 18)

31. What does examination of Exercises 19 through 24 reveal about multiplying by 0.1, 0.01, or 0.001?

LIVING IN THE REAL WORLD

32. What will it cost to fill a 12.8-gallon gas tank if gasoline costs $1.29/gallon?

33. If gasoline costs $1.34/gallon, what will it cost to fill a tank that holds 15.3 gallons?

34. Ground beef costs $3.29/lb. What will a package weighing 1.81 lb cost? Round your answer to the nearest cent.

35. If fresh fish costs $5.99/lb, how much will it cost for a fish that weighs 2.11 lb? Round your answer to the nearest cent.

36. What will three cans of coffee cost if one can costs $4.29?

37. If a phone call costs $0.41 for the first minute and $0.23 for each additional minute, how much will a 12-minute call cost?

38. A phone call costs $0.58 for the first minute and $0.37 for each additional minute. How much will a 16-minute call cost?

39. Thomas earned $5.12/hour for 6 hours on Saturday and overtime pay of $7.68/per hour for 4 hours on Sunday. How much did he earn altogether?

40. Kathy earns $5.40/hr for the first 8 hours and $8.10/hr if she works over 8 hours. How much does she earn if she works 10.5 hours?

41. Water weights 8.34 lb/gallon; how much do 100 gallons weigh?

42. Grey Mountain Community College purchased 1000 pens at a cost of 25.6¢ each. What was the total cost of the pens?

43. Dan bought 2.31 lb of fish priced at $5.99/lb. What was the total cost of the fish?

44. How high is a 1000-sheet stack of paper if each sheet is 0.0017 in. thick?

45. If gas costs 139.9¢ per gallon, how much change will you receive from a $20 bill if you purchase 13.6 gallons?

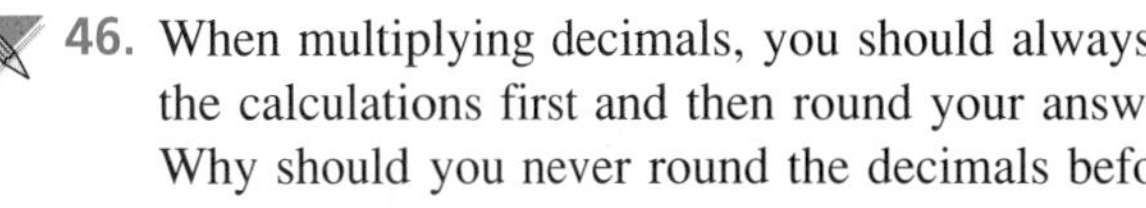

46. When multiplying decimals, you should always do the calculations first and then round your answer. Why should you never round the decimals before you multiply them?

4.5 Division of Decimals

Division by Powers of 10

The easiest numbers to divide by are powers of 10, that is, numbers such as 10, 100, 1000, 10,000, and so on. Suppose that we are given the problem 3.41 ÷ 10. Dividing by 10 is the same as multiplying by $\frac{1}{10}$, or 0.1:

$$3.41 \div 10 = \frac{3.41}{10} = 3.41 \times \frac{1}{10} = 3.41 \times 0.1 = 0.341$$

Now let's divide 3.41 by 1000 using the same procedure:

$$3.41 \div 1000 = \frac{3.41}{1000} = 3.41 \times \frac{1}{1000} = 3.41 \times 0.001 = 0.00341$$

Notice that when we divided by 10 (**one** 0), the net effect was to move the decimal point **one** place to the **left.** When we divided by 1000 (**three** 0s), we moved the decimal point **three** places to the **left.** This rule works in general. When we divide by a power of 10, simply move the decimal point to the **left** the same number of places as there are zeros in the number we are dividing by (the divisor).

When we divide 3.41 by 10,000, we move the decimal point four places to the left (four 0s) and our answer is 0.000341. Summarizing gives us the following rule:

To divide a decimal by a power of 10:

1. Count the number of zeros in the power of 10.
2. Move the decimal point to the **left** this many places.

EXAMPLE 1 Divide: 25.43 ÷ 100.

SOLUTION We move the decimal point two places to the left, since 100 has two 0s:

$$25.43 \div 100 = 0.2543$$

EXAMPLE 2 26.58 ÷ 1000 = 0.02658

Division of Decimals by Whole Numbers

So far, we have restricted our divisors to numbers containing a 1 followed by 0s. Now we consider problems such as 34.4 ÷ 8, which we do as follows:

$$
\begin{array}{r}
4.3 \\
8\overline{)34.4} \\
\underline{32} \\
2\,4 \\
\underline{2\,4} \\
0
\end{array}
$$

Notice that the decimal point in the answer is placed directly above the decimal point in the dividend. This is very important and should not be overlooked.

To divide a decimal by a whole number:

1. Place the decimal point in the quotient directly above the decimal point in the dividend.
2. Divide the numbers as if the decimal point were not there.

EXAMPLE 3 Divide: 6.75 ÷ 25.

SOLUTION

$$
\begin{array}{r}
0.27 \\
25\overline{)6.75} \\
\underline{5\,0} \\
1\,75 \\
\underline{1\,75} \\
0
\end{array}
$$

Again, the decimal point in the answer is directly above the decimal point in the number 6.75, the dividend. ▮▮

If we are asked to round our answer to a specific number of decimal places, the quotient must be carried to one more place in order to round to the desired number of places. Sometimes this means we must attach zeros to the dividend. This is illustrated in the next example.

EXAMPLE 4 Divide 7.1 by 38 and round the quotient to two decimal places.

SOLUTION

$$\begin{array}{r} 0.186 \approx 0.19 \\ 38\overline{)7.100} \leftarrow \text{added two 0s} \\ \underline{3\ 8} \\ 3\ 30 \\ \underline{3\ 04} \\ 260 \\ \underline{228} \\ 32 \end{array}$$

We have attached two 0s on the far right of the dividend so we have three decimal places in the dividend and therefore in the quotient. We then round our answer to two decimal places:

$$0.186 \approx 0.19$$

Division by a Decimal

In a problem like 4.692 ÷ 0.23, the divisor is a decimal. When this happens, it is necessary somehow to change the divisor to a whole number. The problem can also be written as a fraction: $\frac{4.692}{0.23}$. Now, if we multiply both numerator and denominator of the fraction by 100, we get

$$\frac{4.692 \times 100}{0.23 \times 100} = \frac{469.2}{23}$$

As long as we multiply both the numerator *and* the denominator of the fraction by the same number, we will not change the value of the original fraction, only its appearance.

Remember that multiplying by 100 has the effect of moving the decimal point two places to the right. Our new problem now has a whole number in the denominator and we can perform the division as before

$$\begin{array}{r} 20.4 \\ 23\overline{)469.2} \\ \underline{46} \\ 9\ 2 \\ \underline{9\ 2} \\ 0 \end{array}$$

The same result can be obtained in a somewhat easier manner. Our original problem was

$$0.23\overline{)4.692}$$

All we need to do is to move the decimal point in the divisor a sufficient number of places to the right to make it a whole number. Then move the decimal point in the dividend the same number of places to the right. Actually, crossing the decimal point out helps prevent making an error. As you get ready to divide, your problem should look like this:

$$.23.\overline{)4.69.2}$$

Now we can go ahead and divide as we did before:

```
     20.4
23)469.2
   46
     9 2
     9 2
       0
```

To divide a decimal by a decimal:

1. Change the divisor to a whole number by moving the decimal point all the way to the right of the number.
2. Move the decimal point in the dividend to the right the same number of places that you did in the divisor, adding zeros to the dividend if necessary.
3. Divide the decimal as though it were a whole number.
4. Place the decimal point in the quotient directly above the decimal point in the dividend.

EXAMPLE 5 Divide: $8.28 \div 4.6$.

SOLUTION Set the problem up in the familiar way for long division and move the decimal point in both numbers a sufficient number of places to make the divisor a whole number. In this case, we move the decimal point one place, then divide:

```
                     1.8
4.6)8.28 → 4 6.)8 2.8     Move the decimal point one place
                4 6        in both divisor and dividend.
                3 6 8
                3 6 8
                    0
```

EXAMPLE 6 Divide: $0.78584 \div 0.94$.

SOLUTION

```
                       0.836
0.94)0.78584 → 94.)78.584     Move the decimal point two places.
                   75 2
                    3 38
                    2 82
                      564
                      564
                        0
```

EXAMPLE 7 Divide: $351 \div 7.8$.

SOLUTION

$$7.8.\overline{)351.\ \ ?}$$

We move the decimal point one place to the right to make the divisor a whole number, but the decimal point in the dividend is already at the far right. (The decimal point in any whole number is assumed to be immediately after the last digit on the right, even though it is not usually written.) Since we must also move the decimal point one place to the right in the dividend, we must add a zero and then move the decimal point one place to the right. Now we can divide as before.

$$\begin{array}{r} 45. \\ 7.8.\overline{)351.0.} \\ \underline{312} \\ 39\,0 \\ \underline{39\,0} \\ 0 \end{array}$$

Since we do not normally write the decimal point at the right end of a whole number, 45 is our result.

EXAMPLE 8 Divide: $2.698 \div 0.0071$.

SOLUTION

$$0.0071\overline{)2.698} \rightarrow \begin{array}{r} 380. \\ 0071.\overline{)26980.} \\ \underline{213} \\ 568 \\ \underline{568} \\ 0 \end{array} \quad \text{or} \quad 380$$

Place one 0 on the far right of the dividend and move the decimal point 4 places in the divisor and the dividend.

4.5 Exercises

▼ Divide:

1. $24.35 \div 10$
2. $26.98 \div 10$
3. $3.6 \div 10$
4. $7.9 \div 10$
5. $4.26 \div 100$
6. $48 \div 100$
7. $3.6498 \div 1000$
8. $7.2851 \div 1000$
9. $46 \div 10$
10. $75 \div 10$
11. $36 \div 100$
12. $79 \div 100$
13. $6 \div 100$
14. $8 \div 100$
15. $9 \div 1000$
16. $7 \div 1000$

17. 0.31 ÷ 1000

18. 0.46 ÷ 100

▼ Divide:

19. 19.6 ÷ 7

20. 35.4 ÷ 6

21. 32.4 ÷ 9

22. 35.6 ÷ 4

23. 7.44 ÷ 31

24. 35.28 ÷ 63

25. 41.6 ÷ 16

26. 96.6 ÷ 23

27. 138.6 ÷ 33

28. 107.1 ÷ 17

29. 23.22 ÷ 86

30. 50.96 ÷ 91

▼ Divide, rounding all results to two decimal places:

31. $4.4\overline{)2.8}$

32. $1.7\overline{)3.8}$

33. $7.1\overline{)5.8}$

34. $6.3\overline{)2.9}$

35. $0.61\overline{)0.394}$

36. $0.85\overline{)0.844}$

37. 0.453 ÷ 0.231

38. 0.627 ÷ 0.029

39. 17 ÷ 8

40. 15 ÷ 7

41. 0.64 ÷ 0.083

42. 0.52 ÷ 0.074

LIVING IN THE REAL WORLD

43. If Betty Jo earned $117.36 and made $6.52/hr, how many hours did she work?

44. Lilly earns $5.35/hr at her part-time job. How many hours did she work if her paycheck is $90.95?

45. Find the average of 21.6, 30.7, 24.7, 28.6, and 28.5, and round your result to one decimal place.

46. The daily high temperatures in Phoenix for a five-day period in July were 106.4°, 101.8°, 104.2°, 107.1°, and 110.1°. What is the average temperature for the period? Round the result to one decimal place.

47. If you borrow $3614.88 to buy a car, how much will your monthly payments be if you plan to pay the loan off in 24 months?

48. You take out a loan on furniture totaling $725.40. If you wish to pay it off in 12 monthly payments, how much will each payment be?

49. Lawrence traveled 314.3 miles on 11.9 gallons of gasoline. How many miles per gallon did he get? Round your answer to one decimal place.

50. Lois drove from New York to Boston, a distance of 207.4 miles, on a total of 9.2 gallons of gasoline. How many miles per gallon did she get on this trip? Round your answer to the nearest tenth.

51. Babe Ruth, in 22 years of playing baseball, had 2,873 hits in 8,399 times at bat. What was his lifetime batting average? (To find a batting average, divide the number of hits by the number of times at bat.) Round the quotient to the nearest thousandth.

52. Pete Rose, in his 24-year career in major league baseball, had 4,256 hits in 14,053 times at bat (both all-time records). What was Pete Rose's lifetime batting average? (Which is *not* a record.)

53. Scott drove 362.8 miles in 7.5 hours. Compute his average speed by dividing his distance by his time, rounding your result to the nearest tenth.

54. Stock is customarily sold in "round lots" of 100 shares. If an investor pays a total of $5,750 for 100 shares of Amerada Hess Stock, how much does each share cost?

55. In one season, Michael Jordon of the Chicago Bulls basketball team scored 2,633 points in 81 games. What was his average number of points per game, rounded to the nearest tenth of a point?

56. Earvin "Magic" Johnson of the Los Angeles Lakers basketball team scored 1,730 points in 77 games. Calculate his average number of points per game, rounded to one decimal place.

4.6 Interchanging Fractions and Decimals

Converting Fractions to Decimals

Every fraction can be thought of as a numerator divided by a denominator. For instance, the fraction $\frac{1}{4}$ means 1 divided into 4 parts, or, more simply, 1 divided by 4. Similarly, $\frac{3}{8}$ means 3 divided by 8, and $\frac{9}{10}$ means 9 divided by 10, and so on. To perform one of these divisions actually involves doing a long division problem. Let's consider the fraction $\frac{1}{4}$ and divide 1 by 4 using long division. Since there is a decimal point implied immediately after the 1, we write it and then add a couple of zeros after the decimal point, which allows us to do the division. Your work will look like this:

$$\begin{array}{r} 0.25 \\ 4\overline{)1.00} \\ \underline{8} \\ 20 \\ \underline{20} \\ 0 \end{array}$$

We have written the fraction $\frac{1}{4}$ as the decimal .25. It comes as no surprise that if we divide $1 into four parts, we get 25 cents, or $0.25.

Our rule is as follows:

To convert a fraction to a decimal:

Divide the numerator by the denominator. You may find it useful to picture it this way:

$$\text{denominator}\overline{)\text{numerator}}$$

EXAMPLE 1 Write $\frac{4}{5}$ as a decimal.

SOLUTION We use long division to divide 4 by 5. We must remember to write the decimal point after the 4 and add enough zeros after the decimal point to give us the desired number of places in the quotient.

```
   0.8
 5)4.0
   4 0
   ---
     0
```

$$\frac{4}{5} = 0.8$$

EXAMPLE 2 Convert the fraction $\frac{7}{8}$ to a decimal.

SOLUTION

```
   0.875
 8)7.000
   6 4
   ---
     60
     56
     --
      40
      40
      --
       0
```

$$\frac{7}{8} = 0.875$$

So far, all the division problems have come out with no remainder. Let's consider an example where this does not happen. For example, if asked to write the fraction $\frac{1}{3}$ as a decimal, we use long division to divide 1 by 3:

```
   0.3333. . .
 3)1.0000
    9
   --
    10
     9
    --
     10
      9
     --
      1
```

If you work through the problem, you will see that it does not come out with a definite number of decimal places. We keep getting a remainder of 1, so our quotient is an unending string of 3s. To handle this problem, we **round** the quotient to as many decimal places as desired. Rounding to the thousandths place,

$$\frac{1}{3} = 0.333$$

or, more precisely,

$$\frac{1}{3} \approx 0.333$$

(Recall that the symbol ≈ means **approximately equal to.**) It is also common to write the answer

$$\frac{1}{3} = 0.\overline{3}$$

where the line over the 3 means that the 3s will continue if we continue the division. You may see $\frac{1}{3} = 0.333$, but you should not forget that is is really only an approximation.

EXAMPLE 3 Write $\frac{1}{6}$ as a decimal.

SOLUTION Dividing 1 by 6 gives us

$$\begin{array}{r} 0.166\overline{6} \\ 6\overline{)1.0000} \\ \underline{6} \\ 40 \\ \underline{36} \\ 40 \\ \underline{36} \\ 40 \\ \underline{36} \\ 4 \end{array}$$

Again, the line over the last 6 means that the 6s continue indefinitely. Rounding our answer to three decimal places gives us

$$\frac{1}{6} = 0.167 \qquad \text{rounded to the nearest thousandth}$$ ▮▮

If we had been asked to round our result to hundredths, our answer would have been

$$\frac{1}{6} = 0.17$$

If we rounded to ten-thousandths, we would have

$$\frac{1}{6} = 0.1667$$

or, leaving it as a repeating decimal, we have

$$\frac{1}{6} = 0.1\overline{6}$$

The answer can be written with as many decimal places as are needed in a particular situation.

EXAMPLE 4 Convert the fraction $\frac{5}{9}$ to an equivalent decimal. Round your result to the nearest ten-thousandth.

SOLUTION

$$\begin{array}{r} 0.55555 \\ 9\overline{)5.00000} \\ \underline{4\,5} \\ 50 \\ \underline{45} \\ 50 \\ \underline{45} \\ 50 \\ \underline{45} \\ 50 \\ \underline{45} \\ 5 \end{array}$$

$$\frac{5}{9} = 0.5556 \qquad \text{rounded to the nearest ten-thousandth}$$

EXAMPLE 5 Change $6\frac{3}{4}$ to a decimal.

SOLUTION We change the fraction $\frac{3}{4}$ to a decimal and then add it to the 6.

$$\begin{array}{r} 0.75 \\ 4\overline{)3.00} \\ \underline{2\,8} \\ 20 \\ \underline{20} \\ 0 \end{array}$$

Therefore, $6\frac{3}{4} = 6.75$ in decimal form.

Converting Decimals to Fractions

To convert a decimal to its equivalent fraction, we read the decimal as we did in Section 4.1 and write it with a numerator and a denominator, in fraction form. We make use of the place values assigned to numbers occurring after the decimal point to give us the correct denominator.

For example, we read the decimal 0.25 as "twenty-five hundredths," since the digit on the far right, 5, is in the hundredths place. We write it in fraction form as

$$0.25 = \frac{25}{100} = \frac{1}{4}$$

Notice that we reduced the fraction to lowest terms.

To convert a decimal to a fraction:

1. Read the decimal and write it in fraction form;
2. Reduce the fraction to lowest terms.

EXAMPLE 6 Convert 0.8 to a fraction.

SOLUTION We read the decimal as "eight-tenths" and write it as a fraction:

$$0.8 = \frac{8}{10} = \frac{4}{5} \quad \text{reduced to lowest terms}$$

EXAMPLE 7 Convert 6.7 to a mixed number.

SOLUTION We change the decimal part, .7, to a fraction and add it to the whole number part to form a mixed number:

$$.7 = \frac{7}{10}$$

Our result is $6.7 = 6\frac{7}{10}$.

EXAMPLE 8 Convert 8.6 to a mixed number.

SOLUTION

$$\begin{aligned} .6 &= \frac{6}{10} \\ &= \frac{3}{5} \quad \text{reduced to lowest terms.} \\ 8.6 &= 8\tfrac{6}{10} \\ &= 8\tfrac{3}{5} \end{aligned}$$

EXAMPLE 9 Change 0.005 to a fraction.

SOLUTION

$$\begin{aligned} 0.005 &= \frac{5}{1000} \\ &= \frac{1}{200} \end{aligned}$$

4.6 Exercises

Write each of the following as a decimal. If necessary, round the answer to two decimal places.

1. $\frac{1}{5}$
2. $\frac{1}{2}$
3. $\frac{3}{5}$
4. $\frac{3}{4}$
5. $\frac{3}{8}$
6. $\frac{5}{16}$
7. $\frac{2}{7}$
8. $\frac{1}{6}$
9. $\frac{2}{9}$
10. $\frac{5}{6}$
11. $\frac{4}{9}$
12. $\frac{20}{21}$
13. $\frac{13}{32}$
14. $\frac{12}{23}$
15. $6\frac{1}{4}$
16. $7\frac{4}{5}$
17. $8\frac{2}{3}$
18. $4\frac{1}{6}$
19. $23\frac{5}{16}$
20. $28\frac{7}{16}$

Write each of the following decimals as a fraction or a mixed number.

21. 0.3
22. 0.7
23. 0.5
24. 0.9
25. 0.6
26. 0.8
27. 0.45
28. 0.35
29. 2.4
30. 5.8
31. 6.6
32. 4.5
33. 16.25
34. 18.45
35. 0.035
36. 0.025
37. 0.632
38. 0.844
39. 0.0045
40. 0.0075
41. 13.032
42. 29.084
43. 125.05
44. 136.08

LIVING IN THE REAL WORLD

45. The capacity of the gas tank of a Ford Probe is 12.8 gallons. Write 12.8 as a mixed number.

46. A recipe for pie crust calls for $2\frac{1}{4}$ cups of flour. Write this number as a decimal.

47. A dress pattern requires $3\frac{5}{8}$ yards of fabric. Express this number as a decimal.

48. A one-gallon container of laundry detergent is the same as 3.78 liters. Write 3.78 as a mixed number.

49. One foot is approximately the same as 30.48 centimeters. Express 30.48 as a mixed number.

50. On a good day, Jennifer can swim $\frac{3}{4}$ miles in 40 minutes (which is $\frac{40}{60}$ or $\frac{2}{3}$ of an hour). Change the numbers $\frac{3}{4}$ and $\frac{2}{3}$ to decimals.

SUMMARY—CHAPTER 4

The numbers in brackets refer to the section in the text in which each concept is presented.

EXAMPLE

[4.2] **To round a decimal,** first locate the digit in the decimal place to which you wish to round. Then look at the next digit to the right. If it is 5 or greater, increase that digit by 1; if the next digit is less than 5, drop the digits after the decimal place to which you are rounding.

Example [4.2]: $1.0362 = 1.04$ to the nearest hundredth
$4.83 = 4.8$ to the nearest tenth

[4.3] **To add or subtract decimals,** place the numbers in a vertical column with the decimal points under one another. Then add or subtract as you would for whole numbers.

Example [4.3]:

$$\begin{array}{r} 3.67 \\ 0.041 \\ +0.7 \\ \hline 4.411 \end{array}$$

[4.4] **To multiply decimals,** first multiply the numbers without regard to the decimal points. Then count the total number of decimal places in the factors (from the right to the decimal point) and mark off (from right to left) this many places in the answer.

Example [4.4]:

$$\begin{array}{rl} 2.06 & \leftarrow \text{2 decimal places} \\ \times 4.1 & \leftarrow \text{1 decimal place} \\ \hline 206 & \\ 8\,24 & \\ \hline 8.446 & \leftarrow \text{3 decimal places} \end{array}$$

[4.5] **To divide decimals,** move the decimal point in the divisor to the far right of the number. Then move the decimal point in the dividend the same number of places, adding zeros if necessary. Now divide, placing the decimal point in the quotient above the decimal point in the dividend.

Example [4.5]:

$$\begin{array}{r} 5.4 \\ 3.4.\overline{)18.3.6} \\ 17\,0 \\ \hline 1\,3\,6 \\ 1\,3\,6 \\ \hline 0 \end{array}$$

[4.6] **To convert a fraction to a decimal,** divide the numerator by the denominator using long division.

Example [4.6]: $\frac{7}{8} = 8\overline{)7.000}$ giving 0.875

$\frac{5}{6} = 6\overline{)5.000}$ giving $0.83\overline{3}$

$= 0.83$ rounded to the nearest hundredth

[4.6] **To convert a decimal to a fraction,** read the decimal and write it in fraction form. Reduce to lowest terms.

Example [4.6]: $0.35 =$ thirty-five hundredths

$= \frac{35}{100}$

$= \frac{7}{20}$ in lowest terms

REVIEW EXERCISES—CHAPTER 4

The numbers in brackets refer to the section in the text in which similar problems are presented.

[4.1]

1. Write the decimal 0.26 **(a)** as a fraction, **(b)** in words, and **(c)** in expanded notation.
(a) (b) (c)

[4.1]

2. Write the decimal 15.02 **(a)** as a mixed number and **(b)** in words.
(a) (b)

[4.1]

3. Write as a decimal: sixteen and twenty-four thousandths.

[4.1]

4. What place value does the 3 have in the number 26.034?

[4.2]

5. Round 13.681 to the nearest **(a)** tenth; **(b)** hundredth; **(c)** whole number.
(a) (b) (c)

[4.3]

6. $\begin{array}{r} 7.85 \\ 14.007 \\ +\ 2.4 \\ \hline \end{array}$

7. $\begin{array}{r} 4.68 \\ 0.004 \\ +17.09 \\ \hline \end{array}$

[4.3]

8. $\begin{array}{r} 13.262 \\ -\ 9.659 \\ \hline \end{array}$

9. $\begin{array}{r} 18.1 \\ -\ 6.427 \\ \hline \end{array}$

[4.3]

10. What number must be added to 27.834 to make 86?

[4.4]

11. $\begin{array}{r} 1.58 \\ \times\ 6.2 \\ \hline \end{array}$

12. $\begin{array}{r} 0.063 \\ \times\ 0.04 \\ \hline \end{array}$

[4.5]

13. $827.84 \div 10$

14. $827.84 \div 100$

15. $827.84 \div 1000$

16. $61.2 \div 18$

[4.5] Divide, rounding your answers to one decimal place.

17. $0.71\overline{)0.852}$

18. $0.67 \div 0.083$

[4.6] Convert to a decimal, rounding answers to two decimal places if necessary.

19. $\frac{3}{25}$

20. $\frac{5}{7}$

21. $5\frac{3}{8}$

22. $18\frac{3}{7}$

[4.6] Convert to a fraction and reduce to lowest terms.

23. 0.23

24. 0.624

[4.6] Convert to a mixed number. Be sure to reduce to lowest terms.

25. 27.55

26. 12.005

LIVING IN THE REAL WORLD

[4.2-4.5]

27. Michael has a summer job working as grocery clerk. If he earns $208.46 for working 38.25 hours, what is his hourly rate rounded to the nearest cent?

[4.3]

28. Sandy and Donna bought the following refreshments for their party: 10 bottles of soda for $1.09 each, 4 boxes of crackers for $1.89 each, and 2.6 pounds of cheese at $3.45 per pound. How much change from $30.00 did they receive?

CRITICAL THINKING EXERCISES—CHAPTER 4

Tell whether the following are true or false and give reasons why.

1. $\frac{5}{24} = 5\overline{)24.0}$ = 4.8

$$\begin{array}{r} 4.8 \\ 5\overline{)24.0} \\ \underline{20} \\ 4\,0 \\ \underline{4\,0} \\ 0 \end{array}$$

2. 34.949 rounded to the nearest ten is 30.

3. $3.061 = 3 + \frac{6}{10} + \frac{1}{100}$ in expanded notation.

4. Fourteen thousandths is 0.014 written in decimal form.

5. In the 1988 Olympic Games, Florence Griffith-Joyner of the United States won the 100-meter dash in 10.54 seconds. This decimal written as a mixed number is $10\frac{27}{50}$.

▼ Find the error in each of the following:

6. Continental Telephone Co. charges $0.53 for the first minute and $0.29 for each additional minute for a long-distance phone call. How much does an 8-minute call cost?

SOLUTION

$0.53	1st minute
+ 0.29	additional minute
0.82	
× 8	minutes
$6.56	total cost of the call

7.
$$\begin{array}{r} 0.15 \\ 2.3\overline{)3.45} \\ \underline{2\,3} \\ 1\,15 \\ \underline{1\,15} \\ 0 \end{array}$$

8. 26.46 = 26.5 = 27 rounded to the nearest unit.

9. The number 0.027 written in words is twenty-seven hundredths.

10. Subtract 3.45 from 18.
SOLUTION

$$\begin{array}{r} 3.45 \\ -\underline{0.18} \\ 3.28 \end{array}$$

NAME ▮▮▮▮▮ CLASS

ACHIEVEMENT TEST—CHAPTER 4

▼ This test covers the material in Chapter 4.

1. Write 0.073 in words.

2. Write 4.27 in expanded notation.

3. Write 26.07 in words.

4. Write as a decimal: twenty-one and fourteen thousandths.

5. What place value does the 5 have in the number 12.526?

6. Round 12.609 to the nearest **(a)** tenth; **(b)** hundredth; **(c)** whole number.

7. Add:
$$\begin{array}{r} 12.089 \\ 1.4 \\ +\ 3.09 \\ \hline \end{array}$$

8. Subtract 21.06 from 39.

9. Subtract 18.47 from the sum of 22.4, 1.078, and 8.247.

10. Multiply:
$$\begin{array}{r} 1.004 \\ \times 0.036 \\ \hline \end{array}$$

11. Divide: $342.7 \div 100$

12. Divide: $201.6 \div 36$

13. Divide, rounding the quotient to two decimal places.

a. $0.87\overline{)0.937}$

b. $0.4 \div 0.095$

1. ______
2. ______
3. ______
______
4. ______
5. ______
6. (a) (b) (c) ______
7. ______
8. ______
9. ______
10. ______
11. ______
12. ______
13. (a) ______
(b) ______

14. Convert to a decimal, rounding the result to one decimal place.

   a. $\frac{4}{7}$

   b. $5\frac{9}{16}$

15. Convert the decimal 0.385 to a fraction. Reduce to lowest terms.

16. Write 35.06 as a mixed number. Reduce to lowest terms.

17. Sunnyside School District paid $0.67 per box for 100 boxes of paper clips. What was the total cost of the paper clips?

18. Mrs. Anthony is an elementary school teacher who earns $564.23/week. She has $87.33 deducted for federal income tax, $23.12 for Social Security, and $18.92 for state income tax. How much does Mrs. Anthony take home after her deductions have been taken out?

14. (a) ______

(b) ______

15. ______

16. ______

17. ______

18. ______

CHAPTER 5

Ratios, Rates, and Proportions

INTRODUCTION

In this chapter we will learn to compare numbers to form **ratios** and **rates.** We will then use these concepts to form **proportions** and solve a wide variety of practical, everyday problems.

CHAPTER 5—NUMBER KNOWLEDGE

Many people think that a larger size jar of peanut butter is automatically a better buy than a smaller size. Not necessarily! A 16-oz jar of Skippy peanut butter that sells for \$3.39 is not as good a value for your money as a 12-oz jar of the same peanut butter, which sells for \$2.49.

Which of the 5 sizes of soft drinks at Pete's Pizza is the best buy?

Refreshing Beverages

Kids	6 oz.	\$.69
Medium	8 oz.	\$.79
Large	16 oz.	\$.99
Super Size	28 oz.	\$1.09
Pitcher	A 60 oz. Super Value	\$2.49

Pete's Pizza
It's The Best!

5.1 Ratios

Any time that we compare two numbers, we define a **ratio.** For example, we could compare the number 3 to the number 10. This ratio is usually written in one of two ways:

$$3 \text{ to } 10 \qquad \text{or} \qquad \frac{3}{10}$$

The numbers that can be compared using ratios may represent many things. For example, comparing the figure on the left with the one on the right

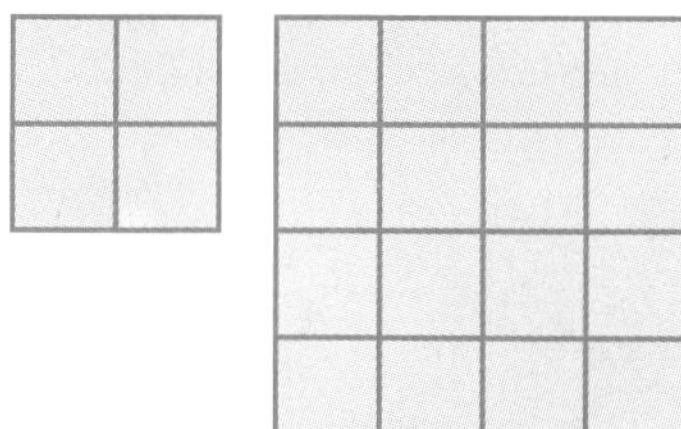

yields the ratio 4 to 16, or $\frac{4}{16}$.

> A **ratio** is the quotient of two quantities of the same kind.
>
> The ratio of a to b is written $\frac{a}{b}$ $\quad (b \neq 0)$

Ratios, like any fractions, can be reduced to lowest terms using the same rules that we learned for writing fractions in lowest terms. The ratio that we are considering here can be written in lowest terms as 1 to 4, or $\frac{1}{4}$. If we look at the figures again, this time looking at the dark outline, we can see that indeed the ratio $\frac{1}{4}$ represents the same relationship between the two figures as the ratio $\frac{4}{16}$.

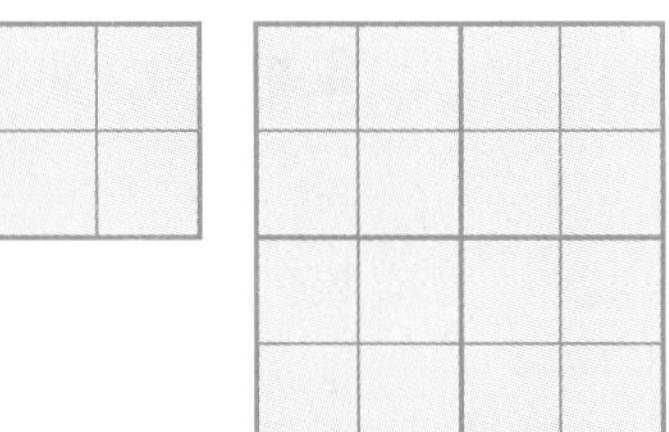

It is also permissible to compare larger quantities with smaller quantities, which gives us a ratio that may be represented by an improper fraction. In our previous example, a comparison of the figure on the right with the one on the left yields the ratio 16 to 4, which in lowest terms is 4 to 1. If we had a room that measured 11 feet by 14 feet, the ratio of the width to the length would be $\frac{11}{14}$, while the ratio of the length to the width would be $\frac{14}{11}$.

EXAMPLE 1 Write the ratio of 12 to 16 as a fraction and reduce it to lowest terms.

SOLUTION

$$\frac{12}{16} = \frac{\overset{3}{\cancel{12}}}{\underset{4}{\cancel{16}}} = \frac{3}{4}$$

EXAMPLE 2 Write the ratio of $4\frac{1}{2}$ to $6\frac{3}{4}$ as a fraction and reduce it to lowest terms.

SOLUTION First we change both mixed numbers to improper fractions:

$$\frac{4\frac{1}{2}}{6\frac{3}{4}} = \frac{\frac{9}{2}}{\frac{27}{4}} \qquad \text{Divide } \frac{9}{2} \text{ by } \frac{27}{4}.$$

$$= \frac{\overset{1}{\cancel{9}}}{\underset{1}{\cancel{2}}} \times \frac{\overset{2}{\cancel{4}}}{\underset{3}{\cancel{27}}} \qquad \text{Invert } \frac{27}{4} \text{ and multiply.}$$

$$= \frac{2}{3}$$

EXAMPLE 3 Write the ratio of 0.06 to 1.8 as a fraction and reduce it to lowest terms.

SOLUTION

$$\frac{0.06}{1.8} = \frac{0.06}{1.8} \times \frac{100}{100} \quad \text{Get rid of the decimal points.}$$

$$= \frac{6}{180} \quad \text{Reduce to lowest terms.}$$

$$= \frac{1}{30}$$

Ratios and Units

Look at the ratio $\frac{2 \text{ feet}}{4 \text{ yards}}$. Your first thought might be to reduce it to $\frac{1}{2}$. However, if we look at a diagram illustrating this ratio, we find that it does not represent the ratio of 1 to 2.

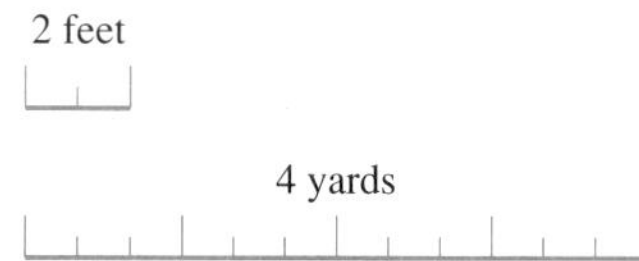

In fact, if we count up the units in the diagram, we find the ratio to be $\frac{2}{12}$, which, when reduced to lowest terms, is $\frac{1}{6}$. This example illustrates one of the most important things about ratios.

> In order to reduce a ratio containing units, the two quantities must always be expressed in the same units.

In our example, the ratio of $\frac{2 \text{ feet}}{4 \text{ yards}}$ must be written as $\frac{2 \text{ feet}}{12 \text{ feet}}$ first. In other words, when simplifying a ratio, it should always compare feet to feet, inches to inches, pounds to pounds, and so on.

The ratio of 8 inches to 2 feet must be written using only one set of units. Since 2 feet is the same as 24 inches, the correct ratio is $\frac{8 \text{ inches}}{24 \text{ inches}} = \frac{1}{3}$.

EXAMPLE 4 Write the ratio of one dime to one quarter as a fraction and reduce it to lowest terms.

SOLUTION Remember that both parts of a ratio must have the same units. We change both the dime and the quarter to cents, giving us

$$\frac{10 \text{ cents}}{25 \text{ cents}} = \frac{\overset{2}{\cancel{10}} \cancel{\text{cents}}}{\underset{5}{\cancel{25}} \cancel{\text{cents}}} = \frac{2}{5}$$

Note that we "cancelled" the cents, since they were the same units. This procedure is done frequently in science. Our final answer, $\frac{2}{5}$, has no units at all. We call it a **pure number.** ▮▮

EXAMPLE 5 Write the ratio of $\frac{1}{2}$ hour to $\frac{3}{4}$ hour as a fraction and reduce it to lowest terms.

SOLUTION Since the units are the same, they cancel and we have

$$\frac{\frac{1}{2}\ \text{hour}}{\frac{3}{4}\ \text{hour}} = \frac{\frac{1}{2}}{\frac{3}{4}} \qquad \frac{1}{2} \text{ is divided by } \frac{3}{4}.$$

$$= \frac{1}{\not{2}_1} \cdot \frac{\not{4}^2}{3} \qquad \text{Invert } \frac{3}{4} \text{ and multiply.}$$

$$= \frac{2}{3}$$

ALTERNATE SOLUTION We could have changed the units from hours to minutes, which would have eliminated the fractions.

$$\frac{\frac{1}{2}\ \text{hour}}{\frac{3}{4}\ \text{hour}} = \frac{30\ \text{minutes}}{45\ \text{minutes}} \qquad \text{Divide numerator and denominator by 15.}$$

$$= \frac{2}{3} \qquad \text{The result is the same as before.}$$

▮▮

EXAMPLE 6 A building 120 feet high casts a shadow 40 feet long. Express this as a ratio.

SOLUTION

$$\frac{120\ \text{feet}}{40\ \text{feet}} = \frac{\overset{3}{\not{120}}\ \not{\text{feet}}}{\underset{1}{\not{40}}\ \not{\text{feet}}} = \frac{3}{1}$$

▮▮

STOP

A 1 in the denominator in the case of fractions would not normally be written. However, in the case of ratios we keep it. Our answer in Example 6 means that the building is in a ratio of 3 to 1 with its shadow, or that the building is three times as high as its shadow is long.

5.1 Exercises

▼ Write the ratio of the figure on the left to the figure on the right as a fraction. Reduce the ratio to lowest terms if necessary.

1.

2. 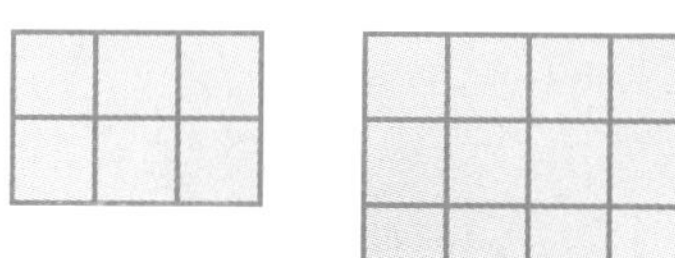

3. 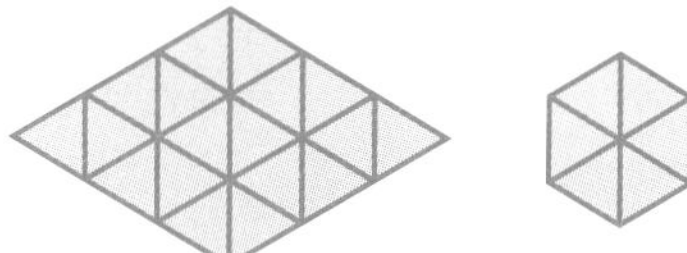

4. 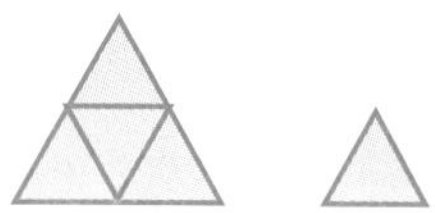

5. 

6. 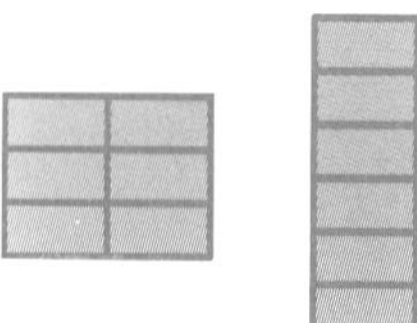

▼ Write each of the ratios in lowest terms.

7. $\frac{3}{12}$

8. $\frac{4}{22}$

9. 3 to 15

10. 30 to 90

11. $\frac{30}{10}$

12. $\frac{105}{15}$

13. 39 to 52

14. 26 to 78

15. $\frac{6}{1000}$

16. $\frac{11}{1100}$

17. $\frac{17}{6}$

18. $\frac{29}{11}$

19. $\frac{360}{40}$

20. $\frac{240}{40}$

21. $\frac{2}{3}$ to $\frac{8}{7}$

22. $\frac{8}{5}$ to $\frac{4}{25}$

23. $\frac{\frac{9}{15}}{\frac{27}{25}}$

24. $\frac{\frac{11}{4}}{\frac{22}{8}}$

25. $\frac{3\frac{1}{3}}{7\frac{1}{2}}$

26. $\frac{4\frac{2}{3}}{6\frac{1}{4}}$

27. $\frac{6\frac{3}{8}}{\frac{2}{3}}$

28. $\frac{9\frac{1}{5}}{\frac{7}{8}}$

29. $\frac{0.04}{0.12}$

30. $\frac{0.25}{0.75}$

31. $\frac{0.6}{2.4}$

32. $\frac{0.9}{3.6}$

33. $\dfrac{0.34}{17}$

34. $\dfrac{0.64}{16}$

35. $\dfrac{\frac{3}{4}}{0.15}$

36. $\dfrac{\frac{1}{4}}{0.75}$

▼ Express the following as ratios. Remember to use only one set of units in a ratio. Reduce to lowest terms where possible.

37. 3 eggs to 1 dozen eggs

38. 2 hours to 15 minutes

39. 5 inches to 2 feet

40. 1 yard to 1 foot

41. 1 dime to 1 quarter

42. 40 minutes to 1 hour

43. 4 nickels to 2 dollars

44. 2 dozen to 6

45. $\frac{1}{2}$ hour to $1\frac{1}{4}$ hours

46. 2 feet 2 inches to 1 yard

5.2 Rates

So far in our dealing with ratios, either there were no units at all or the units were the same or could be made the same. For example:

$$\frac{2}{3}, \quad \frac{2 \text{ inches}}{5 \text{ inches}}, \quad \frac{3 \text{ feet}}{6 \text{ inches}}$$

A **rate,** on the other hand, is the comparison of two quantities each of a different type. For example, $\frac{\text{miles}}{\text{hour}}$ (miles per hour) or $\frac{\text{dollars}}{\text{hour}}$ (dollars per hour) are rates that are probably familiar to you. A rate is generally written as a fraction and the units in both numerator and denominator are stated.

EXAMPLE 1 Gail traveled 120 miles in 3 hours. Express this as a rate in lowest terms.

SOLUTION

$$\frac{120 \text{ miles}}{3 \text{ hours}} = \frac{40 \text{ miles}}{1 \text{ hour}} \quad \text{or} \quad 40 \frac{\text{miles}}{\text{hour}}$$

If the denominator of a rate is 1, it is not usually written and we read the answer as 40 miles per hour. Rates in which the denominator is 1 are called **unit rates.**

EXAMPLE 2 Don earns \$226 in a 40-hour work week. Express this as a unit rate in decimal form. (This is actually his rate of pay.)

SOLUTION

$$\frac{226 \text{ dollars}}{40 \text{ hours}} = 5.65 \frac{\text{dollars}}{\text{hour}}$$

or \$5.65 per hour.

In the case of dollars, a decimal rounded to two places is the most common way of expressing the quotient.

Unit Pricing

Unit pricing is an example of unit rates that has become increasingly popular in grocery stores. It gives the consumer a basis on which to compare prices of different sizes or brands of an item.

EXAMPLE 3 If milk costs \$2.28 per gallon, \$2.18 for a half-gallon, or \$0.63 for a quart ($\frac{1}{4}$ gallon), which size provides the most milk for your money?

SOLUTION

1 gallon $\quad \dfrac{\$2.28}{1 \text{ gallon}} = \dfrac{228¢}{1 \text{ gallon}} = 228¢ \text{ per gallon}$

$\dfrac{1}{2}$ gallon $\quad \dfrac{\$1.18}{\frac{1}{2} \text{ gallon}} = \dfrac{118¢}{0.5 \text{ gallon}} = 236¢ \text{ per gallon}$

$\dfrac{1}{4}$ gallon $\quad \dfrac{\$0.63}{\frac{1}{4} \text{ gallon}} = \dfrac{63¢}{0.25 \text{ gallon}} = 252¢ \text{ per gallon}$

The one-gallon container has the lowest unit price, which means it provides the most milk for your money. ▮▮

5.2 Exercises

▼ Write the following comparisons as unit rates.

1. 364 miles on 14 gallons
2. 132 dollars in 22 hours
3. 143 points in 11 games
4. \$7.74 for 6 cans
5. 432 miles in 9 hours
6. 100 meters in 10 seconds
7. $6\frac{1}{3}$ tons on 19 acres
8. 180 feet in 20 seconds
9. 170 words in 5 minutes
10. 135 feet in 15 minutes

▼ Write the following comparisons as unit rates, rounding your answer to the nearest tenth.

11. 26 gallons in 5 hours
12. 175 miles in 4 hours
13. 190 feet in 85 seconds
14. 100 yards in 9.8 seconds
15. 190 words in 6 minutes

16. 262 miles in $5\frac{1}{2}$ hours

17. 725 square feet with 3 gallons

18. 137 yards in 31 carries

19. 12.6 grams in 4.1 liters

20. 14.5 pounds in 6 quarts

24. Tracy got 36 hits in 124 times at bat. Express this as a unit rate in decimal form rounded off to the nearest thousandth.

25. Nicole is comparing different brands of canned tuna. Brand A contains 6 oz and costs 89¢. Brand B contains $5\frac{1}{2}$ oz and costs 79¢. Find the unit price of each to determine which size is the better buy.

26. Mr. Francisco wants to know which is the best value, a 20-oz container of dish detergent that costs $1.39 or a 12-oz container of the same detergent that costs 89¢.

LIVING IN THE REAL WORLD

21. In 5 minutes, Maryanna can type 316 words with no errors; what is her typing rate?

22. Maryanna can take 714 words of dictation in 11 minutes. What is her dictation rate?

23. George shot 190 points in 26 games last season. How many points did he shoot per game?

27. What is the difference between a ratio and a rate?

28. What is a unit rate?

29. Of what value is unit pricing to consumers? How can you make use of unit pricing when grocery shopping?

5.3 Proportion

A **proportion** is an equation consisting of two equal ratios. That is, if $\frac{a}{b}$ and $\frac{x}{y}$ are two ratios, then the equation $\frac{a}{b} = \frac{x}{y}$ is a proportion. This proportion can also be written $a:b = x:y$. The first and last terms, a and y, are called the **extremes** and the middle two terms, b and x, are called the **means.**

There is a very important relationship between the means and the extremes in a proportion: The product of the means *equals* the product of the extremes. For example, if $\frac{a}{b} = \frac{x}{y}$, then $a \cdot y = b \cdot x$. This process is commonly called cross-multiplication.

Cross-multiplication rule

In a proportion, the product of the means equals the product of the extremes.

$$\frac{a}{b} = \frac{c}{d} \qquad \frac{a}{b} \times \frac{c}{d} \qquad a \cdot d = b \cdot c$$

Cross-multiplication provides an easy way to tell if a statement is a proportion or not.

EXAMPLE 1 Does $\frac{7}{12} = \frac{21}{36}$?

SOLUTION To find out, we cross-multiply: $\frac{7}{12} \times \frac{21}{36}$.

$$7 \cdot 36 = 252 \qquad \text{and} \qquad 12 \cdot 21 = 252$$

Since the product of the means, $12 \cdot 21$, and the product of the extremes, $7 \cdot 36$, both equal 252, the statement $\frac{7}{12} = \frac{21}{36}$ is a proportion. ❚❚

EXAMPLE 2 Does $\frac{2}{3} = \frac{12}{16}$?

SOLUTION Cross-multiplying gives the product of the means as $3 \cdot 12 = 36$ and the product of the extremes as $2 \cdot 16 = 32$. Since $36 \neq 32$, we conclude that $\frac{2}{3} \neq \frac{12}{16}$, and we say that the statement $\frac{2}{3} = \frac{12}{16}$ is not a proportion. ❚❚

5.3 Exercises

▼ Replace the question mark in each of the following by = if the statement is a proportion and by ≠ if the statement is not a proportion.

1. $\frac{3}{4} ? \frac{9}{12}$
2. $\frac{2}{3} ? \frac{10}{15}$
3. $\frac{2}{3} ? \frac{7}{10}$
4. $\frac{3}{4} ? \frac{12}{15}$
5. $\frac{5}{2} ? \frac{20}{8}$
6. $\frac{20}{5} ? \frac{100}{25}$
7. $\frac{6}{8} ? \frac{9}{12}$
8. $\frac{9}{5} ? \frac{45}{20}$
9. $\frac{1}{12} ? \frac{36}{3}$
10. $\frac{5}{15} ? \frac{3}{1}$

11. $\frac{3.6}{0.12} \ ? \ \frac{72}{24}$

12. $\frac{4.2}{5.2} \ ? \ \frac{2.1}{2.6}$

13. $\frac{\frac{2}{5}}{\frac{5}{8}} \ ? \ \frac{\frac{16}{50}}{\frac{1}{2}}$

14. $\frac{\frac{3}{5}}{\frac{1}{2}} \ ? \ \frac{\frac{7}{8}}{\frac{1}{4}}$

15. $\frac{\frac{9}{16}}{72} \ ? \ \frac{\frac{1}{4}}{32}$

16. $\frac{2\frac{1}{3}}{6} \ ? \ \frac{1\frac{1}{8}}{\frac{3}{4}}$

▼ Given the proportion $\frac{3}{4} = \frac{12}{16}$, answer the following:

17. Are you allowed to reverse the two ratios? That is, does $\frac{3}{4} = \frac{12}{16}$ mean the same as $\frac{12}{16} = \frac{3}{4}$?

18. Are you allowed to invert one of the ratios? That is, does $\frac{3}{4} = \frac{12}{16}$ mean the same as $\frac{4}{3} = \frac{12}{16}$?

19. Are you allowed to invert both ratios in a proportion? That is, does $\frac{3}{4} = \frac{12}{16}$ mean the same as $\frac{4}{3} = \frac{16}{12}$?

5.4 Solving Proportions

We shall now put our newfound knowledge of ratio and proportion to work solving problems for us. Proportions are frequently used to solve problems in which three out of four parts of a statement are known and we have to find a value for the missing part that will make the statement a true proportion.

For example, if on a map 2 inches represents 50 miles, how many miles are represented by 5 inches? If we represent the distance we are trying to find by a variable, say *d* for distance, the proportion looks like this:

$$\frac{2}{50} = \frac{5}{d}$$

As you can see, three parts of the proportion are given and we are being asked to find the fourth. To solve for the variable *d*, we will make use of the fact that the product of the means equals the product of the extremes in a proportion. In other words, we cross-multiply, which gives us the equation

$$2d = 5 \cdot 50 \qquad \text{or} \qquad 2d = 250$$

Dividing both sides of the equation by 2 gives us

$$\frac{\cancel{2}d}{\cancel{2}} = \frac{\overset{125}{\cancel{250}}}{\cancel{2}} \qquad \text{or} \qquad d = 125$$

To check to see if the value we found for *d* is correct, we substitute it back into the original proportion: $\frac{2}{50} = \frac{5}{125}$. Cross-multiplication gives us $2 \cdot 125 = 5 \cdot 50$, or $250 = 250$, so $d = 125$ yields a true proportion.

EXAMPLE 1 Solve the following proportion for x: $\frac{3}{4} = \frac{x}{28}$.

SOLUTION We cross-multiply and solve the resulting equation as follows:

$$\frac{3}{4} = \frac{x}{28}$$

$$4 \cdot x = 3 \cdot 28 \quad \text{Cross-multiply.}$$

$$4x = 84$$

$$\frac{\cancel{4}x}{\cancel{4}} = \frac{\overset{21}{\cancel{84}}}{\cancel{4}} \quad \text{Divide both sides by 4.}$$

$$x = 21$$

Check: Is $\frac{3}{4} = \frac{21}{28}$ a proportion? Since $4 \cdot 21 = 84$ and $3 \cdot 28 = 84$, it is a proportion. ✓

EXAMPLE 2 Find the value for h that makes the proportion $\frac{h}{8} = \frac{21}{6}$ true.

SOLUTION

$$\frac{h}{8} = \frac{21}{6}$$

$$6 \cdot h = 8 \cdot 21 \quad \text{Cross-multiply.}$$

$$6h = 168$$

$$\frac{\cancel{6}h}{\cancel{6}} = \frac{\overset{28}{\cancel{168}}}{\cancel{6}} \quad \text{Divide both sides by 6.}$$

$$h = 28$$

Check: Does $\frac{28}{8} = \frac{21}{6}$?

$$6 \cdot 28 = 8 \cdot 21$$

$$168 = 168 \quad ✓$$

EXAMPLE 3 Solve $\frac{30}{40} = \frac{18}{x}$ for x.

SOLUTION

$$\frac{30}{40} = \frac{18}{x}$$

$$30 \cdot x = 40 \cdot 18 \quad \text{Cross-multiply.}$$

$$30x = 720$$

$$\frac{\cancel{30}x}{\cancel{30}} = \frac{\overset{24}{\cancel{720}}}{\cancel{30}} \quad \text{Divide both sides by 30.}$$

$$x = 24$$

You may have wondered if the ratio $\frac{30}{40}$ in Example 3 could have been reduced before solving the proportion. Actually, this would simplify our work, as you will see in the next example.

EXAMPLE 4 Solve $\frac{30}{40} = \frac{18}{x}$ for x.

SOLUTION This time we simplify the ratio $\frac{30}{40}$ to $\frac{3}{4}$ before cross-multiplying.

$$\frac{30}{40} = \frac{18}{x}$$

$$\frac{3}{4} = \frac{18}{x}$$

$$3 \cdot x = 4 \cdot 18 \qquad \text{Cross-multiply.}$$

$$3x = 72$$

$$\frac{\cancel{3}x}{\cancel{3}} = \frac{\overset{24}{\cancel{72}}}{\cancel{3}} \qquad \text{Divide both sides by 3.}$$

$$x = 24$$

EXAMPLE 5 Solve the proportion $\frac{\frac{1}{4}}{y} = \frac{4\frac{1}{2}}{7}$.

SOLUTION Change $4\frac{1}{2}$ to the improper fraction $\frac{9}{2}$ and cross-multiply:

$$\frac{\frac{1}{4}}{y} = \frac{\frac{9}{2}}{7}$$

$$\frac{9}{2} \cdot y = \frac{1}{4} \cdot 7 \qquad \text{Cross-multiply}$$

$$\frac{\frac{9}{2} \cdot y}{\frac{9}{2}} = \frac{\frac{7}{4}}{\frac{9}{2}} \qquad \text{Divide both sides by } \frac{9}{2}.$$

$$\frac{\cancel{2}}{\cancel{9}} \cdot \frac{\cancel{9}}{\cancel{2}} y = \frac{7}{\underset{2}{\cancel{4}}} \cdot \frac{\overset{1}{\cancel{2}}}{9} \qquad \text{This is the same as multiplying by } \frac{2}{9}.$$

$$y = \frac{7}{18}$$

Check: $\frac{\frac{1}{4}}{\frac{7}{18}} = \frac{\frac{9}{2}}{7}$

$$\frac{\overset{1}{\cancel{9}}}{2} \cdot \frac{7}{\underset{2}{\cancel{18}}} = \frac{7}{4} \quad \text{and} \quad \frac{1}{4} \cdot 7 = \frac{7}{4} \quad ✓$$

5.4 Exercises

▼ Solve the following proportions for the variable and check:

1. $\frac{3}{x} = \frac{4}{8}$

2. $\frac{1}{5} = \frac{y}{20}$

3. $\frac{6}{75} = \frac{4}{w}$

4. $\frac{5}{4} = \frac{30}{h}$

5. $\frac{m}{8} = \frac{75}{12}$

6. $\frac{8}{a} = \frac{20}{250}$

7. $\frac{10}{3} = \frac{s}{99}$

8. $\frac{x}{6} = \frac{3.5}{10.5}$

9. $\frac{k}{52} = \frac{14}{13}$

10. $\frac{0.09}{1} = \frac{18}{t}$

11. $\frac{5.2}{2.6} = \frac{m}{14}$

12. $\frac{\frac{2}{3}}{\frac{5}{8}} = \frac{6}{n}$

13. $\frac{y}{6} = \frac{5}{7}$

15. $\frac{\frac{1}{2}}{x} = \frac{\frac{3}{4}}{\frac{7}{8}}$

14. $\frac{2\frac{1}{2}}{x} = \frac{4\frac{1}{4}}{2\frac{2}{3}}$

5.5 Solving Practical Problems Using Proportions

Let's return for a moment to our problem involving the map: If, on a map, 2 inches represents 50 miles, how many miles are represented by 5 inches? The most important thing to understand about the problem is that the ratio of inches to miles does not change, regardless of how many inches we consider. As before, we let the unknown number of miles be represented by a variable, *d*. This gives us the proportion

$$\frac{2 \text{ inches}}{50 \text{ miles}} = \frac{5 \text{ inches}}{d \text{ miles}} \qquad \text{Method 1}$$

Labels on units help to ensure that the proportion is set up correctly. The units must be in the same order on each side of the proportion, in this case, inches to miles = inches to miles. If each ratio involves the same types of units, for example,

$$\frac{2 \text{ inches}}{5 \text{ inches}} = \frac{50 \text{ miles}}{d \text{ miles}} \qquad \text{Method 2}$$

we must be sure that the numerators correspond to each other and the denominators correspond to each other. In this example the numerators correspond to each other. In this example the numerators correspond, 2 inches to 50 miles, as do the denominators, 5 inches to *d* miles. An *incorrect* proportion would be

$$\frac{2 \text{ inches}}{50 \text{ miles}} = \frac{d \text{ miles}}{5 \text{ inches}} \quad \text{or} \quad \frac{2 \text{ inches}}{5 \text{ inches}} = \frac{d \text{ miles}}{50 \text{ miles}} \qquad \text{Set up incorrectly}$$

In the first example, we have inches to miles = miles to inches, which is incorrect because the units are not in the same order on each side of the equation. In the second example, the values in the numerators and denominators do not correspond to each other. In the original problem, 2 inches does not correspond to *d* miles, nor does 5 inches correspond to 50 miles.

Once we are certain that the proportion is set up correctly, we can drop the labels, since they are not necessary in the solution of the proportion and indeed may even be confusing. Solving the proportion for the missing term, *d,* is then as before:

$$\frac{2}{50} = \frac{5}{d}$$

$$2 \cdot d = 5 \cdot 50 \quad \text{Cross-multiply.}$$

$$2d = 250$$

$$\frac{\overset{1}{\cancel{2}}d}{\underset{1}{\cancel{2}}} = \frac{\overset{125}{\cancel{250}}}{\underset{1}{\cancel{2}}} \quad \text{Divide both sides by 2.}$$

$$d = 125$$

Our answer is: 5 inches on the map represents 125 miles.

EXAMPLE 1 If a car can go 112 miles on 7 gallons of gasoline, how far can it travel on 11 gallons of gasoline?

SOLUTION We represent the unknown distance by a variable; let's use *m,* for miles. Now we set up the proportion, being careful to keep the units in the right order: 7 gallons corresponds with 112 miles, and 11 gallons corresponds with the unknown distance, *m* miles. The proportion is

$$\frac{7 \text{ gallons}}{112 \text{ miles}} = \frac{11 \text{ gallons}}{m \text{ miles}}$$

Now we drop the labels and solve:

$$\frac{7}{112} = \frac{11}{m}$$

$$7 \cdot m = 11 \cdot 112$$

$$7m = 1232$$

$$\frac{\overset{1}{\cancel{7}}m}{\underset{1}{\cancel{7}}} = \frac{\overset{176}{\cancel{1232}}}{\underset{1}{\cancel{7}}}$$

$$m = 176 \text{ miles}$$

The car can travel 176 miles on 11 gallons of gasoline. ❙❙

EXAMPLE 2 If Christine scores 99 points in the first 6 basketball games that she plays, how many points can she expect to score in a 26-game season if she continues to score at the same rate?

SOLUTION We let x = the number of points for the season and set up the proportion:

$$\frac{99 \text{ points}}{6 \text{ games}} = \frac{x \text{ points}}{26 \text{ games}}$$

$$6 \cdot x = 99 \cdot 26 \qquad \text{Drop the units and cross-multiply}$$

$$\frac{\overset{1}{\cancel{6}}x}{\underset{1}{\cancel{6}}} = \frac{\overset{429}{\cancel{2574}}}{\underset{1}{\cancel{6}}}$$

$$x = 429 \text{ points}$$

Christine can expect to score about 429 points.

EXAMPLE 3 Under certain conditions, 1 unit of oxygen reacts with 2 units of hydrogen. How many units of oxygen will react with 23 units of hydrogen under the same conditions?

SOLUTION We represent the unknown quantity of oxygen by x and set up by the proportion. One unit of oxygen corresponds to 2 units of hydrogen, and x units of oxygen correspond to 23 units of hydrogen, giving us the following proportion:

$$\frac{1 \text{ oxygen}}{2 \text{ hydrogen}} = \frac{x \text{ oxygen}}{23 \text{ hydrogen}}$$

Now we drop the units and solve the proportion.

$$\frac{1}{2} = \frac{x}{23}$$

$$2x = 23$$

$$\frac{\overset{1}{\cancel{2}}x}{\underset{1}{\cancel{2}}} = \frac{23}{2}$$

$$x = \tfrac{23}{2} \text{ or } 11\tfrac{1}{2} \text{ units of oxygen}$$

5.6 Exercises

LIVING IN THE REAL WORLD

▼ Solve each of the following practical problems using proportions.

1. If you can drive 180 miles in 4 hours, how long will it take you to drive 450 miles at the same rate?

2. A certain drug is administered according to body weight. If 2 cubic centimeters (cc) of the drug are given for each 35 pounds of body weight, how many cc of the drug should be given to a person weighing 160 pounds?

3. If a recipe calls for 3 cups of sugar and serves 8 people, how many cups of sugar are needed if the recipe is increased to serve 20?

4. If 3 pairs of curtains require a total of 10 yards of material, how many yards of material are required to make 4 pairs of curtains?

5. A 50-pound bag of fertilizer will cover 5000 square feet of land. How much land can be covered by a 70-pound bag?

6. Two units of oxygen combine with one unit of carbon under certain conditions. How many units of carbon will combine with 123 units of oxygen under these conditions?

7. The scale on a drawing is 1 inch represents 6 feet. What are the dimensions of a house that measures $6\frac{1}{2}$ inches by 5 inches on the drawing? (**Hint:** separate into two problems.)

8. If a car can travel 126 miles on 8 gallons of gas, how much gas will be required to make a trip of 550 miles?

9. A recent survey showed that 14 of every 15 families in a certain city own at least one television set. If there are 2768 families in this city, how many of them own one television set or more?

10. A businessman earned $136 on a $2000 investment in 1 year. How much would he have earned if he had invested $6750?

11. If there are 5 milligrams of a substance in a solution of 150 milliliters, how many milligrams of the substance would you find in 500 milliliters of the solution?

12. A business spends 14 cents of every dollar it takes in for advertising. If the total sales for a week average $3260, how much is spent per week for advertising?

13. If you earned $89 in 2 days, how much would you expect to make if you worked for 5 days at the same rate of pay?

14. 87 divided by what number is the same as 27 divided by 6?

15. If you can average 450 miles of driving per day, how many days will it take to make a trip of 2415 miles from New York City to Phoenix, Arizona?

16. A weekly paycheck of $360 has $79.20 withheld for federal taxes. How much will you have earned before taxes by the time a total of $1346.40 has been deducted?

17. On a map, 1 inch represents a distance of 150 miles. How many inches represent 2450 miles?

18. A basketball player scores 168 points in 8 games. How many points will she score in 20 games if she continues to score at the same rate?

19. If 6 ounces of a cereal contain 2.5 grams of sugar, how many grams of sugar will 16 ounces of the cereal contain?

20. If you can drive your car 148 miles in 4 hours, how far will you travel at the same rate in 10 hours?

21. If avocados cost 3/$2.00, approximately how much will 2 avocados cost?

22. If 25 grams of a breakfast cereal contain $\frac{1}{2}$ gram of fat, how many grams of fat will be contained in 90 grams of the cereal?

23. If the tax on a home worth $85,000 is $1850, how much is the tax on a home valued at $100,000 if taxed at the same rate?

24. On a map, $\frac{1}{2}$ inch represents a distance of 140 miles. How many miles do $3\frac{3}{4}$ inches represent?

25. A time study shows that a worker can assemble 9 items in 30 minutes. How many items can a worker be expected to assemble at the same rate in a 7-hour shift?

26. A jogger can run $2\frac{1}{2}$ miles in $22\frac{1}{2}$ minutes. How many minutes will it take her to run 8 miles if she continues to run at the same rate?

SUMMARY—CHAPTER 5

The numbers in brackets refer to the section in the text in which each concept is presented.

[5.1] A **ratio** is a comparison of two numbers.

EXAMPLES

[5.1] 3 to 7 or $\frac{3}{7}$

[5.1] To **reduce a ratio to lowest terms,** the units must be the same.

[5.1] $\frac{3 \text{ inches}}{2 \text{ feet}} = \frac{3 \text{ inches}}{24 \text{ inches}} = \frac{1}{8}$

[5.2] A **rate** is a comparison of two quantities each of a different type. A rate in which the denominator is 1 is called a **unit** rate.

[5.2] $\frac{90 \text{ miles}}{2 \text{ hours}} = \frac{45 \text{ miles}}{1 \text{ hour}}$

or 45 miles per hour

[5.3] A **proportion** is a statement that two ratios are equal.

[5.3] $\frac{2}{3} = \frac{8}{12}$

[5.3] In any proportion, the product of the means equals the product of the extremes. This is called the **cross-multiplication rule.**

[5.3] $\frac{2}{3} = \frac{8}{12}$

$$\underbrace{3 \cdot 8}_{\text{means}} = \underbrace{2 \cdot 12}_{\text{extremes}}$$

$$24 = 24$$

[5.4] If any one of the four parts of a proportion is unknown, it can be *solved* by using cross-multiplication.

[5.4] $\frac{3}{2} = \frac{x}{6}$

$2x = 18$

$x = 9$

REVIEW EXERCISES—CHAPTER 5

The numbers in brackets refer to the section in the text in which similar problems are presented.

▼ [5.1]

1. Write as a fraction the ratio of the area of the figure on the left to that of the figure on the right. Reduce to lowest terms if necessary.

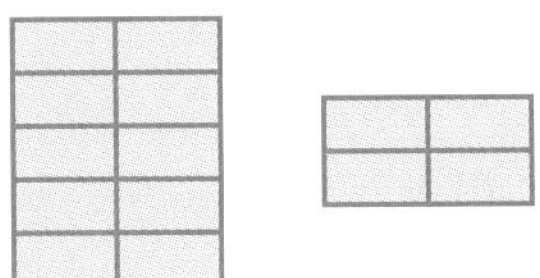

▼ [5.2]

2. Write each of the following ratios in lowest terms.

(a) $\frac{16}{20}$ **(b)** $\frac{2\frac{3}{4}}{5\frac{1}{2}}$ **(c)** $\frac{51}{11}$

(d) $\frac{\text{3 nickels}}{\text{6 dimes}}$ **(e)** $\frac{2\frac{1}{2}\text{ hours}}{1\frac{1}{2}\text{ hours}}$

▼ [5.3]

3. Write each of the following comparisons as a unit rate.

(a) 153 dollars in 17 hours

(b) 336 words in 8 minutes

(c) 137 miles in 3 hours (round your answer to two decimal places)

▼ [5.3]

4. Replace the question mark by = or ≠.

(a) $\frac{36}{5} \;?\; \frac{79.2}{11}$ **(b)** $\frac{2}{3} \;?\; \frac{4}{9}$

(c) $\frac{2\frac{1}{2}}{5} \;?\; \frac{3\frac{3}{4}}{7\frac{1}{2}}$

▼ [5.4]

5. Solve and check:

(a) $\frac{2}{7} = \frac{x}{42}$ **(b)** $\frac{9}{y} = \frac{3}{7}$

(c) $\frac{3.2}{5.6} = \frac{1.4}{t}$

LIVING IN THE REAL WORLD

▼ [5.5]

6. If 20 ounces of a breakfast cereal contains 11 grams of sugar, how many ounces of sugar will 3 grams of cereal contain?

7. If you can drive your car 220 miles in 5 hours, how long will it take you to drive 396 miles if you travel at the same rate?

8. Ben can type 270 words in 6 minutes. How many words can he type in 20 minutes if he continues to type at the same rate?

CRITICAL THINKING EXERCISES—CHAPTER 5

▼ Tell whether each of the following is true or false and give reasons why.

1. $\frac{\text{3 feet}}{\text{12 yards}} = \frac{1}{4}$

2. $\frac{\text{4 nickels}}{\text{2 quarters}} = \frac{\text{20 cents}}{\text{50 cents}} = \frac{2}{5}$

3. In his career, Reggie Jackson hit 534 home runs and struck out 2,459 times. Write the ratio of strikeouts to home runs.

SOLUTION:

$$\frac{534}{2{,}459}$$

4. The ratio $\frac{0.6}{2.4}$ in lowest terms is $\frac{\overset{1}{0.\not{6}}}{\underset{0.4}{\not{2.4}}} = \frac{1}{0.4}$.

5. A 10-ounce package of frozen corn costs 89¢. The unit price is $\frac{10 \text{ ounces}}{89¢} = 11.2¢/\text{ounce}$ rounded to one decimal place.

▼ Find the error in each of the following:

6. The ratio $\frac{6 \text{ feet}}{3 \text{ feet}}$ in lowest terms is $\frac{\overset{2}{\not{6}} \not{\text{feet}}}{\underset{1}{\not{3}} \not{\text{feet}}} = \frac{2}{1} = 2$.

7. Does $\frac{6}{8} = \frac{12}{9}$ form a proportion?

SOLUTION:

$6 \cdot 12 = 72$ and $8 \cdot 9 = 72$,

so $\frac{6}{8} = \frac{12}{9}$ is a proportion.

8. Solve the proportion $\frac{2}{5} = \frac{x}{20}$.

SOLUTION:

$$2 \cdot x = 5 \cdot 20$$
$$2x = 100$$
$$x = 50$$

9. Kevin McEnroe earns \$420 per week and has \$92 deducted for federal taxes and social security. How much will he have earned by the time \$736 has been deducted?

SOLUTION:

$$\frac{\$420}{\$92} = \frac{\$736}{x}$$

$$420 \cdot x = 92 \cdot 736$$

$$420x = 67{,}712$$

$$\frac{\not{420}x}{\not{420}} = \frac{67{,}712}{420}$$

$$x = \$161.22$$

10. The ratio of men to women at Southern University is 2 to 3. If there are 3,600 students at Southern, how many of them are men?

SOLUTION:

$$\frac{2}{3} \cdot 3{,}600 = \frac{2}{\underset{1}{\not{3}}} \cdot \overset{1{,}200}{\not{3{,}600}} = 2{,}400 \text{ men}$$

NAME ▮▮▮▮▮ CLASS

ACHIEVEMENT TEST—CHAPTER 5

This test covers the material in Chapter 5.

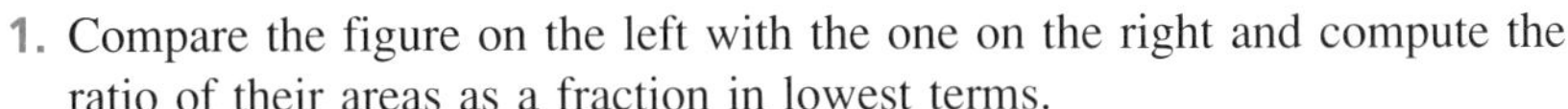

1. Compare the figure on the left with the one on the right and compute the ratio of their areas as a fraction in lowest terms.

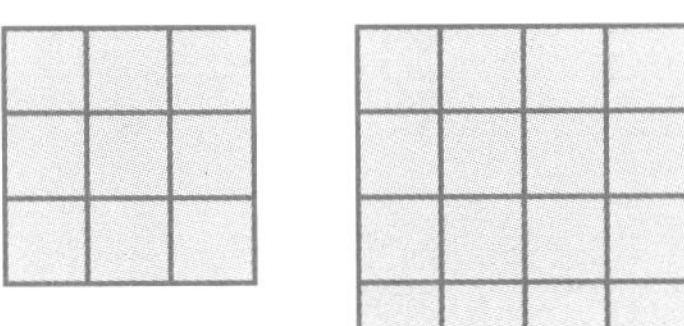

2. Reduce each ratio to lowest terms.

   a. $\frac{26}{65}$ b. $\frac{\text{9 inches}}{\text{3 feet}}$

3. Write as a unit rate:

   a. $448 for 28 square yards of carpet.

   b. 312 miles on 11.7 gallons of gasoline (round your answer to one decimal place)

4. Replace the question mark by $=$ or $\neq$.

   a. $\frac{5}{6} ? \frac{35}{42}$ b. $\frac{17}{6} ? \frac{12}{34}$ c. $\frac{2\frac{3}{5}}{7\frac{4}{5}} ? \frac{\frac{2}{3}}{2}$

5. The scale on a drawing is 1 inch represents 8 feet. What are the dimensions of a room that measures 3 inches by $2\frac{3}{4}$ inches on the drawing?

6. The property tax on a $65,000 house is $700. How much will the tax be on a $95,000 home that is taxed at the same rate?

1. ______

2. a. ______ b. ______

3. a. ______ b. ______

4. a. ______ b. ______ c. ______

5. ______

6. ______

7. Gary drove 168 miles in $3\frac{1}{2}$ hours. How far will he drive in 8 hours if he travels at the same rate?

7. ______________________

8. Don is comparing different sizes of a particular soft drink. A 12-ounce bottle costs 75¢ and a 16-ounce bottle of the same drink costs 99¢. Find the unit price of each and determine which is the best value.

8. ______________________

____

____

____

CHAPTER **6**

Percent

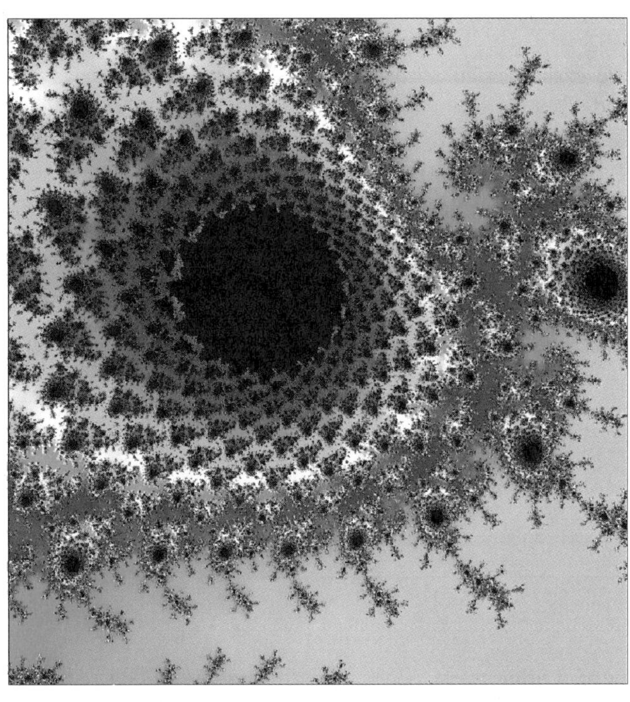

INTRODUCTION

Previously we studied ratios and proportions. In this chapter we will investigate an example of a proportion called the **percent** proportion. Percent is one of the most widely used applications of mathematics and we will use it to solve many different kinds of real-world problems.

NUMBER KNOWLEDGE—CHAPTER 6

Sportcasters frequently say that an athlete is "giving 110%." What exactly does that mean?

If your Spanish test is worth 140 points and you want to get a B, which is 80%, how many points must you earn?

Lou, a used car salesman, sold two cars for $2,000 each. On the first car he earned a profit of 25% of the cost, and on the second he sold it at a loss equal to 25% of the cost. His wife, Lucille, in an effort to cheer him up, remarked, "Things could be worse, at least you broke even." "Nonsense," said Lou, "I lost over $250 today." Who was right, Lou or Lucille?

6.1 The Meaning of Percent

Many everyday things are expressed in terms of percent and percentage. Taxes, interest rates on loans and savings accounts, efficiency of machines, salesmen's commissions, mark-ups and mark-downs on merchandise, and grades on students' test papers are but a few of the relations that are normally expressed in percent.

When we studied fractions, we read $\frac{31}{100}$ as thirty-one **hundredths,** which we interpreted to mean 31 parts out of 100. Similarly, when we studied decimals we read 0.31 as thirty-one **hundredths.** Now we are going to explore a third way of writing **hundredths.** It is written using the symbol "%" and is read **percent.**

> Percent means per hundred.

Thus 31% is another way of writing thirty-one hundredths, or $\frac{31}{100}$.

In a similar manner, we have the following:

$$6\% = \frac{6}{100}$$

$$50\% = \frac{50}{100}$$

$$100\% = \frac{100}{100}$$

$$315\% = \frac{315}{100}$$

$$4.6\% = \frac{4.6}{100}$$

We can also make the following statements:

1. 87 out of 100 means 87%.
2. Thirty-one hundredths means 31%.
3. A quarter is 25 hundredths of a dollar, or 25% of a dollar.

The definition, percent means per hundred, should be remembered well, because we will use it frequently in the following sections.

Changing a Percent to a Decimal

We have just learned that percent means per hundred. Thus 31% means $\frac{31}{100}$. This is indeed fortunate because, as we learned in the unit on fractions, it is quite simple to divide any number by 100. We just move the decimal point 2 places to the left. For example, $\frac{31}{100} = 0.31$.

If you have trouble remembering this rule, you can always perform the long division and obtain the same answer.

$$\begin{array}{r} 0.31 \\ 100\overline{)31.00} \\ \underline{30\ 0} \\ 1\ 00 \\ \underline{1\ 00} \\ 0 \end{array}$$

Summarizing, then, we have

$$31\% = \frac{31}{100} = 0.31$$

and we have established a relationship among a percent, a fraction, and a decimal. This is an important relationship, because when we deal with problems that involve percent, we shall usually find it necessary to write percents as equivalent decimals.

Stated formally, our rule is this:

To change a percent to a decimal:

Remove the % symbol and move the decimal point two places to the **left.**

EXAMPLE 1 Change 42% to a decimal.

SOLUTION We drop the % symbol and move the decimal point two places to the left. Remember, there is a decimal point implied, but not written, immediately after the 2 in the number 42.

$$42\% \rightarrow 0.42. \rightarrow 0.42$$

EXAMPLE 2 Change 275% to a decimal.

SOLUTION We drop the percent symbol and move the decimal point two places to the left:

$$275\% = 2.75$$

EXAMPLE 3 Change 1.5% to a decimal.

SOLUTION

$$1.5\% = 0.015$$

EXAMPLE 4 Change $\frac{3}{4}\%$ to a decimal.

SOLUTION Since $\frac{3}{4} = 0.75$, we have

$$\frac{3}{4}\% = 0.75\% = 0.0075$$

EXAMPLE 5 Change 0.092% to a decimal.

SOLUTION

$$0.092\% = 0.00092$$

STOP

A common error is to move the decimal point in the wrong direction when changing from a percent to a decimal or from a decimal to a percent. It helps to remember that "percent" means "per hundred" or "divided by 100." Since dividing by 100 will always yield a smaller number, changing a percent to a decimal will always make the number part smaller.

$$75\% = \frac{75}{100} = 0.75$$

0.75 is smaller than 75

Changing a Decimal to a Percent

To write a decimal as a percent, we reverse the process we have just been using. Our rule is as follows:

To change a decimal to a percent:

Move the decimal point two places to the **right** and add a % symbol.

EXAMPLE 6 Write the decimal 0.29 as an equivalent percent.

SOLUTION

$$0.29 \rightarrow 0.29. \rightarrow 29\%$$

EXAMPLE 7 Express the decimal 1.34 as a percent.

SOLUTION

$$1.34 = 134\%$$

EXAMPLE 8 Write 0.4 as a percent.

SOLUTION

$$0.4 = 0.40 = 40\%$$

Notice that we had to add a zero before moving the decimal point to the right two places.

6.1 Exercises

▼ Write each percent as a decimal.

1. 56%
2. 23%
3. 49%
4. 77%
5. 4.6%
6. 3.2%
7. 7.6%
8. 14.9%
9. 0.05%
10. 0.08%
11. 0.32%
12. 0.47%
13. $23\frac{1}{2}\%$
14. $56\frac{1}{2}\%$
15. $\frac{1}{8}\%$
16. $\frac{1}{4}\%$
17. 340%
18. 560%
19. 139%
20. 123%

▼ Write each decimal as a percent.

21. 0.31
22. 0.49
23. 0.85
24. 0.73
25. 1.25
26. 2.15
27. 2.04
28. 1.08
29. 0.6
30. 0.9
31. 0.7
32. 0.1
33. 0.063
34. 0.004
35. 0.0091
36. 0.0063
37. 0.0325
38. 0.0575
39. 0.876
40. 0.942
41. What does "percent" mean?
42. Why do you move the decimal point two places to the left when changing from a percent to a decimal?
43. Why do you move the decimal point two places to the right when changing from a decimal to a percent?

6.2 Changing Fractions to Percents and Percents to Fractions

Changing a Fraction to a Percent

It is sometimes useful to express a given fraction as a percent. Actually, this does not require any new knowledge on our part, only putting together some things that we learned previously.

Suppose that we are asked to write the fraction $\frac{7}{8}$ as an equivalent percent. We do this by writing the fraction as an equivalent decimal and then writing the decimal as an equivalent percent.

First, to write $\frac{7}{8}$ as a decimal, we use long division to divide the numerator 7 by the denominator 8:

$$\begin{array}{r} 0.875 \\ 8\overline{)7.000} \\ \underline{6\,4} \\ 60 \\ \underline{56} \\ 40 \\ \underline{40} \\ 0 \end{array}$$

So, we have $\frac{7}{8} = 0.875$ and we have written the fraction as an equivalent decimal. We learned in the last section how to write a decimal as an equivalent percent by moving the decimal point two places to the right:

$$\frac{7}{8} = 0.875 = 87.5\%$$

The rule is this:

To change a fraction to a percent:

Write the fraction as a decimal and then change the decimal to a percent.

EXAMPLE 1 Express the fraction $\frac{1}{4}$ as an equivalent percent.

SOLUTION We must change the fraction to a decimal and then change the decimal to a percent. To write $\frac{1}{4}$ as a decimal, we use long division to divide 1 by 4:

$$\begin{array}{r} 0.25 \\ 4\overline{)1.00} \\ \underline{8} \\ 20 \\ \underline{20} \\ 0 \end{array}$$

or $\frac{1}{4} = 0.25$. Then we write 0.25 as a percent, giving us

$$\frac{1}{4} = 0.25 = 25\%$$

EXAMPLE 2 Write the fraction $\frac{2}{5}$ as a percent.

SOLUTION First we write $\frac{2}{5}$ as an equivalent decimal. We divide 2 by 5:

$$\begin{array}{r} 0.4 \\ 5\overline{)2.0} \\ \underline{2.0} \\ 0 \end{array}$$

or $\frac{2}{5} = 0.4$. Then we have $0.4 = 0.40 = 40\%$. Our final answer is $\frac{2}{5} = 40\%$.

EXAMPLE 3 Write $\frac{5}{8}$ as a percent.

SOLUTION We write $\frac{5}{8}$ as a decimal:

$$\begin{array}{r} 0.625 \\ 8\overline{)5.000} \\ \underline{4\,8} \\ 20 \\ \underline{16} \\ 40 \\ \underline{40} \\ 0 \end{array}$$

This gives us

$$\frac{5}{8} = 0.625 = 62.5\%$$

EXAMPLE 4 The fraction $\frac{1}{3}$ equals what percent?

SOLUTION We write $\frac{1}{3}$ as a decimal:

$$3\overline{)1.0^{1}0^{1}0} = 0.33\overline{3}$$ Divide 1 by 3 using short division.

Recall that the line over the last 3 means that the 3s keep repeating as we keep dividing. Therefore

$$\frac{1}{3} = 0.33\overline{3} = 33.\overline{3}\% \quad \text{or} \quad 33\tfrac{1}{3}\%$$

or 33.3% rounded to one decimal place. ▮▮

EXAMPLE 5 Write $\frac{9}{40}$ as a percent.

SOLUTION We write $\frac{9}{40}$ as a decimal:

$$\begin{array}{r} 0.225 \\ 40\overline{)9.000} \\ \underline{8\,0} \\ 1\,00 \\ \underline{80} \\ 200 \\ \underline{200} \\ 0 \end{array}$$

Therefore $\frac{9}{40} = 0.225 = 22.5\%$. ▮▮

Changing a Percent to a Fraction

Our next task in this section is to write a given percent as an equivalent fraction. This is exactly the reverse process of the operation we just discussed. Again, the process is an application of the definition of percent: *percent means per hundred.*

Suppose that we are given 9% and asked to write it as an equivalent fraction. Remembering that percent means per hundred, we write 9% as the fraction $\frac{9}{100}$.

Whenever possible, we reduce the fraction to lowest terms. In this case, $\frac{9}{100}$ is already in simplest form, so we are finished:

$$9\% = \frac{9}{100}$$

The rule is as follows:

To change a percent to a fraction:

Drop the percent symbol, divide the number by 100, and reduce to lowest terms.

EXAMPLE 6 Write 40% as an equivalent fraction.

SOLUTION From the meaning of percent, we have $40\% = \frac{40}{100}$. Now we reduce to lowest terms:

$$\frac{40}{100} = \frac{4}{10} = \frac{2}{5} \quad \text{or} \quad 40\% = \frac{2}{5}$$

EXAMPLE 7 Express 34.5% as an equivalent fraction.

SOLUTION

$$34.5\% = \frac{34.5}{100}$$

To get rid of the decimal point in the fraction, we multiply numerator and denominator by 10, giving us $\frac{345}{1000}$, which reduces to $\frac{69}{200}$.

EXAMPLE 8 Express 125% as a fraction.

SOLUTION

$$125\% = \frac{125}{100} = \frac{5}{4} \quad \text{or} \quad 1\frac{1}{4}$$

Our answer is $125\% = \frac{5}{4}$ or $1\frac{1}{4}$. Either the improper fraction or the mixed number is acceptable.

EXAMPLE 9 Write 6% as a fraction.

SOLUTION

$$6\% = \frac{6}{100} = \frac{3}{50}$$

EXAMPLE 10 Change 0.25% to a fraction reduced to lowest terms.

SOLUTION

$$0.25\% = \frac{0.25}{100} = \frac{25}{10000} = \frac{1}{400}$$

EXAMPLE 11 Change 12.5% to a fraction reduced to lowest terms.

SOLUTION

$$12.5\% = \frac{12.5}{100} = \frac{125}{1000} = \frac{1}{8}$$

6.2 Exercises

▼ Write each fraction as a percent. Round your answer to the nearest tenth of a percent.

1. $\frac{4}{5}$
2. $\frac{3}{4}$
3. $\frac{5}{8}$
4. $\frac{2}{5}$
5. $\frac{3}{8}$
6. $\frac{7}{20}$
7. $\frac{4}{25}$
8. $\frac{6}{25}$
9. $\frac{5}{16}$
10. $\frac{7}{16}$
11. $1\frac{1}{2}$
12. $1\frac{3}{4}$
13. $1\frac{3}{5}$
14. $1\frac{3}{10}$
15. $1\frac{7}{10}$

▼ Write each percent as a fraction or mixed number; reduce to lowest terms.

16. 50%
17. 75%
18. 20%
19. 40%
20. 15%
21. 9%
22. 7%
23. 8%
24. 6%
25. 12.5%
26. 25.5%
27. 5.75%
28. 3.25%
29. 175%
30. 200%
31. 180%
32. 160%
33. 0.3%
34. 0.7%
35. 0.09%
36. 0.07%

▼ Use equivalent forms to fill in the table.

	Percent	Decimal	Fraction
37.	25%		
38.		0.04	
39.		0.32	
40.			$\frac{1}{8}$
41.	17.5%		
42.			$\frac{4}{5}$
43.		0.116	
44.			$\frac{4}{19}$
45.	0.008%		
46.	122%		
47.			$\frac{6}{6}$
48.	$7\frac{1}{2}\%$		
49.		0.001	

LIVING IN THE REAL WORLD

50. Janet earns $\frac{1}{8}$ of her college expenses by working in the summer. What percent of her expenses is this?

51. Only 22% of the population in the United States eats breakfast on a regular basis. Write this percent as a decimal.

52. A blouse is marked down 20%. What fraction does this represent?

53. One-sixth of the students at Piedmont Community College take some math course in their freshman year. What percent does this represent? (Round your answer to the nearest tenth of a percent.)

54. In 1992, studies showed that 16.5% of the population over the age of 65 was living below the poverty level. What fraction of the population is this? Reduce to lowest terms.

55. **a.** One-eighth of the students in Mrs. Rubinstein's English class claim that they need ten or more hours of sleep per night. What percent of her class is this?

b. If you sleep ten hours a night, what fraction of your day do you spend sleeping?

6.3 The Percent Proportion

Consider Figure 6.1, in which is pictured a 50-gallon drum that contains 30 gallons of oil. The **amount** that is in the drum is 30 gallons. The total capacity of the drum is 50 gallons, which we will call the **whole.** What fraction of the drum is filled with oil? It is $\frac{30}{50} = \frac{3}{5}$, so we can say that the drum is $\frac{3}{5}$ full. It would also be reasonable to ask what **percent** this represents. Since $\frac{3}{5} = 0.60 = \frac{60}{100} = 60\%$, the drum is 60% full. This example gives us a proportional relationship between these three quantities: amount (A), the whole (W), and percent (P). We have

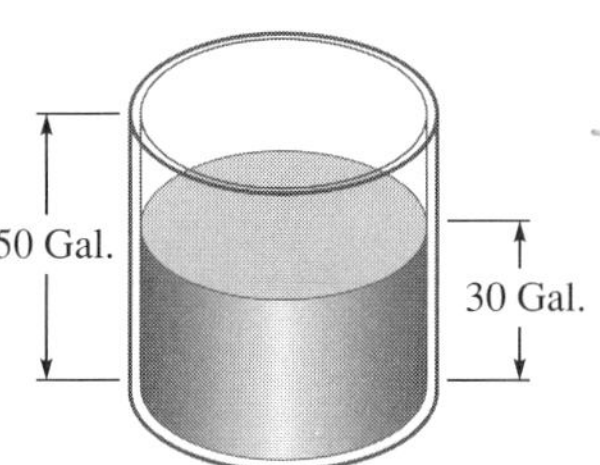

FIGURE 6.1.

$$\frac{30}{50} = \frac{60}{100} \quad \text{or} \quad \frac{\text{amount}}{\text{whole}} = \frac{\text{percent}}{100}$$

More simply,

$$\frac{A}{W} = \frac{P}{100}$$

This is called the **percent proportion.** Note that there are three variables in this proportion: A, W, and P. If we know the value of any *two* of these variables, we can use our knowledge of solving proportions to find the value of the third variable.

EXAMPLE 1 Suppose that we want to find P in the preceding example, where $A = 30$ gallons and $W = 50$ gallons.

SOLUTION We substitute the known values into the percent proportion.

$$\frac{A}{W} = \frac{P}{100}$$

$$\frac{30}{50} = \frac{P}{100}$$

$$50 \cdot P = 30 \cdot 100 \quad \text{Cross multiply.}$$

$$P = \frac{30 \cdot 100}{50} = 60$$

Since we are looking for percent, we add a % symbol to our answer. The drum is 60% full. ▮▮

EXAMPLE 2 Solve for W in the percent proportion when $A = 3.5$ and $P = 70$.

SOLUTION

$$\frac{A}{W} = \frac{P}{100}$$

$$\frac{3.5}{W} = \frac{70}{100}$$

$$(70)(W) = (3.5)(100)$$

$$W = \frac{350}{70} = 5$$

▮▮

6.3 Exercises

▼ Solve for the unknown term in the percent proportion, $\frac{A}{W} = \frac{P}{100}$.

1. Find A when $P = 25$ and $W = 200$.
2. If $P = 60$ and $A = 4.8$, find W.
3. Find A when $W = 12$ and $P = 30$.
4. Find W when $P = 26$ and $A = 3$.
5. If $A = 7$ and $W = 35$, find P.
6. Find A when $W = 46$ and $P = 9$.
7. If $A = 2.8$ and $W = 4.8$, find P.
8. Find P when $W = 36$ and $A = 54$.
9. If $A = 26$ and $W = 6.5$, find P.
10. Find W when $A = 12$ and $P = 136$.
11. Find P when $A = 15$ and $W = 6$.
12. If $P = 100$ and $A = 40$, find W.
13. Find A when $P = 100$ and $W = 135$.
14. Find A when $P = 110$ and $W = 120$.

6.4 Solving Percent Problems; Identifying the Parts of the Percent Proportion

As we have mentioned before, one of the most important steps in solving a mathematical problem is translating from the English language into the language of mathematics. Before we can solve the percent proportion, we must first identify its parts: A, W, and P.

The easiest of these to identify is percent, P; since it is always written along with the word "percent" or the symbol "%."

EXAMPLE 1 Identify P in each of the following.

(a) Find the number that is 17% (P) of 25 — $P = 17\%$

(b) 12% (P) of what number is 24? — $P = 12\%$

(c) 9 is what percent (P) of 26%? — P is unknown.

The whole (W) almost always follows the words "percent of" or "% of" and can usually be identified this way. You should also remember that W means "the whole thing" or "100 percent."

EXAMPLE 2 Identify W in each of the following.

(a) Find the number that is 17% **of** 25 (W). — $W = 25$

(b) 12% **of** what number (W) is 24? — W is unknown.

(c) 9 is what **percent of** 26 (W)? — $W = 26$

The amount (A) represents some percent of the whole and is easily identified as the part of the percent proportion that remains after P and W have been identified. To illustrate this, we'll look at the same examples again.

EXAMPLE 3

(a) Find the number (A) that is 17% (P) of 25 (W). — A is unknown.

(b) 12% (P) of what number (W) is 24 (A)? — $A = 24$

(c) 9 (A) is what percent (P) of 26 (W)? — $A = 9$

Solving Percent Problems

Now that the parts of the percent proportion have been identified, we are in a position to solve problems for the unknown quantity. Let's look at the same three examples again.

EXAMPLE 4 Find the number that is 17% of 25.

SOLUTION We identify the parts of the percent proportion, P, W, and A, as before:

Find the number (A) that is 17% (P) of 25 (W).

Now we substitute the known values in the percent proportion and solve for the unknown quantity A:

$$\frac{A}{W} = \frac{P}{100}$$

$$\frac{A}{25} = \frac{17}{100}$$

$$(100)(A) = (17)(25)$$

$$100A = 425$$

$$A = \frac{425}{100} = 4.25$$

▮▮

EXAMPLE 5 12% of what number is 24?

SOLUTION We identify P, W, and A:

12% (P) of what number (W) is 24 (A)?

Now we substitute in the percent proportion:

$$\frac{A}{W} = \frac{P}{100}$$

$$\frac{24}{W} = \frac{12}{100}$$

$$12W = 2400$$

$$W = \frac{2400}{12} = 200$$

▮▮

EXAMPLE 6 9 is what percent of 26?

SOLUTION We identify P, W, and A:

9 (A) is what percent (P) of 26 (W)?

Next we substitute in the percent proportion:

$$\frac{A}{W} = \frac{P}{100}$$

$$\frac{9}{26} = \frac{P}{100}$$

$$26P = 900$$

$$P = \frac{900}{26} = 34.6\% \qquad \text{rounded to the nearest tenth}$$

Many times problems involving percent are given as application problems. The methods we just discussed are extremely helpful in identifying the necessary parts of the application problem, even though the application problem must sometimes be restated to fit our needs. Study the following examples carefully; the time spent will be worthwhile.

EXAMPLE 7 In a class of 36 students, 4 were absent. What percent of the class was absent?

SOLUTION Let's restate the problem in one sentence and identify P, W, and A. We know that P is the percent and W follows the words "percent of," so W is "a class of 36" or just 36. Note that W represents the entire class, or 100% of the class. The remaining part is the amount A, which is 4.

A		P		W	
4	students absent is what	percent	of a class of	36	?

$$\frac{A}{W} = \frac{P}{100}$$

$$\frac{4}{36} = \frac{P}{100}$$

$$36P = 400$$

$$P = \frac{400}{36} = 11.1\% \qquad \text{rounded to the nearest tenth}$$

EXAMPLE 8 If your salary is \$246.30 per week and you received a 9% increase, how much more money will you receive each week?

SOLUTION We restate the problem and identify P, W, and A.

P		W		A	
9%	of	\$246.30	is	what number	?

$$\frac{A}{W} = \frac{P}{100}$$

$$\frac{A}{\$246.30} = \frac{9}{100}$$

$$100A = 2216.7$$

$$A = \frac{2216.7}{100} = \$22.17 \qquad \text{rounded to the nearest cent}$$

6.4 Exercises

▼ Solve the following percent problems using the method described in the preceding examples.

1. Sixteen is what percent of 48?

2. What number is 15% of 326?

3. Find 0.002% of 31.6.

4. What percent of 45,000 is 150?

5. Find 308% of 41.6.

6. Twenty-five percent of what number is 35?

7. Find the percent of 72 that is 12.

8. Find the percent of 12 that is 72.

9. Find 86% of 23.

10. Twelve is 36% of what number?

11. Fifteen percent of what number is 27?

12. What number is equal to 0.04% of 36?

13. Four percent of a number is 0.72. Find the number.

14. Fifty is what percent of 250?

15. One hundred and fifteen percent of what number is 80.5?

16. What percent of 42 is 21?

LIVING IN THE REAL WORLD

▼ In each of the following, rewrite each problem as a single sentence before solving.

17. A house is worth $45,500. How much would the owner receive if it were destroyed by fire and it was insured for 80% of its value?

18. A light bulb manufacturer estimates that about 0.73% of the light bulbs that he ships are defective. How many bulbs would be defective in a shipment of 500,000?

19. If the interest rate on a charge card is $1\frac{1}{4}$% on the unpaid balance per month and your unpaid balance this month is $312, how much interest will you pay this month?

20. If a professional baseball team wins 78 of the 114 games that it plays, what percent of its games has it won? What percent has it lost?

21. If your rent of $320 per month is increased by 8%, how much rent will you pay next month?

22. If your salary is going to be increased by 10% next year and then reduced by 10% the following year, how will your final salary compare with your present salary? Why? (**Hint:** Start with a $100/week salary and see what happens.)

23. If a car dealer pays $900 for a used car and wants to make a profit of 20% of the cost, what should the sticker price of the car be?

24. If there are 612 males and 468 females in the freshman class of Sebastian College, what percent of the class is male? What percent is female?

25. If the dropout rate at Crest College is 35% in the freshman year, how many students of 750 who start will finish the year?

26. How much money must you invest at a rate of 10% per year to earn interest of $10,000 per year?

27. If a coat is reduced from $100 to $85 and a suit is reduced from $180 to $165, which was reduced by the greater percent?

28. A certain bronze alloy is 52.3% copper. How many pounds of copper are contained in 600 pounds of the bronze alloy?

29. If you receive 88% on a math exam containing 25 problems, how many did you get correct?

30. If you get 80% on an economics test that has 30 exercises on it, how many did you get wrong?

31. The owner of a ladies' boutique pays $50 for a dress from her wholesaler. She wants to mark it up by 80% of her cost. Find 80% of the $50 cost and add this result to the $50 cost price to find the retail price. Next find 180% of the cost price of $50. You should get the same result. Why? Which method is easier?

32. Using the method discovered in Exercise 31, find the selling price for each of the following items.

Item	Cost	Markup	Selling price
Shirt	$16	50%	
Ring	$120	200%	
Toothbrush	$1.04	15%	
Textbook	$31.75	20%	
Automobile	$20,108	14%	

SUMMARY—CHAPTER 6

The numbers in brackets refer to the section in the text in which each concept is presented.

EXAMPLES

[6.1] **Percent** means per hundred.

[6.1] 47% means $\frac{47}{100}$.

[6.1] **To change a percent to a decimal,** remove the % symbol and move the decimal point two places to the left.

[6.1] 47% = 0.47
312% = 3.12
7.9% = 0.079

[6.1] **To change a decimal to a percent,** move the decimal point two places to the right and add a % symbol.

[6.1] 0.36 = 36%
1.58 = 158%
0.062 = 6.2%

[6.2] **To change a fraction to a percent,** write the fraction as a decimal and then change the decimal to a percent.

[6.2] $\frac{1}{8} = 0.125 = 12.5\%$

[6.2] **To change a percent to a fraction,** drop the percent sign, divide the number by 100, and reduce to lowest terms.

[6.2] $13\% = \frac{13}{100}$

[6.3] The **percent proportion** is $\frac{A}{W} = \frac{P}{100}$, where ***P*** is the **percent** and is written with the symbol "%" or the word "percent"; ***W*** is the **whole** and usually follows the words "percent of"; and ***A*** is the **amount** and is identified as the remaining part after P and W have been found.

[6.3] $\frac{A}{W} = \frac{P}{100}$

P		W		A
8%	of	what number	is	10 ?

$P = 8$
W is unknown
$A = 10$

[6.4] **To solve a percent problem,** identify P, W, and A, substitute them into the percent proportion, and solve the proportion for the unknown variable.

[6.4] $\frac{A}{W} = \frac{P}{100}$

$\frac{10}{W} = \frac{8}{100}$

$8W = 1000$

$W = \frac{1000}{8} = 125$

REVIEW EXERCISES—CHAPTER 6

The numbers in brackets refer to the section in the text in which similar problems are presented.

[6.1] Write each percent as a decimal.

1. 18%
2. 6.4%
3. 0.05%
4. $18\frac{3}{4}\%$
5. 312%

[6.1] Write each decimal as a percent.

6. 0.41
7. 2.85
8. 0.9
9. 0.0093
10. 0.315

[6.2] Write each fraction as a percent. Round answers to the nearest tenth.

11. $\frac{3}{5}$
12. $\frac{7}{8}$
13. $1\frac{4}{5}$
14. $1\frac{2}{7}$

[6.2] Write each percent as a fraction or a mixed number, reduced to lowest terms.

15. 60%
16. 35%
17. 14.5%
18. 0.04%

[6.2] Use equivalent forms to fill in the table.

	Percent	Decimal	Fraction
19.	30%		
20.		0.08	
21.			$\frac{3}{8}$
22.	18.75%		
23.		1.24	
24.			$\frac{2}{9}$

[6.3–6.4] Solve the following percent problems using the percent proportion.

25. What percent of 48 is 36?

26. Find 16% of 350.

27. Thirty percent of what number is 57?

[6.3–6.4] Rewrite each word problem as a single sentence and solve.

28. Mr. Fredericks was given a 6% salary increase on his salary of $30,500. How much was his raise in pay?

29. Ron Pullins receives 15% commission on the total amount of his sales. This week he received $570 in commissions. What was the total amount of his sales?

30. Denise received a $25 increase in her weekly salary. If her original salary was $400 per week, what percentage increase does this represent?

31. If you are charged $32.94 sales tax on a $549.00 dinette set, what percent sales tax is this?

CRITICAL THINKING EXERCISES—CHAPTER 6

Tell whether each of the following is true or false and state why.

1. $0.062\% = 6.2$

2. $125\% = 1.25$

3. $\frac{3}{4} = 0.75\%$

4. $0.35\% = \frac{35}{100} = \frac{7}{20}$

5. George Shepard correctly answered 20 questions on his exam and got a score of 74%. There were 25 questions on the test.

Find the error in each of the following:

6. What number is equal to 6% of 24?

SOLUTION

$$\frac{A}{W} = \frac{P}{100}$$

$$\frac{24}{W} = \frac{6}{100}$$

$$6W = \frac{2400}{6}$$

$$W = 400$$

7. Writing 0.4 as a percent gives us 0.004%.

8. Lou, a used car salesman, sold two cars for $2000 each. On the first car he earned a profit of 25% of the cost, and on the second he sold it at a loss equal to 25% of the cost. His wife, Lucille, in an effort to cheer him up, remarked, "Things could be worse, at least you broke even."

NAME ▮▮▮▮▮ CLASS

ACHIEVEMENT TEST—CHAPTER 6

▼ Write each percent as a decimal.

1. 14% 1. ______
2. 3.5% 2. ______
3. $14\frac{1}{4}\%$ 3. ______

▼ Write each decimal as a percent.

4. 0.38 4. ______
5. 0.045 5. ______
6. 0.2 6. ______

▼ Write each fraction as a percent. Round each decimal to the nearest thousandth.

7. $\frac{4}{5}$ 7. ______
8. $\frac{8}{7}$ 8. ______
9. $1\frac{3}{5}$ 9. ______

▼ Write each percent as a fraction or a mixed number, reduced to lowest terms.

10. 85% 10. ______
11. 25.5% 11. ______
12. 0.06% 12. ______

▼ Solve the following percent problems using the percent proportion.

13. How much is 28% of 425? 13. ______
14. What percent of 72 is 12? 14. ______
15. Seventy percent of what number is 231? 15. ______

▼ Rewrite each word problem as a single sentence and solve.

16. Lisa's weekly salary is $380. Her deductions amount to $72. What percent does this represent? Round your answer to the nearest tenth of a percent.

16. ____________________

17. Monticello High School won 70% of the games that it played. If they played 30 games, how many did they win?

17. ____________________

18. Steve went on a diet and lost 32 pounds. If this represents 20% of his original weight, how much did he weigh before he went on the diet? How much does he weigh now?

18. ____________________

NAME ▮▮▮▮▮ CLASS

CUMULATIVE REVIEW—CHAPTERS 4, 5, 6

1. Write 0.34 as a fraction in lowest terms. 1. ______

2. In the number 5.341, what place value does the 3 have? 2. ______

3. Round 51.863 to the nearest tenth. 3. ______

4. Add: 3.041 + 18.11 + 6.524. 4. ______

5. Subtract: $\begin{array}{r} 66.423 \\ -18.258 \\ \hline \end{array}$ 5. ______

6. Multiply: $\begin{array}{r} 3.24 \\ \times 0.62 \\ \hline \end{array}$ 6. ______

7. Divide, rounding your answer to the nearest tenth: $0.42\overline{)0.9743}$ 7. ______

8. Write $\frac{3}{16}$ as a decimal. 8. ______

9. Write the ratio of 2 dimes to 3 quarters in lowest terms. 9. ______

10. Megan earned $136 in 17 hours. Write as a unit rate. 10. ______

11. Solve for x: $\frac{3}{x} = \frac{7}{42}$. 11. ______

12. An 18-oz box of cereal contains 4 oz of sugar. How many ounces of sugar will 3 oz of the cereal contain? 12. ______

13. Write $5\frac{1}{2}\%$ as a decimal.

14. Write 0.342 as a percent.

15. Write $\frac{3}{16}$ as a percent.

16. Write 45% as a fraction, reduced to lowest terms.

17. Find 30% of 360.

18. What percent of 65 is 13?

19. Forty percent of what number is 104?

20. Lori Godard received a 4% increase in her salary of $33,500. How much was her increase in pay? What is her new salary?

13. ____________________

14. ____________________

15. ____________________

16. ____________________

17. ____________________

18. ____________________

19. ____________________

20. ____________________

CHAPTER 7

Descriptive Statistics

INTRODUCTION

A city police department wants to know if there is any connection between the number of burglaries committed and the day of the week on which they occur. Do more burglaries happen on a Saturday night than on a Tuesday night? If so, they may decide to increase the number of patrols on Saturday as a preventative measure.

The bits of information that the police department will gather for their study are called **data.** The people who collect and analyze this data are called **statisticians** and they are trained in the study of **statistics.** Methods for describing and presenting such data fall under the heading of **descriptive statistics;** this is the subject of this chapter.

NUMBER KNOWLEDGE—CHAPTER 7

What Do You Know about Car Liability Insurance?

Automobile liability insurance protects you, as a driver, against injury to other people or damage to their property. The limits of how much your insurance company will pay for bodily injury are given using numbers like 10/20, which means they are responsible for paying no more than $10,000 for each injured person and a total maximum of $20,000 for any accident.

Suppose you are involved in an accident in which two people are injured, and they are awarded $15,000 and $4,000, respectively, for their injuries. Your insurance company will pay the first person only $10,000, the limit of their liability, and the full $4,000 to the second injured person. You are responsible for paying the additional $5,000 to the first person.

Higher coverages, such as 25/50, 100/300, or even 500/1,000 (that's $500,000 per person and $1,000,000 per accident), are available for a relatively small increase in premium payments.

Suppose you injure someone in an accident and there is some reason to believe that you're at fault. If the injured person sues you for $100,000 to cover his medical expenses and your coverage is 10/20, to what extent is your insurance company liable? If your coverage is 100/300, to what extent is your insurance company liable? Do you think that the amount of coverage that you have will affect the amount of effort and money they will put forth on your behalf to defend you?

You should always ask your insurance agent for an explanation of policy limitations and relative costs. Only then can you make a wise decision.

7.1 Measures of Central Tendency

Mean or Average

If you receive test scores of 70, 85, 80, and 93 during a semester, what is your average test score for the term? It's a problem that no doubt interests all of us. To find the **mean** or **average** of a set of quantities, we add all of them together and divide by the total number of quantities. In this case we add all of the test scores and divide by 4, since there were 4 tests:

$$\text{average} = \frac{70 + 85 + 80 + 93}{4} = \frac{328}{4} = 82$$

The average score is 82. This single number that represents or characterizes the data is called a **statistic.** Frequently this number tends to be near the center of the data and is called a *measure of central tendency.*

$$\text{mean or average} = \frac{\text{the sum of all the quantities}}{\text{the number of quantities}}$$

EXAMPLE 1 Find the average grade if a student's test scores are 61, 70, 86, 68, and 85.

SOLUTION

$$\text{average} = \frac{\text{sum of scores}}{\text{number of scores}} = \frac{61 + 70 + 86 + 68 + 85}{5} = \frac{370}{5} = 74$$

EXAMPLE 2 Find the average of the following quantities: 136, 125, 147.

SOLUTION

$$\text{average} = \frac{136 + 125 + 147}{3} = \frac{408}{3} = 136$$

EXAMPLE 3 Find the average weight of the offensive line of a football team if the players weigh 261 pounds, 275 pounds, 245 pounds, 310 pounds, and 244 pounds.

SOLUTION

$$\text{average} = \frac{261 + 275 + 245 + 310 + 244}{5} = \frac{1{,}335}{5} = 267 \text{ pounds}$$

Grade Point Average (GPA) • • At most colleges, students earn what's called a *grade point average,* which depends both on what grade they earn in the course and also on how many credits the course is worth.

EXAMPLE 4 Last semester Monica earned the grades given in the following table. Calculate her grade point average, assuming grade values as given.

Course	Grade	Credits
English	B	3
College Algebra	C	4
Spanish I	A	3
History	A	3
Phys. Ed.	C	1

GRADE VALUES

A = 4

B = 3

C = 2

D = 1

F = 0

SOLUTION To find Monica's GPA, we average what are called *quality points.* Quality points are found by multiplying grade values by the number of credits that the course is worth.

COURSE	GRADE VALUES	×	CREDITS	=	QUALITY POINTS
English	3	×	3	=	9
College Algebra	2	×	4	=	8
Spanish I	4	×	3	=	12
History	4	×	3	=	12
Phys. Ed.	2	×	1	=	2
			14		43

Finally, to calculate Monica's GPA, we divide the number of quality points by the number of credits. This number is usually rounded to the nearest hundredth.

$$\text{GPA} = \frac{\text{quality points}}{\text{credits}} = \frac{43}{14} = 3.07$$

Monica's grade point average for the semester was 3.07.

Median

A second measure of central tendency of data is called the **median.** Suppose five students earned grade point averages of 2.4, 3.1, 1.9, 3.8, and 2.1. We arrange the numbers in order from smallest to largest.

1.9, 2.1, 2.4, 3.1, 3.8

The middle number, 2.4, is called the median.

To find the median of a collection of data:

Arrange the data in order by size and then find the middle of the data. If there is an odd number of items of data, it will be the middle item. If there is an even number of items of data, average the two middle numbers.

EXAMPLE 5 Find the median of the following set of numbers:

76, 43, 85, 68, 65, 87, 49, 58, 61

SOLUTION First we arrange the data in order:

43, 49, 58, 61, 65, 68, 76, 85, 87

↑
median

EXAMPLE 6 Salaries for a small company are as follows: $18,000; $23,000; $31,000; $16,000; $28,000; $42,000. Find the median salary.

SOLUTION We first arrange the salaries in order:

$16,000 $18,000 $23,000 $28,000 $31,000 $42,000

Since there is an even number of salaries (6), there is no middle number and we must average the two middle salaries to find the median:

$$\text{median} = \frac{23{,}000 + 28{,}000}{2} = \frac{51{,}000}{2} = \$25{,}500$$

Mode

The last measure of central tendency that we will consider is called the **mode** and it is defined to be the number that occurs most frequently in a set of data. For example, in the set of data 4, 2, 9, 11, 9, 3, 9, 5, 7, 6, the number 9 occurs most frequently (three times) and therefore the mode is 9.

The **mode** of a set of data is the number that occurs most often. If no number occurs more than once, there is no mode.

EXAMPLE 7 Find the mode of the following set of data:

26, 31, 31, 35, 35, 35, 38, 41, 41, 45, 49

SOLUTION The mode is 35, since it occurs three times, more often than any other number.

EXAMPLE 8 William Villanueva received the following quiz grades in Psychology: 10, 8, 9, 8, 7, 6, 8, 7, 10, and 7. Find the mode.

SOLUTION Since the numbers 7 and 8 each occur three times, there are two modes, 7 and 8. We call such a set of data *bimodal*.

EXAMPLE 9 The following high temperatures were recorded over a period of seven days in Allentown, PA:

68° 59° 63° 67° 54° 71° 70°

What is the mode for this set of data?

SOLUTION Since no number occurs more than once, we say that the set of data has *no mode*.

EXAMPLE 10 Find (a) the mean, (b) the median, and (c) the mode for the following set of exam scores:

73, 81, 81, 78, 92

SOLUTION **(a)** $\text{mean} = \frac{73 + 81 + 81 + 78 + 92}{5} = 81$

(b) To find the median, we arrange the scores in order by size:

73, 78, 81, 81, 92

The median is the middle score, 81.

(c) The mode is 81, since it occurs more than any other score.

7.1 Exercises

▼ Find the mean of each set of data.

1. 3, 7, 5, 8, 4, 9
2. 56, 48, 37, 28, 41
3. 73, 82, 85, 92, 88
4. 16, 21, 14, 19, 23, 15, 11
5. 142, 139, 156, 187
6. 7, 10, 8, 5, 7, 11

▼ Find the median of each set of data.

7. 26, 28, 31, 25, 29

8. 128, 130, 114, 121, 133

9. 16, 24, 18, 21, 18, 21, 15

10. 63, 78, 79, 64, 63, 67, 74

11. 32, 34, 32, 41, 30, 38

12. 121, 114, 118, 120, 121, 116

▼ Find the mode for each set of data.

13. 6, 8, 11, 8, 9, 5, 8, 4

14. 26, 26, 19, 26, 20, 20, 21, 26

15. 16, 19, 16, 13, 21, 19, 12, 24

16. 3, 1, 5, 2, 2, 6, 7, 5, 8

17. 6, 9, 20, 14, 8, 10, 17

18. 116, 123, 104, 111, 119

▼ Find the mean, median, and mode of each of the following sets of data.

19. 10, 15, 20, 5, 15

20. 68, 85, 82, 82, 73

21. 13, 25, 32, 18, 27, 18

22. 9, 11, 14, 16, 12, 10, 14, 19

23. 2.7, 3.5, 1.6, 3.2, 4.6

24. 6.9, 8.2, 6.9, 4.3, 8.8

LIVING IN THE REAL WORLD

25. If you receive scores of 78, 86, 66, 92, and 83 on your history exams during the semester, what is your average score?

26. The attendance in Mr. Smiley's speech class for five days was 26, 31, 24, 24, and 25 students. What was the average daily attendence?

27. Dawn Oretsky did the following number of sit-ups each day for a seven-day period: 23, 27, 38, 36, 35, 35, and 29. Find the mean, the median, and the mode.

28. Mr. Cummins recorded the following grades for ten students in his astronomy class: 83, 88, 72, 96, 75, 75, 91, 93, 64, and 79. Find the mean, median, and mode of the ten scores.

29. Using radar, the state police measured the following speeds in m.p.h. of fifteen cars on a state highway: 62, 60, 58, 62, 63, 67, 57, 54, 64, 62, 65, 56, 51, 62, and 64. Find the mean, median, and mode of the data.

30. A consumer research group priced paper towels in eight stores as follows: $1.39, $1.37, $1.39, $1.39, $1.49, $1.35, $1.39, and $1.55. Find the mean, median, and mode of the data.

▼ In Exercises 31–36, calculate the grade point average (GPA). Assume grade values of A = 4, B = 3, C = 2, D = 1, and F = 0. Round final results to the nearest hundredth.

31.

Grade	Credits
C	3
B	3
B	4
A	4

32.

Grade	Credits
A	3
B	4
C	3
C	3
D	2

33.

Grade	Credits
D	3
B	3
B	3
C	3
C	3

34.

Grade	Credits
A	4
A	3
B	3
B	3
B	1

35.

Course	Grade	Credits
History	A	3
Physics	F	4
Calculus	C	4
English	B	3

36.

Course	Grade	Credits
Phys. Ed.	B	1
Health	B	2
Sociology	C	3
English	B	3
Typing	A	3

Janet Morrison earned the following grades. Use this information to answer Exercises 37–40.

Class	Credits	Grade
Psychology	3	A
Algebra	3	B
Chemistry	4	C
English	3	C

37. What is Janet's grade point average?

38. If Janet had received a B in English, what would her grade point average have been?

39. If Janet had received Bs in both English and Chemistry, what would her grade point average have been?

40. What would Janet's grade point average have been if she had received Cs in all of her courses?

Last semester Ming Wang received the following grades in his courses. Use this information to answer Exercises 41–43.

Class	Credits	Grade
English	3	C
Algebra	3	B
Physics	4	D
History	3	B
Keyboarding	1	A

41. Calculate Ming's grade point average

42. If Ming had earned a C in Physics instead of a D, what would his grade point average have been?

43. If Ming had received a B in English instead of a C, by how much would he have increased his GPA?

44. In order to pass a math course, an overall average of 70 is required. If you scored grades of 65, 70, and 70 on your first three exams, what is the lowest grade you can receive on your fourth exam and still pass the course?

45. When finding the average of a list of numbers, will the result always lie between the smallest and largest numbers in the list? Why or why not?

46. In the town of Poorich, there are five residents: the mayor, who earns \$199,992 annually, and four workers, who each earn \$2 per year. Calculate the mean, median, and mode for the five salaries. The four workers decide to apply for a federal poverty grant. Which measures of central tendency would be most beneficial to use on their applications and which would they like to omit?

7.2 Tables

A good way of presenting data in a clear and easily understood way is by using a table.

EXAMPLE 1 A combination of cold and wind make you feel colder than the actual temperature of the air. Table 7.1 indicates the wind-chill effect at various temperatures and wind speeds.

TABLE 7.1. WIND-CHILL FACTOR

		Temperature (°F)							
		35	30	25	20	15	10	5	0
Wind Speed (mph)	5	33	27	21	16	12	7	0	−5
	10	22	16	10	3	−3	−9	−15	−22
	15	16	9	2	−5	−11	−18	−25	−31
	20	12	4	−3	−10	−17	−24	−31	−39
	25	8	1	−7	−15	−22	−29	−36	−44
	30	6	−2	−10	−18	−25	−33	−41	−49
	35	4	−4	−12	−20	−27	−35	−43	−52
	40	3	−5	−13	−21	−29	−37	−45	−53
	45	2	−6	−14	−22	−30	−38	−46	−54

Source: National Weather Service

(a) How cold does it feel when the temperature is 25° F and the wind is blowing at 30 miles per hour?

(b) When the temperature is 30° F, how cold does it feel if there is a wind of 40 mph?

(c) What combination of temperature and wind will give an effective temperature of −3?

SOLUTION **(a)** We locate 25 along the top of Table 7.1 and then move down that column to the row with 30 at the left of the table. At the intersection of this column and row, we find the number −10, indicating that it will *feel* like the temperature is 10 degrees below zero.

(b) The intersection of 30° F and 40 mph gives a wind-chill effect of −5. That is, a temperature of 30° F with a wind of 40 mph feels the same as a temperature of −5° on a windless day.

(c) From Table 7.1, we can see that 15° F with a wind of 10 mph or 25° F with a wind of 20 mph will both produce a wind-chill effect of −3.

EXAMPLE 2 Table 7.2 presents data that will allow us to calculate the cost of a college education from 1970 to 1991.

TABLE 7.2 AVERAGE COLLEGE COSTS: 1970–1991

	Tuition and Required Fees			Board Rates			Dormitory Charges		
	All institutions	2-yr. colleges	4-yr. universities	All institutions	2-yr. colleges	4-yr. universities	All institutions	2-yr. colleges	4-yr. universities
Public:									
1970	$ 323	$ 178	$ 427	$ 511	$ 465	$ 540	$ 369	$ 308	$ 395
1980	583	355	840	867	894	898	715	572	749
1990	1,356	756	2,035	1,635	1,581	1,728	1,513	962	1,561
1991	1,454	824	2,159	1,691	1,594	1,767	1,612	1,050	1,658
Private									
1970	1,533	1,034	1,809	561	546	608	436	413	503
1980	3,130	2,062	3,811	955	924	1,078	827	769	999
1990	8,147	5,196	10,348	1,948	1,811	2,339	1,923	1,663	2,411
1991	8,772	5,570	11,379	2,074	1,989	2,470	2,063	1,744	2,654

Source: U.S. Dept. of Education

(a) What was the average tuition for a public two-year college in 1970?

(b) How much did dormitory charges increase between 1980 and 1990 at public four-year universities?

(c) In 1990, what was the difference in tuition between private two-year and four-year schools?

SOLUTION **(a)** In the intersection of the second column and the first row of Table 7.2, we find $178.

(b) The dormitory charge in 1980 was $749 and in 1990 it was $1,561. The increase was $1,561 − $749 = $812. As you can see, it more than doubled in ten years.

(c) Tuition costs in 1990 for private two-year schools was $5,196 and for private four-year schools it was $10,348. The difference was $10,348 − $5,196 = $5,152.

7.2 Exercises

Use Table 7.3 to answer questions 1–8.

TABLE 7.3. HEAT INDEX

Relative Humidity	Temperature (°F) 70	75	80	85	90	95	100
0%	64	69	73	78	83	87	91
10%	65	70	75	80	85	90	95
20%	66	72	77	82	87	93	99
30%	67	73	78	84	90	96	104
40%	68	74	79	86	93	101	110
50%	69	75	81	88	96	107	120
60%	70	76	82	90	100	114	132
70%	70	77	85	93	106	124	144
80%	71	78	86	97	113	136	
90%	71	79	88	102	122		
100%	72	80	91	108			

Source: National Weather Service

This table shows how high humidity together with high temperature reduces the body's ability to cool itself. This index tells how the temperature will *feel* to you at various combinations of temperature and relative humidity.

1. What temperature will it feel like to you when the actual temperature is 80° F and the relative humidity is 90%?

2. How hot will it feel if the temperature is 90° F and the relative humidity is 80%?

3. What temperature will it feel like if it is 100° F with a relative humidity of 10%?

4. If it is 90° F and the humidity is 50%, what will the temperature feel like to you?

5. If the actual temperature is 85° F and stays the same but the humidity increases from 60% to 80%, what increase in temperature will you sense has taken place?

6. At a constant temperature of 90° F, if the humidity increases from 70% to 90%, what temperature increase will your body feel has taken place?

7. If the relative humidity stays constant at 70% and the temperature increases from 70° F to 80° F, a change of 10°, what increase in temperature will you feel has taken place?

8. Assume a constant relative humidity of 80%. If the temperature increases 15° over the course of the day, from 75° to 90°, what increase in temperature will your body sense has taken place?

Use Table 7.4 to answer Exercises 9–16. This table shows attendance from May through September for the four High County parks.

TABLE 7.4. MONTHLY ATTENDANCE AT HIGH COUNTY PARKS

Park	May	June	July	Aug	Sept
Windy Hill	364	566	819	863	314
Maple Road	284	414	776	719	212
West End	303	412	669	667	319
Stone Mtn.	312	624	807	833	321

9. How many people visited West End Park in August?

10. What was the number of visitors for the month of May at Stone Mtn. Park?

11. What was the total number of visitors at all four parks during the month of July?

12. How many people visited Windy Hill Park during the five-month period from May through September?

13. How much did attendance increase from June to July at West End Park?

14. What was the decrease in the number of people who visited Stone Mtn. Park from August to September?

15. How much did total attendance change for the four parks from May to June?

16. At which park did attendance double from May to June?

▼ Use Table 7.5 to answer Exercises 17–26. This table gives a comparison of nutritional values for two breakfast cereals.

TABLE 7.5. NUTRITIONAL VALUES OF CEREALS PER 1-OZ SERVING

	Cornflakes		Raisin bran	
	Without milk	With $\frac{1}{2}$ cup skim milk	Without milk	With $\frac{1}{2}$ cup skim milk
Calories	100	140	120	160
Protein	2 g	6 g	3 g	7 g
Carbohydrates	24 g	30 g	30 g	36 g
Fat	0 g	0 g	1 g	1 g
Cholesterol	0 mg	0 mg	0 mg	0 mg
Sodium	290 mg	350 mg	200 mg	260 mg
Potassium	35 mg	240 mg	230 mg	430 mg

17. Which of the two cereals contains the least number of calories per serving?

18. How many grams (g) of carbohydrates does 1 oz of raisin bran with $\frac{1}{2}$ cup of skim milk contain?

19. Which cereal provides more protein?

20. What is the difference in the amount of sodium contained in the two cereals?

21. How many calories does $\frac{1}{2}$ cup of skim milk contain?

22. How many grams (g) of protein does $\frac{1}{2}$ cup of skim milk contain?

23. How many milligrams (mg) of sodium does 2 oz of cornflakes without milk contain?

24. Which cereal would be best for someone on a low-cholesterol diet?

25. Which cereal contains the least amount of potassium?

26. Why would cornflakes with skim milk be a good choice for a person on a low-fat, low-cholesterol diet?

7.3 Bar Graphs, Line Graphs, and Pictographs

Graphs

One way of communicating data quickly and clearly is by presenting it in a graph.

Bar Graphs • • A **bar graph** is a convenient way to show comparisons between large and small quantities of data. They are easy to read and provide a quick means of presenting information in an accurate way. They can be drawn either vertically or horizontally.

EXAMPLE 1 This bar graph shows the number of calories used in an hour by a 150-lb person doing various kinds of exercise.

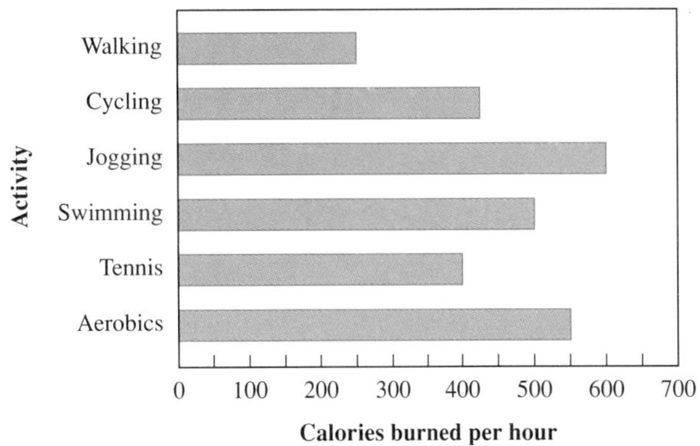

(a) How many calories per hour will be used by Mary Jo, a tennis player who weighs about 150 lb?

(b) Which of the exercises listed burns the least number of calories per hour?

(c) Which exercise burns about 500 calories per hour?

SOLUTION **(a)** If we go to the right end of the bar labeled "Tennis" and look down to the scale at the bottom of the graph, we read about 400 calories.

(b) The shortest bar, representing *walking,* burns the least number of calories per hour, about 250.

(c) We locate 500 on the scale at the bottom and then go upward until we reach a bar that ends at that point. We find that *swimming* burns about 500 calories per hour.

Line Graphs • • **Line graphs,** at a glance, show changes and trends of data over a period of time.

EXAMPLE 2 The following line graph gives average typing test scores, in words per minute (wpm), over a six-week period of instruction at Midtown Business School.

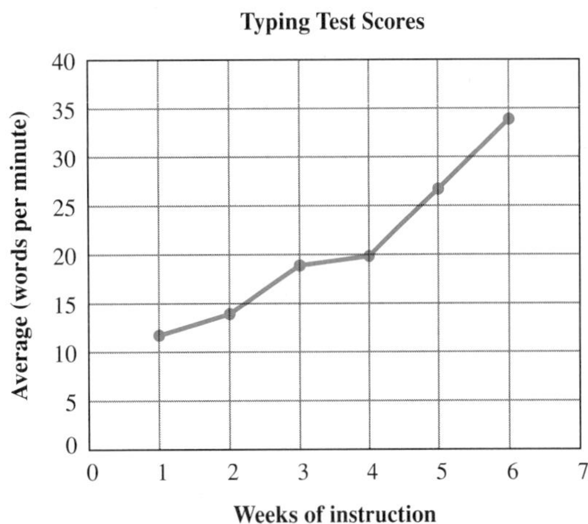

(a) What was the average score at the end of the fourth week?

(b) Between which two weeks was the increase the least?

(c) What trend does the graph show?

SOLUTION **(a)** The average score at the end of the fourth week was about 20 wpm.

(b) Between weeks 3 and 4, the graph is the "flattest," which indicates that the increase is the least.

(c) The average score always increases ove the entire six-week period of instruction. ■ ■

Pictographs • • A **pictograph** is actually a *picture graph* and uses a picture to represent a specific amount, rather than numbers to present data. Usually a key is given that tells the reader how much each picture or figure represents.

EXAMPLE 3 The pictograph shows how many degrees were awarded, by academic area, for a recent graduating class at Ocean County College.

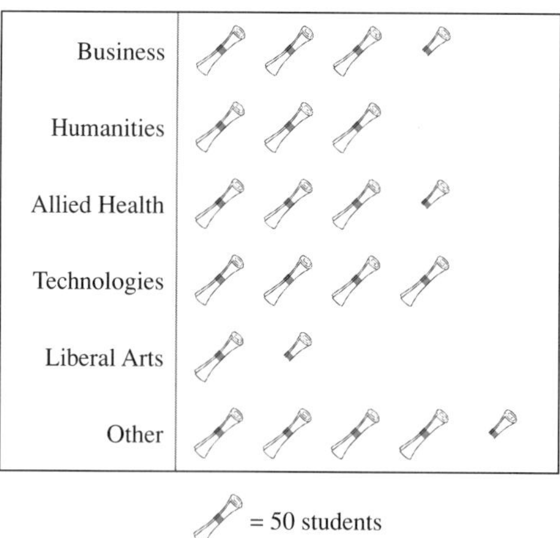

(a) How many students graduated in the humanities?

(b) How many diplomas were awarded in the business area?

(c) How many more students graduated in technologies than in the allied health fields?

(d) Which two areas of study in the graph have the same number of graduates?

SOLUTION **(a)** Since there are three figures representing the number of humanities graduates and each figure represents 50 students, there are (3)(50) = 150 humanities graduates.

(b) There are $3\frac{1}{2}$ diploma figures following business, so we multiply $3\frac{1}{2}$ or $\frac{7}{2}$ by 50, which gives us 175 business graduates.

(c) There are four symbols representing technologies and three and a half symbols representing allied health, a difference of one-half. So $\frac{1}{2}$ (50) = 25 more students graduated in the technologies than in the allied health fields.

(d) Allied health and business each have $3\frac{1}{2}$ figures, so there are $\left(3\frac{1}{2}\right)(50) = \left(\frac{7}{2}\right)(50) = 175$ graduates in both areas.

EXAMPLE 4 Make a pictograph showing the sales of new homes for Nation Wide Real Estate Co. for the first six months of the year. Let each picture of a house represent 100 sales and show the following data.

- Jan.— 300 sales
- Feb.—250 sales
- Mar.—200 sales
- Apr.—400 sales
- May—600 sales
- Jun.— 550 sales

SOLUTION The smallest number of sales is 200 (Mar.), which requires two house figures, and the largest is 600 (May), which requires six house symbols. This gives some idea of how large our graph must be.

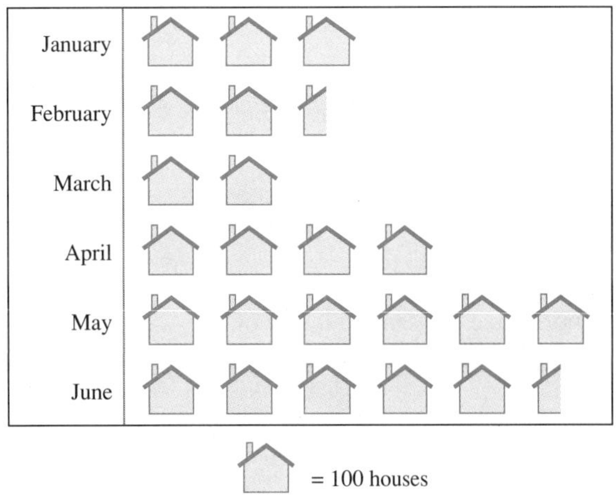

7.3 Exercises

▼ This bar graph shows how many students have declared a major in each of six areas of study at Eastern State University

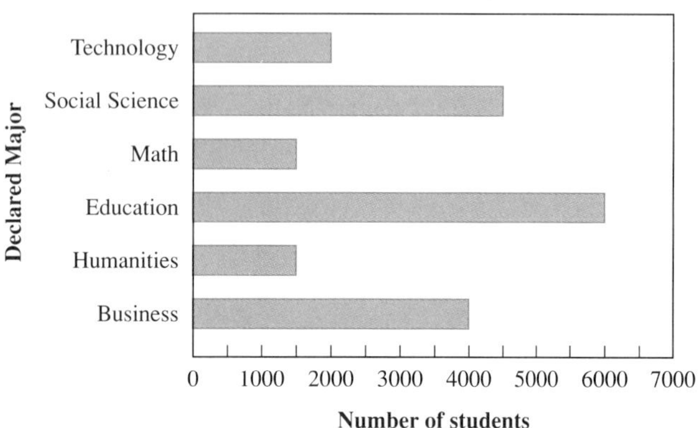

1. Which field of study has the greatest number of students?

2. How many students declared business as their major?

3. How many social science majors are there?

4. Which major has the fewest number of students?

5. Which major has four times as many students as math does?

6. How many more education majors are there than business majors?

7. Which major has only half as many students as there are business majors?

8. Which two majors have the same number of students?

▼ The following vertical bar graph shows the depths, at the deepest point, of each of the Great Lakes.

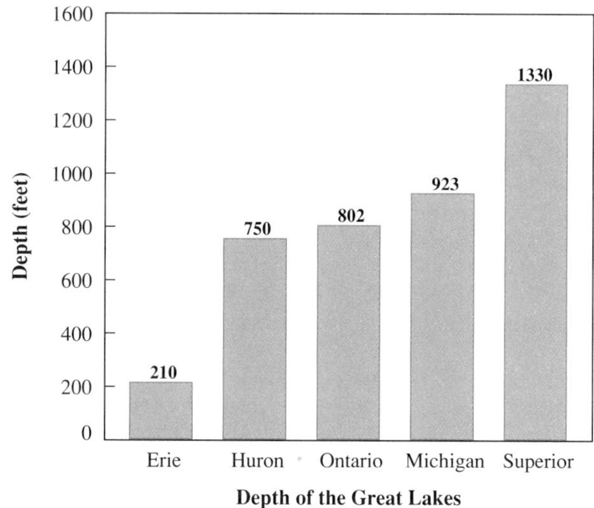

9. Which is the deepest of the Great Lakes?

10. Which two lakes are closest in depth?

11. What is the difference in depth between the shallowest and deepest Great Lakes?

12. How much deeper is Lake Michigan than Lake Erie?

▼ The following line graph shows the number of inches of rainfall over a six-year period.

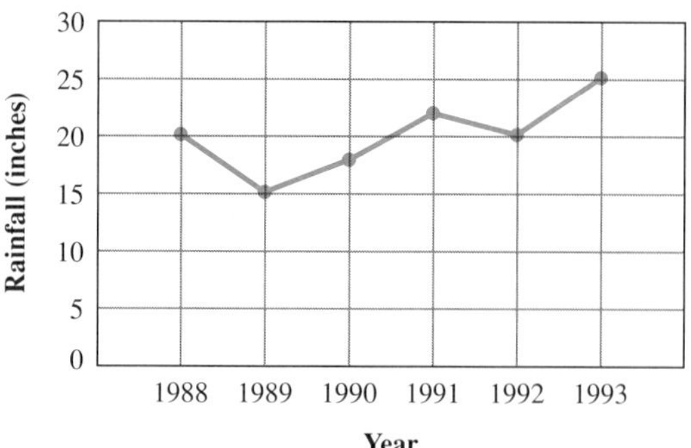

13. In which year did the most rainfall occur?

14. In which year did the least rainfall occur?

15. How many inches of rain fell in the year 1989?

16. Approximately how much more rainfall occurred in 1991 than in 1990?

17. Do you see any upward or downward trend over the given six-year period?

18. What is the difference in rainfall between the years in which the least and the most rain fell?

▼ The following two line graphs show both the high and the low temperatures for a one-week period of time.

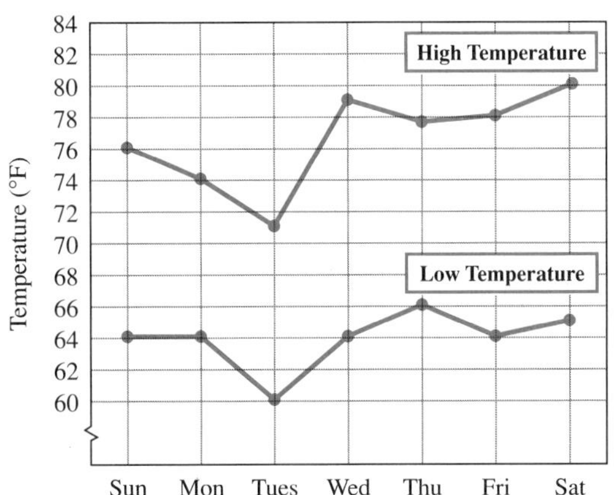

19. On what day did the lowest temperature occur?

20. What was the highest temperature during the week?

21. Between which two days did the high temperature increase the greatest amount?

22. On what day was the difference between the high and low temperatures the greatest?

23. What is the maximum difference in the high temperature over the seven-day period?

24. Do the high- and low-temperature lines tend to follow each other or not?

▼ This pictograph shows the car sales for each of the five salespersons at Honest John's Car Co.

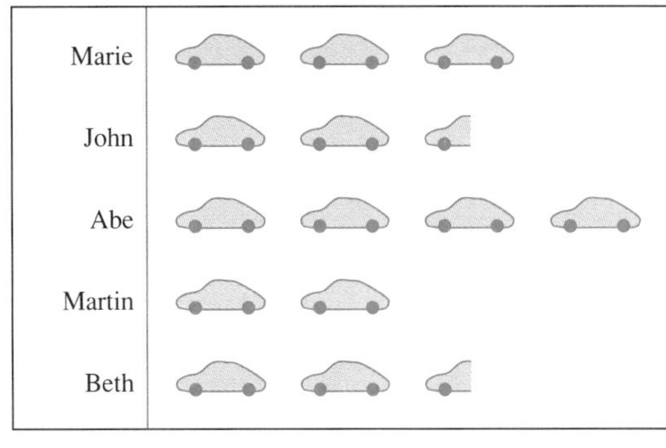

= 10 cars

25. Who sold the most number of cars? How many did he or she sell?

26. Who sold the least number of cars? How many did he or she sell?

27. How many cars were sold altogether, by all five salespersons?

28. Which two people sold the same number of cars?

29. How many more cars did Abe sell than John?

30. Who sold half as many cars as Abe?

▼ This pictograph shows the amount of earnings at each of five branches of County Savings Bank.

County Savings Bank Earnings – by Branch

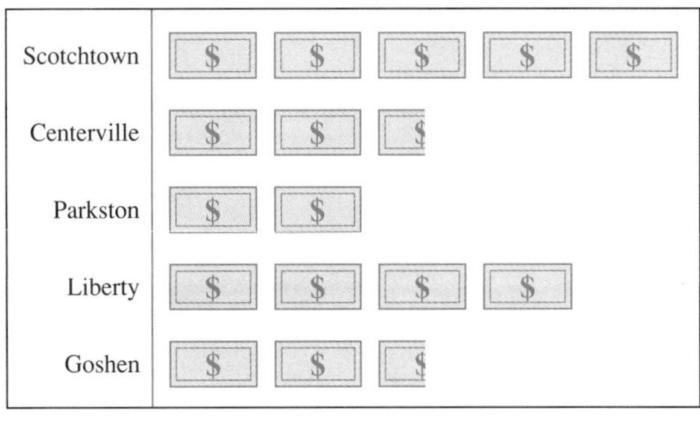

= $1,000,000

31. How much did the Parkston branch earn?

32. Which branch earned the least? How much did it earn?

33. Which two branches earned the same amount?

34. How much more did the Liberty branch earn than the branch in Goshen?

35. How much did all five branches earn altogether?

36. What was the difference between the most profitable branch and the least profitable branch?

37. Centerville and Goshen combined earned the same amount as which branch?

38. Parkston earned exactly half as much as which branch?

7.4 Circle Graphs

A *circle graph* is used to show how some quantity is divided up into parts or percents. Each part is represented by a pie-shaped piece called a *sector* of the circle. The whole circle always represents 100% of the total amount being divided up.

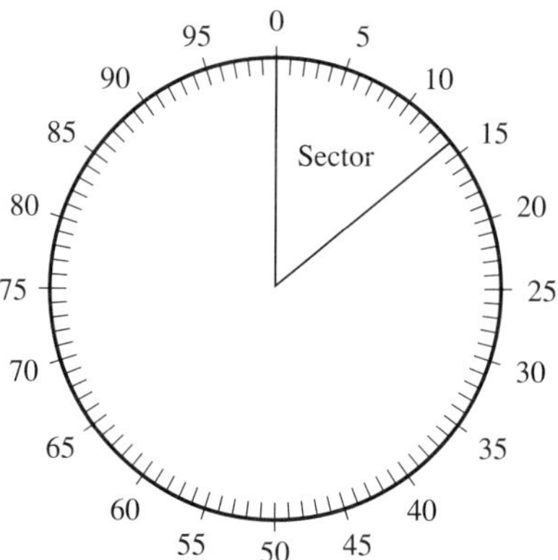

EXAMPLE 1 The sources of revenue for Smartmore Community College are given by the following circle graph.

Revenues – Smartmore Community College

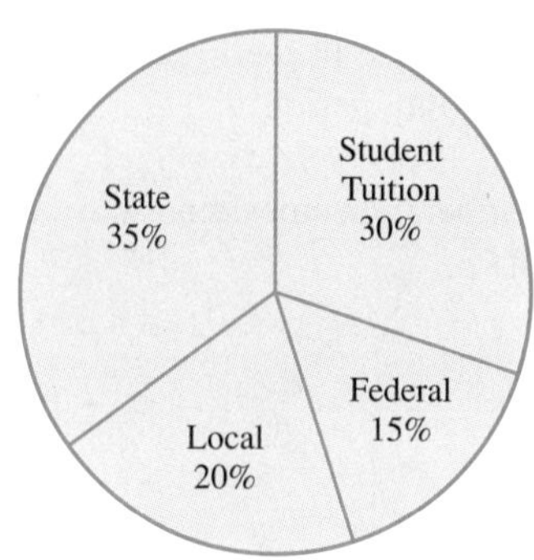

(a) Which source accounts for the greatest percent of revenue?

(b) What percentage do the federal and state governments supply in total?

(c) If the total revenue received is $4,000,000, how much of this is supplied by student tuition?

SOLUTION **(a)** The state contributes 35%, the greatest amount.

(b) The state contributes 35% and the federal government pays 15% for a total of 50%.

(c) Student tuition accounts for 30% of the total $4,000,000 revenue, giving us

$$(0.30)(4{,}000{,}000) = \$1{,}200{,}000 \qquad \text{from student tuition}$$

EXAMPLE 2 The following circle graph shows how the Assante family spends their monthly income.

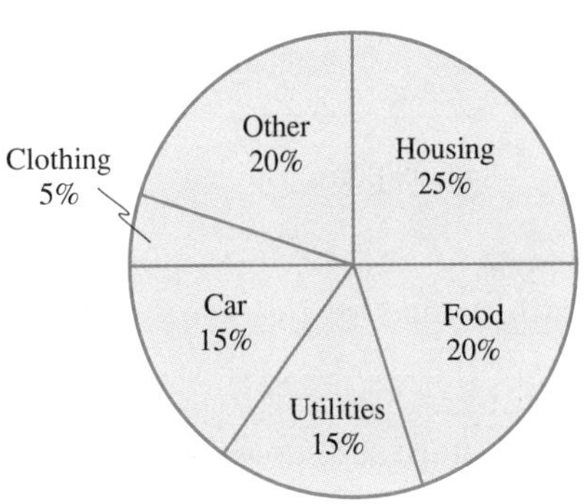

(a) What is the sum of all of the percentages in the graph?

(b) For which item does the Assante family spend the largest percentage of their income?

(c) Which two categories are equal in amount spent?

(d) Altogether, what percent of family income is spent for food and housing?

(e) If the monthly income is $3000, how much is spent for car expenses?

SOLUTION **(a)** 25% + 20% + 15% + 15% + 5% + 20% = 100%

(b) They spend the most for housing (25%).

(c) Food and "other" are both 20%, and car and utilities are both 15%.

(d) Food + housing = 20% + 25% = 45%

(e) Car costs amount to 15% of the monthly income, so (0.15)(3000) = $450 is spent on car expenses.

7.4 Exercises

Use the accompanying circle graph to answer Exercises 1–6. The graph represents the typical college expenses for a student at Tricounty State College.

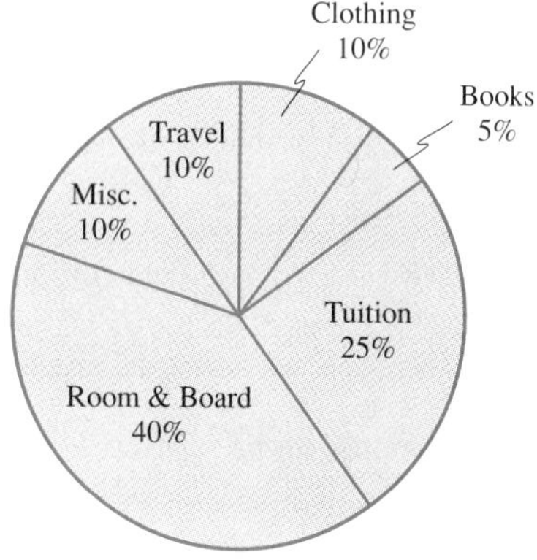

1. Which category is the largest part of the student's expenses?

2. What percent of the students expenses are for tuition?

3. Which three items each account for 10% of expenses?

4. What is the total percent of expenses that are spent for tuition and books?

5. If the total expenses amount to \$6000, how much is spent for tuition?

6. How much does the student spend for room and board if the total expenses amount to \$6000?

Use the following circle graph to answer Exercises 7–12. A research company surveyed 400 people at a mall to see what kind of shampoo they preferred. The brands were labeled *A, B, C, D,* and *E.*

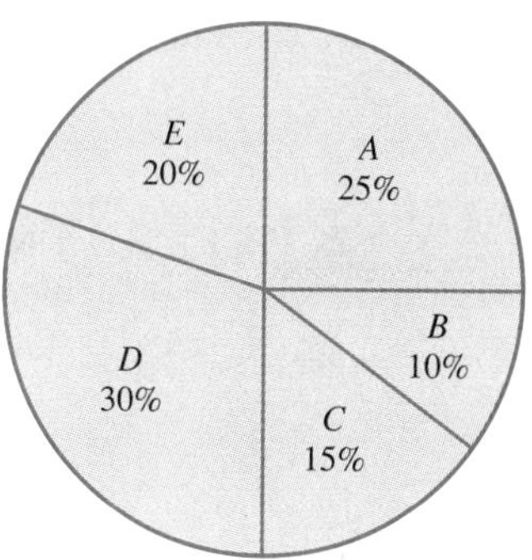

7. Which brand of shampoo was preferred by most people?

8. What percentage of the people surveyed preferred brand *B*?

9. Of the 400 people surveyed, how many of them liked brand *A*?

10. How many people in the survey preferred brand *E*?

11. What brand is preferred two-to-one over brand *B*?

12. What brand is preferred twice as often as brand *C*?

SUMMARY—CHAPTER 7

The numbers in brackets refer to the section in the text in which each concept is presented.

[7.1] **Mean** or **average:** $= \dfrac{\text{the sum of all the quantities}}{\text{the number of quantities}}$

[7.1] The **median** is the middle number in a set of data after the data has been arranged in order by size. If there is an even number of numbers in the data, average the two middle numbers.

[7.1] The **mode** is the number that occurs most frequently in a set of data.

[7.2] The example at the right refers to the following **table:**

NUTRITIONAL VALUES OF CEREAL (PER 1 OZ SERVING)

	Oat Loops	Toasties	Honey Flakes
Calories	100	110	190
Protein	2 g	2 g	1 g
Carbohydrates	20 g	30 g	28 g
Fat	0 g	2 g	8 g
Sodium	240 mg	180 mg	330 mg

[7.3] The example at the right refers to the following **bar graph,** which shows the number of appliances sold in a week at South-side sales Appliance Store.

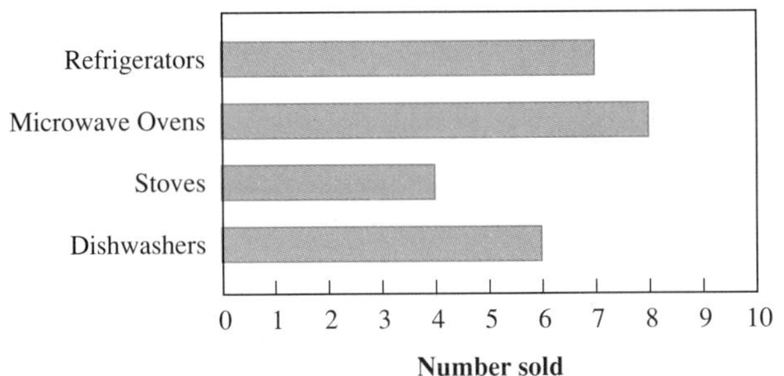

EXAMPLES

[7.1] The average of 4, 11, 8, 7, and 10 is

$$\frac{4 + 11 + 8 + 7 + 10}{5} = \frac{40}{5} = 8$$

[7.1] Find the median of 3, 6, 9, 4, 11, 15, and 7. Put the data in order:

3, 4, 6, 7, 9, 11, 15

The middle number, 7, is the median.

[7.1] The mode of 4, 6, 7, 6, 5, 3, 6, 1, and 6 is 6.

[7.2] How many calories are there in

a. 1 oz of Honey Flakes?

b. 3 oz of Honey Flakes?

SOLUTION

a. 190 from the table

b. (3)(190) = 570 calories

[7.3]

a. How many microwave ovens were sold?

b. How many more refrigerators were sold than stoves?

c. Which appliance had the least number sold?

SOLUTION

a. 8

b. $7 - 4 = 3$

c. stoves (4)

[7.3] The example at the right refers to the following **line graph,** which shows sales results for a dress shop over a six-month period.

EXAMPLES

[7.3]

a. What were sales for May? $8,000

b. How much did sales increase from Feb. to April?

$$\$7{,}000 - \$4{,}000 = \$3{,}000$$

c. What is the sales trend over the entire six months? Sales increased steadily over the whole six-month period.

[7.3] The example at the right refers to the following **pictograph,** which shows boat sales at Swinging Bridge Marina.

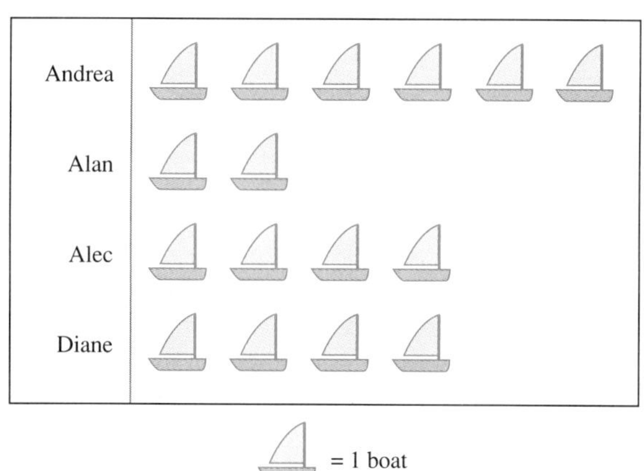

[7.3]

a. Who sold the greatest number of boats? Andrea sold 6.

b. Which two people sold the same number of boats? Alec and Diane both sold 4.

c. How many more boats did Andrea sell than Alan? 5

[7.4] The example at the right refers to the following **circle graph,** which shows how a student's final grade is calculated.

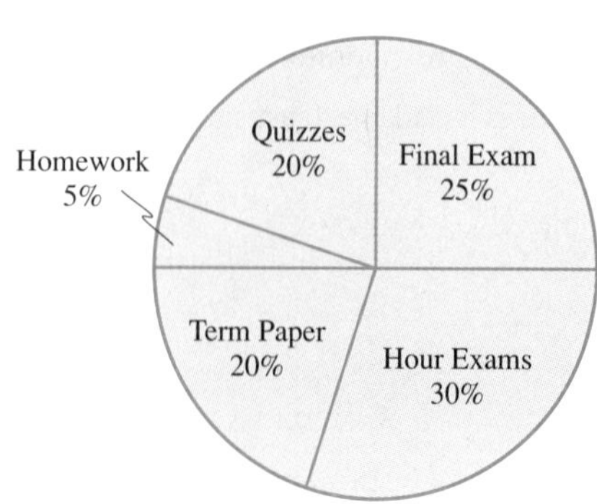

[7.4]

a. Which activity contributes the most toward the final grade? Hour exams, 30%

b. Which two activities are equally weighted? Quizzes and the term paper both count 20%.

c. What portion of the final grade does the final exam make up? 25%

REVIEW EXERCISES—CHAPTER 7

The numbers in brackets refer to the section in the text in which similar problems are presented.

[7.1]

1. Find the mean of the numbers 23, 26, 22, 28, 32, and 37.

[7.1]

2. If you bowl games of 136, 145, and 109, what is your average for the three games?

[7.1]

3. Find the median for the following set of data: 16, 12, 14, 18, 26, 17.

[7.1]

4. Eight men record their cholesterol levels: 168, 206, 231, 204, 191, 170, 158, and 216. What is the median level?

[7.1]

5. What is the mode of the numbers 164, 166, 161, 164, 166, 162, 169, 158, 165, and 164?

[7.1]

6. Mel Cohen earned the following quiz grades in his Spanish class: 8, 10, 9, 10, 7, 7, 8, 7, 4, 8, 8, 10, 6, 8, and 8. What is the mode?

[7.1]

7. Calculate the grade point average (GPA) for the following list of grades. Assume that A = 4, B = 3, C = 2, D = 1, and F = 0

Grade	Credits
B	4
C	3
A	3
A	4
B	1

[7.1]

8. The speeds of ten cars were clocked on radar as follows: 68, 54, 62, 62, 62, 58, 57, 71, 63, and 62. Find the mean, median, and mode.

[7.2]

9. The following table gives the number of employees for each division by shift and by gender at Genco Electronics Corp.

	Day Shift		Night Shift	
Department	Male	Female	Male	Female
Sales	9	7	0	0
Personnel	6	11	2	1
Assembly	24	31	9	23
Shipping	6	1	3	2

(a) How many women from the day shift work in sales?

(b) How many women altogether work in personnel?

(c) How many more men on the day shift work in assembly than in shipping?

(d) How many employees are in the personnel department in total?

(e) How many women work on the day shift?

(f) How many employees work on the night shift?

▼ [7.3]

10. The following bar graph shows the total sales (in dollars) for one week at each of four fast food restaurants:

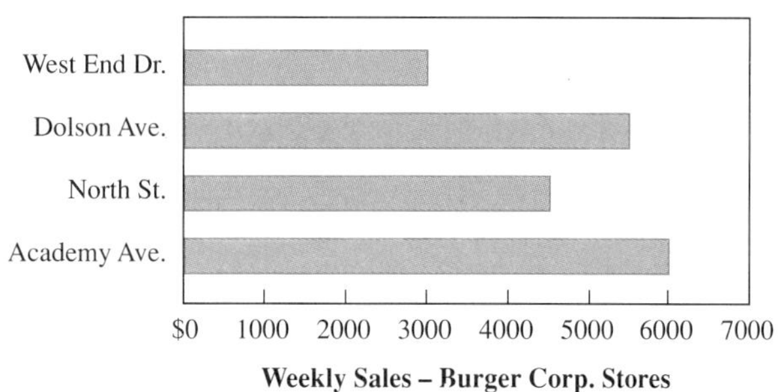

(a) What were the sales for the week at the North St. restaurant?

(b) Which location had the greatest sales?

(c) How much more money did the Dolson Ave. store bring in than the West End Dr. location?

(d) In which location were sales twice as great as West End Dr?

(e) What were the total sales for all four restaurants?

▼ [7.3]

11. The following line graph shows the number of points that the North Central High School basketball team has scored for their first ten games.

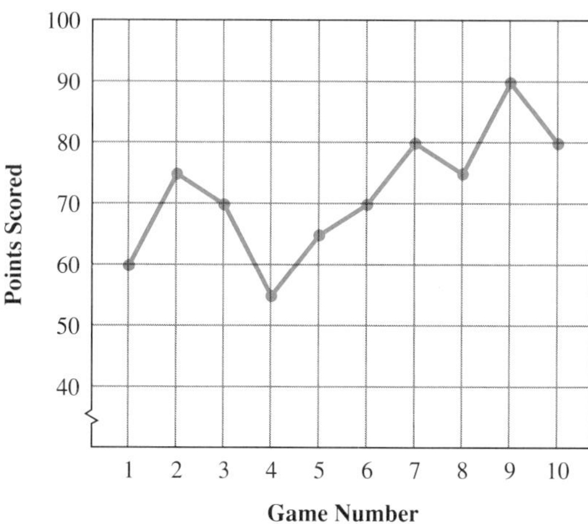

(a) In which game did they score the most points? How many did they score?

(b) In which game did they score the lowest number of points? How many did they score?

(c) How many more points did the team score in the eighth game than in the fourth game?

(d) Has there been a general upward or downward trend in the scoring over the first ten games?

▼ [7.3]

12. The following pictograph shows the number of soccer games won by various teams in the Under-Ten Soccer League.

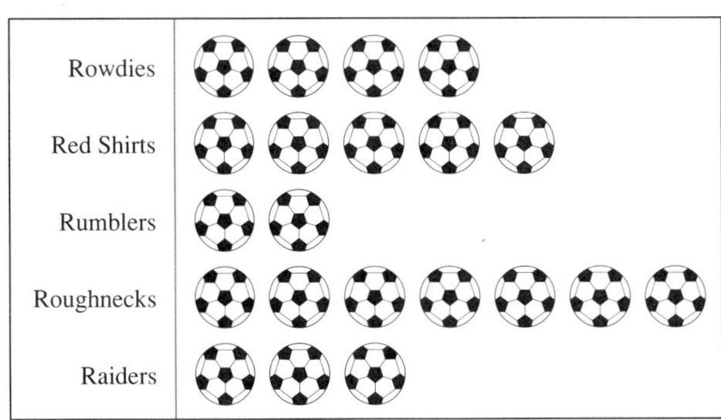

(a) How many games have the Red Shirts won?

(b) Which team has won the fewest games?

(c) Which team has won exactly twice as many games as the Rumblers?

(d) Which team is in third place?

(e) How many games have been played in the league so far?

▼ [7.4]

13. The following circle graph shows the types of property sold at Houston Realtors.

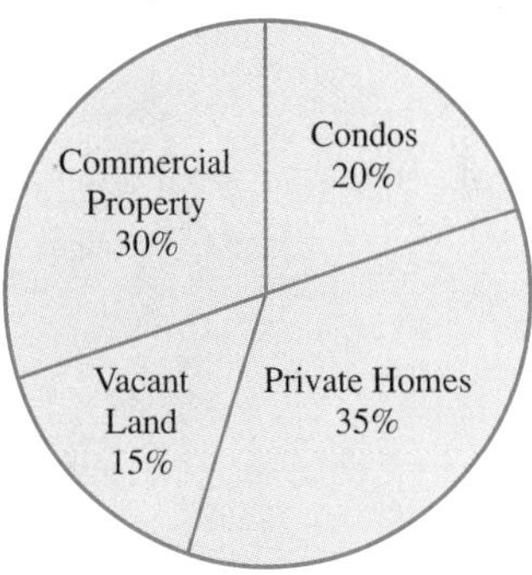

(a) Which type of property accounts for the greatest portion of sales?

(b) What percentage of total sales does commercial property and vacant land account for altogether?

(c) It was the company's goal to have the total of condo and commercial property sales account for half of their sales. Did they reach their goal?

(d) If the company sells a total of 160 properties, how many of them will be private homes?

(e) If the company sells a total of 160 properties, how many condos will they have sold?

CRITICAL THINKING EXERCISES—CHAPTER 7

1. Julian received grades of 70 and 80 on his first two exams. His average so far is $\frac{70 + 80}{2} = \frac{150}{2} = 75$. If he earns an 85 on his next exam, he reasons that his average for the three exams will be the average of the first two exams, 75, and the third exam, 85, or $\frac{75 + 85}{2} = \frac{160}{2} = 80$. Find the error in Julian's reasoning.

2. The mean of twelve numbers is 7. What is the sum of the twelve numbers?

3. Professor Rickard decides to add 10 points to each student's grade on their algebra exam. What does this do to the average test grade for the class?

4. Give an example of a set of seven numbers in which the median is less than the mean.

5. Find a set of five numbers in which the mean and the mode are the same but the median is different.

NAME ▮▮▮▮▮ CLASS

ACHIEVEMENT TEST—CHAPTER 7

1. What is the mean of the data 38, 42, 51, 61, and 43? 1. ______

2. Find the median for the set of numbers 36, 41, 35, 41, 37, and 40. 2. ______

3. What is the mode of the following group of grade point averages: 2.6, 3.4, 1.9, 2.5, 2.6, 2.7, 2.9, 3.9, 2.6, and 3.1. 3. ______

4. Jack wants to make the wrestling team and so he goes on a diet. He weighs himself every morning for seven days and records the following weights (in pounds): 165, 161, 162, 161, 160, 159 and 159. Find the mean, median, and mode. 4. ______

______

5. Calculate the grade point average (GPA) from the following table. Assume that A = 4, B = 3, C = 2, D = 1, and F = 0 5. ______

Course	Grade	Credits
Calculus	B	4
French	B	3
Physics	A	3
Chemistry	C	4
English	B	3

6. Table 7.6 shows the number of calories used per minute for various activities, by body weight.

TABLE 7.6. CALORIES USED PER MINUTE

	Weight in Pounds		
Activity	**100**	**150**	**200**
Walking (3 mph)	2.7	4.0	5.4
Walking (4 mph)	3.9	5.8	7.8
Swimming (crawl)	5.8	8.7	11.6
Running (11-min mile)	6.1	9.1	12.2
Running (8-min mile)	9.4	14.1	18.8

a. How many calories are used per minute by a 150-lb person who runs 11-minute miles?

b. How many more calories per minute does a 200-lb person use by walking at 4 mph rather than at 3 mph?

c. How many calories are used by a 100-lb woman who swims steadily for 20 minutes?

7. The following bar graph shows the number of books, by category, sold at the Scholars' Book Store last week.

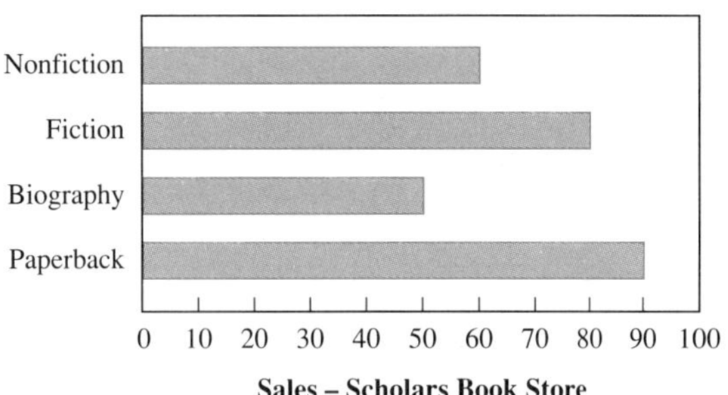

a. How many paperback books were sold?

b. How many more fiction books were sold than nonfiction books?

c. How many books were sold altogether?

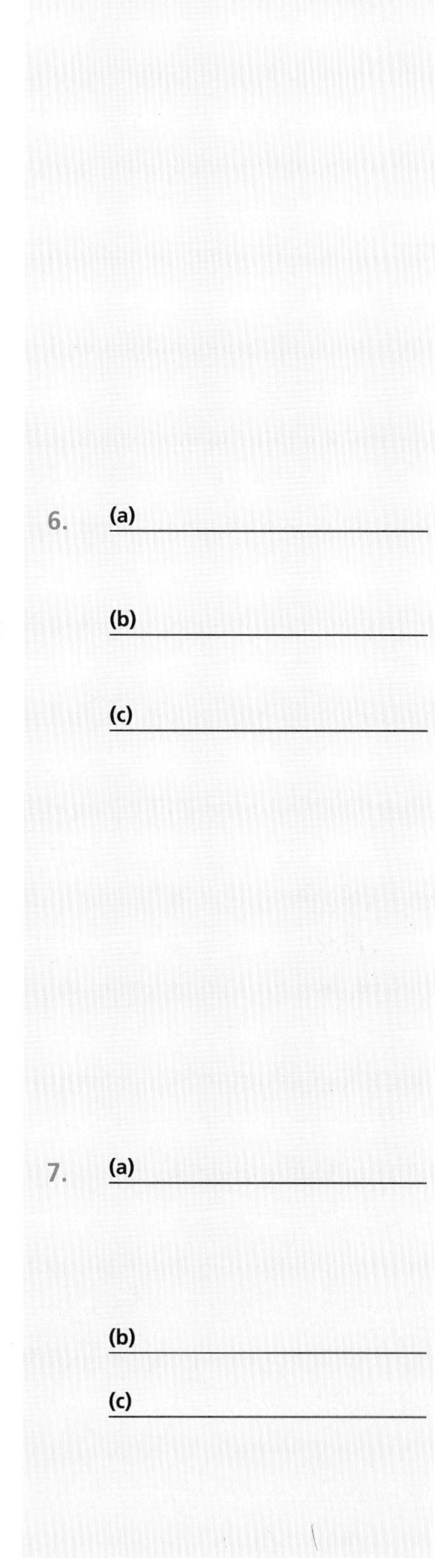

8. The following line graph shows the hourly temperature readings between midnight and 8:00 A.M. on June 1 in New Paltz, New York.

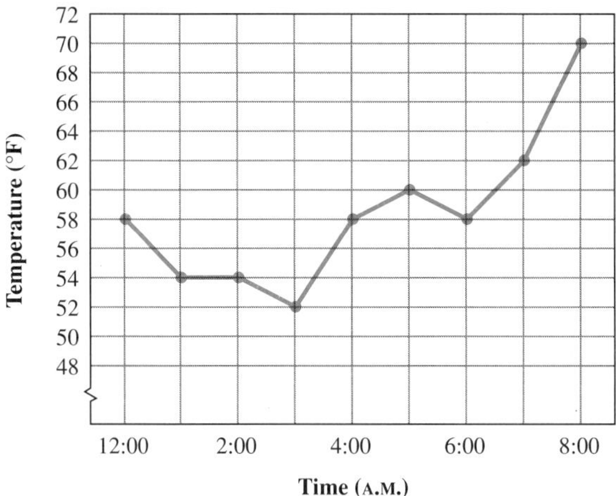

a. What was the temperature at 4:00 A.M.?

b. What is the difference between the highest and lowest temperature readings?

c. Between which two hours did the temperature increase the most?

d. At what time was the temperature the lowest?

e. Between which two hours did the temperature stay the same?

9. The following pictograph represents how many students graduated with various majors from Hardford College.

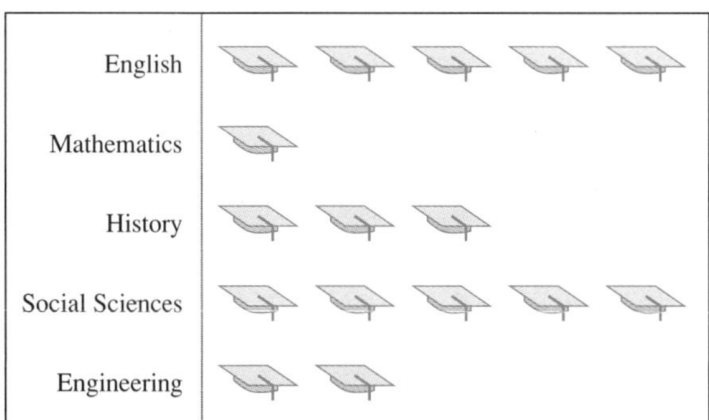

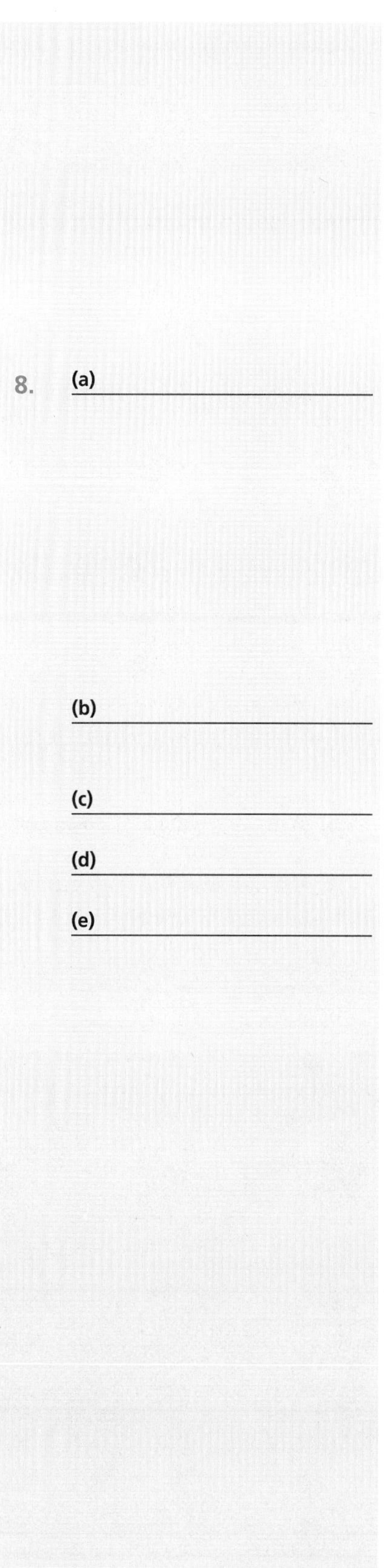

a. How many graduates majored in mathematics?

b. Which two majors had the same number of graduates?

c. How many more students graduated in English than in history?

10. The following circle graph shows which activities were preferred by customers at Best-Fit Health Club.

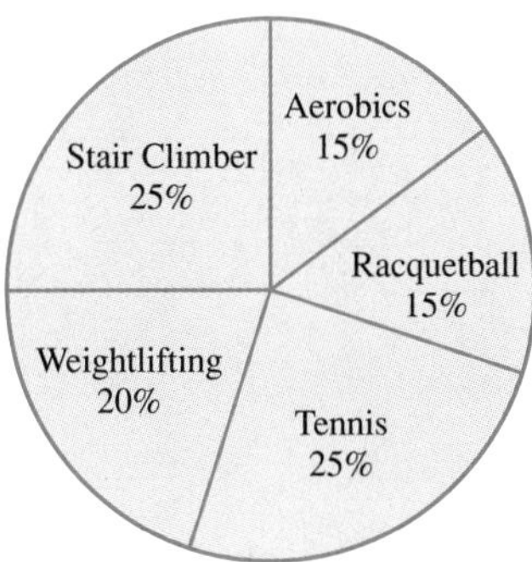

a. Which activity is most popular?

b. What percentage of customers prefer tennis?

c. If there are 2800 club members, how many of them prefer weightlifting?

9. (a) ____________
(b) ____________
(c) ____________

10. (a) ____________
____
(b) ____________
(c) ____________

CHAPTER 8

Signed Numbers—The Language of Algebra

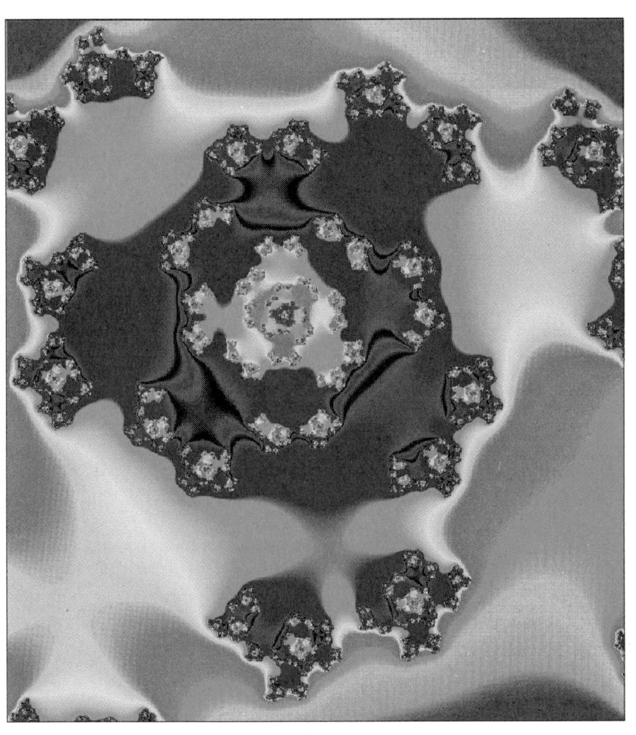

INTRODUCTION

We learn algebra in order to solve problems in the real world. The signed numbers are the building blocks of algebra. In this chapter you will learn about the meaning of signed numbers, some of their properties, how to combine them using six basic operations, and how to "punctuate" them using grouping symbols.

This knowledge will enable us to move on to the business of learning more about algebra.

NUMBER KNOWLEDGE—CHAPTER 8

Consider an ordinary checkerboard that is eight squares by eight squares, as in the following diagram.

A little thought should convince you that you can cover the entire checkerboard with dominos, each of which covers exactly two squares.

Suppose you were to remove the upper left square and the lower right square from the checkerboard. Do you think that you can still cover the remaining squares with dominos?

The solution is in the answer section in the back of the book, but spend some time trying to solve the problem on your own before you look at the answer.

8.1 Introduction to Signed Numbers

The first numbers we see in arithmetic are the **counting numbers:** 1, 2, 3, 4, 5, 6, and so on. When we *add* two counting numbers we always get a counting number for an answer, for example, $6 + 2 = 8$, and $15 + 12 = 27$.

Now let's see what happens when we *subtract* one counting number from another.

$15 - 7 = 8$ A counting number

$10 - 6 = 4$ A counting number

$3 - 8 = ?$

Can such a subtraction be performed and, if so, what does the answer look like? To answer the question, consider the case where the temperature is 3 degrees above zero and then drops 8 degrees. The new temperature is 5 degrees below zero, or negative 5 degrees.

This example of temperature above and below zero nicely illustrates the concept of positive and negative numbers. These positive and negative numbers, along with zero, are called the **signed numbers,** or the **integers.** This set, or group, of signed numbers is often looked at in terms of the *number line.* To make a number line, we select a point on a line and label it zero. Then we place the positive numbers equally spaced to the right of zero and place a positive sign (+) in front of each of them. The negative numbers are equally spaced to the left of zero and a negative sign (−) is placed in front of each.

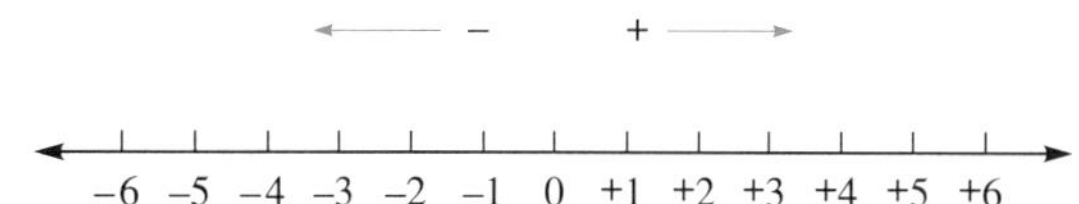

It is common practice to omit the + sign in front of positive numbers; thus a number with neither a + sign nor a − sign in front of it is positive. For example, 7 is +7 and 34 is +34.

Signed numbers are the language of algebra, and we must become skilled in their use.

8.1 Exercises

Locate each of the following numbers on the number line.

1. −3
2. −1
3. +5
4. +6
5. −4
6. 5

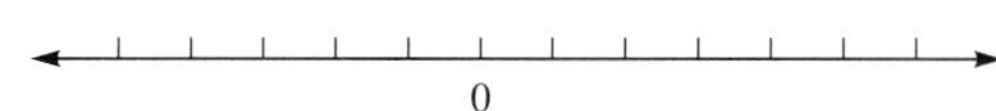

LIVING IN THE REAL WORLD

Represent each quantity by a signed number.

7. You lose $5.00 in a bet.
8. You receive a bill for $23.50.
9. A certain mountain is 7,288 feet above sea level.
10. Death Valley lies 280 feet below sea level.
11. The year is 1990 A.D.
12. Euclid lived around 300 B.C.
13. A football team is penalized 15 yards.
14. You get a bonus of $500.
15. You are penalized $25 for a late payment.
16. The boiling point of water is 100° C above zero.
17. A stock goes down $2\frac{1}{4}$ points.
18. Mt. Everest is 29,028 feet above sea level.
19. The Marianas Trench in the Pacific Ocean is 36,198 feet below sea level.
20. The temperature in Albany, New York, on Christmas Day in 1900 was 16° F below zero.
21. You overdraw your checking account by $38.55.
22. Mt. Kilimanjaro, the highest peak in Africa, is 19,340 feet above sea level.

23. Write down five numbers from your own experiences that can be represented by signed numbers.

8.2 Absolute Value

Before we can add signed numbers we must first learn about a concept called **absolute value.** The absolute value of a signed number is the number of units that it is located from zero on the number line. For example, +3 is 3 units away from zero on the number line, so the absolute value of $+3$ is 3, and it is written $|+3| = 3$.

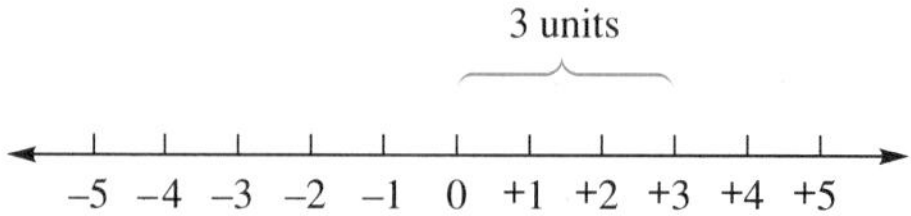

What about the absolute value of -3? Since it is also 3 units away from zero, its absolute value is also 3, which is written $|-3| = 3$.

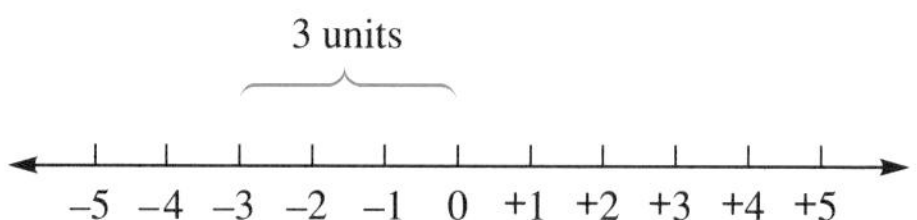

EXAMPLE 1

(a) $|+3| = 3$ is read, "The absolute value of $+3$ is 3."

(b) $|-3| = 3$ is read, "The absolute value of -3 is 3."

(c) $|-14| = 14$ is read, "The absolute value of -14 is 14."

(d) $|0| = 0$ is read, "The absolute value of zero is zero."

EXAMPLE 2

(a) $|-2| + |+6| = 2 + 6 = 8$ Find the absolute value before adding.

(b) $|-6| + |-8| = 6 + 8 = 14$

8.2 Exercises

Find each of the following absolute values:

1. $|-8|$
2. $|+5|$
3. $|-4|$
4. $|-6|$
5. $|-1462|$
6. $|+46|$
7. $|0|$
8. $|+34|$
9. $|+6|$
10. $|-62|$
11. $|-1.7|$
12. $|+3.8|$
13. On the Kansas Turnpike, town *A* is at milepost 79 and town *B* is at milepost 94. How far is it from town *A* to town *B?* How far is it from town *B* to town *A?*

Find the value of each of the following.

14. $|-3| + |+2|$
15. $|-6| + |+4|$
16. $|-7| + |+6|$
17. $|+2| + |-18|$
18. $|3| + |-5|$
19. $|-7| + |14|$
20. $|23| + |17|$
21. $|-23| + |-17|$
22. $|-16| - |-5|$
23. $|16| - |5|$
24. $|-17| - |+2|$
25. $|13| - |-4|$

8.3 Greater Than and Less Than

On the number line, as we move to the right the numbers get larger, continuing on and on in such a way that there is no largest integer. Similarly, as we move to the left of zero on the number line, the negative numbers have smaller values. Therefore, there is also no smallest negative integer.

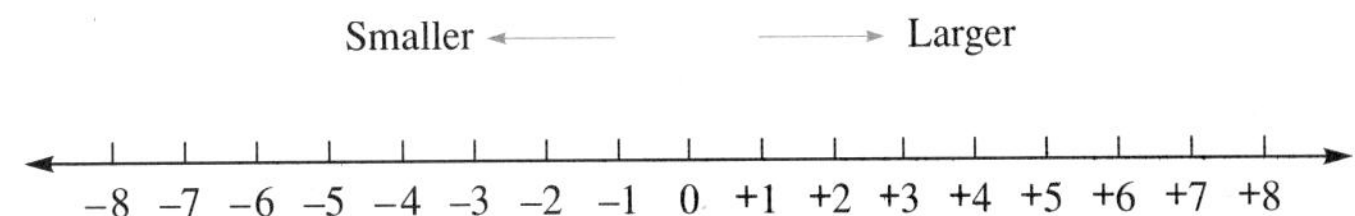

From this we can see that when any two numbers on the number line are compared, the one lying to the right of the other must always be larger. For example, $+6$ is greater than -2 since $+6$ lies to the right of -2 on the number line. This is usually written

$$+6 > -2$$

and is read "$+6$ is greater than -2."

We can also write

$$-2 < +6$$

which is read "-2 is less than $+6$."

EXAMPLE 1

(a) $6 < 18$ is read "6 is less than 18."

(b) $-3 > -8$ is read "-3 is greater than -8."

(c) $4 > -2$ is read "4 is greater than -2."

(d) $-7 < 4 < 6$ is read "-7 is less than 4 and 4 is less than 6."

Notice that $4 < 8$ and $8 > 4$ really mean the same thing, even though we read them differently.

The symbol

$>$ is read "is greater than"

and

$<$ is read "is less than."

STOP You can remember which way the symbol goes by noticing that the small side of the symbol is on the side of the smaller number and the larger side of the symbol is found on the side of the larger number.

small side	$<$	large side
smaller number		larger number

EXAMPLE 2 Which is greater, $|+6|$ or $|-8|$?

SOLUTION

$$|+6| = 6$$
$$|-8| = 8$$

Since 8 is greater than 6, we know that $|-8|$ is greater than $|+6|$ and we write

$$|-8| > |+6|$$

8.3 Exercises

Indicate which of the following are true and which are false.

1. $-6 < -5$
2. $3 > -4$
3. $7 > 14$
4. $6 < -7$
5. $3 > -4$
6. $-17 > -6$
7. $4 < -8.6$
8. $3.6 < -1.4$
9. $-784 < -326$
10. $-\frac{7}{8} < -\frac{1}{4}$
11. $-7\frac{1}{2} > \frac{2}{3}$
12. $\frac{5}{16} > -\frac{1}{8}$
13. $|-8| > |-9|$
14. $|-2| > -10$
15. $|+6| < |-10|$
16. $|-4| > 0$
17. $|-16| < 0$
18. $|-2| < |+2|$

Insert $>$ or $<$ between each pair of numbers to make a true statement.

19. 17 14
20. -4 12
21. 23 -5
22. -7 -4
23. 46 49
24. $13\frac{3}{4}$ $-11\frac{1}{2}$
25. 0 14
26. $\frac{5}{8}$ $-2\frac{1}{3}$
27. 586 -12
28. 0 -4
29. 314 -8.9
30. $-24\frac{2}{3}$ -25
31. $|-8|$ $|+4|$
32. $|-16|$ $|+9|$
33. -9 $|-6|$
34. -4 $|-12|$
35. $|-14|$ 0
36. 0 $|+4|$

LIVING IN THE REAL WORLD

37. Suppose that you ran out of gas at milepost 67 on the Pennsylvania Turnpike. Your map shows a town at milepost 81 and a town at milepost 56. Which town is your closest source of gasoline?

38. If the temperature in Chicago on January 5 is -11 degrees and on January 6 the temperature is $+18$ degrees, which temperature is closer to zero degrees?

8.4 Adding Signed Numbers

To add signed numbers, we again use the number line. To add a positive number we move to the right; to add a negative integer we move to the left. We always start at zero.

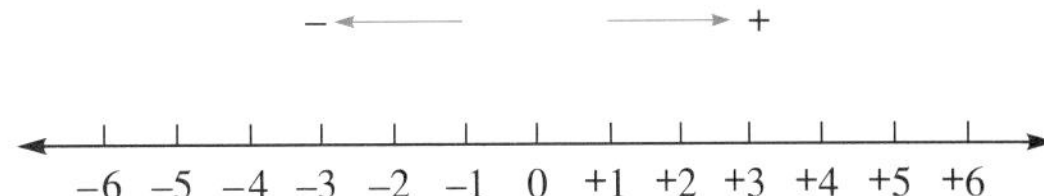

Let's start with an example that we already know the answer to.

EXAMPLE 1 On the number line, 4 + 2 is added by starting at zero, moving four spaces to the right to represent +4, and then moving two more spaces to the right to represent +2. We end up at +6, which is the answer: 4 + 2 = 6.

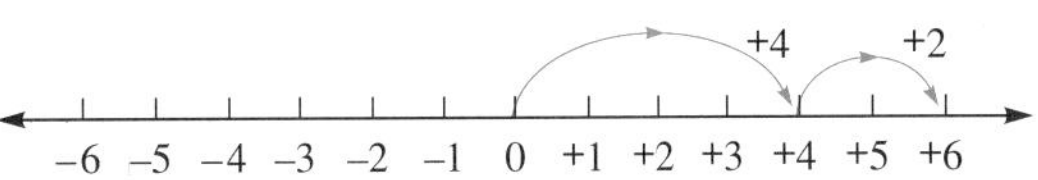

EXAMPLE 2 The temperature outside drops 6 degrees (−6) in an hour and then rises 4 degrees (+4) the next hour. What is the total change in temperature over the two-hour period?

SOLUTION The problem, when written in symbols, looks like this: (−6) + (+4). The answer will be negative, because we move farther to the left (6 units) than we do to the right (4 units). How much farther? We move two units farther, because the difference between $|-6|$ and $|+4|$ is 2. So (−6) + (+4) = −2, or the temperature dropped 2 degrees.

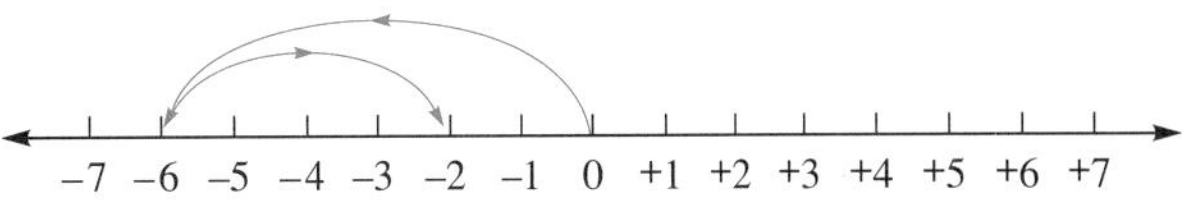

EXAMPLE 3 Let's add 5 + (−2). Starting at zero, we first move 5 units to the right and then 2 units to the left. The answer is +3 or 3, so 5 + (−2) = 3.

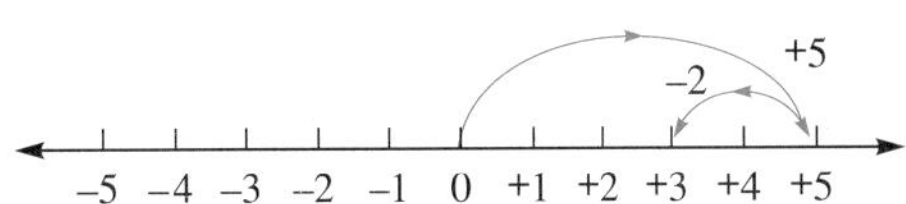

EXAMPLE 4 Calculate $-3 + (-4)$. A move of 3 units to the left plus another move of 4 units to the left leaves us at -7, the answer: $-3 + (-4) = -7$.

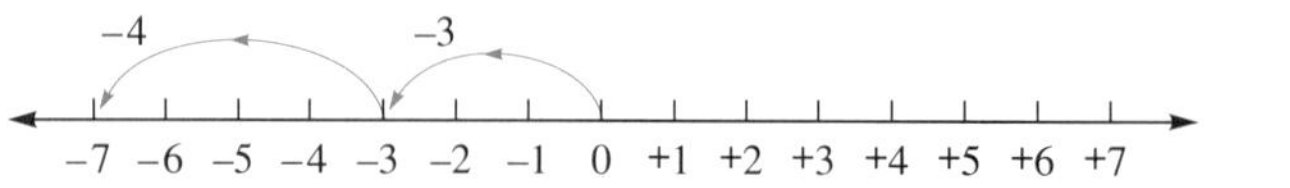

EXAMPLE 5 Show that $-6 + 2 = -4$.

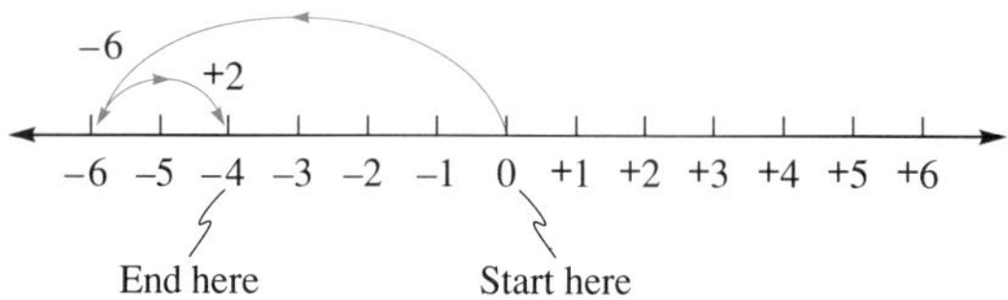

The sign + or − tells us which direction to move in, and the size of the number (its absolute value) tells us how far to move.

The number line provides a good visual explanation of what is taking place when we add signed numbers. However, it is cumbersome to draw a number line everytime we are asked to add two signed numbers, particularly if the numbers are very large. Therefore we must establish a rule that provides us with the same result as using the number line.

To add signed numbers:

1. If both numbers have the **same** sign, **add** the absolute values and keep the common sign for the answer.
2. If the numbers have **different** signs, find the **difference** in their absolute values and take the sign of the number with the larger absolute value for the answer.

EXAMPLE 6 $3 + 4 = 7$

Since 3 and 4 have the same sign, we add their absolute values and keep the same sign (+).

EXAMPLE 7 $-4 + 1 = -3$

Since -4 and $+1$ have different signs, we find the difference between their absolute values, 4 and 1, which is 3, and give it the sign of the larger absolute value.

EXAMPLE 8 $-6 + (-5) = -11$

Both -6 and -5 have the same sign, so we add their absolute values ($6 + 5 = 11$) and keep the common sign (−).

EXAMPLE 9 $2 + (-6) = -4$
Since 2 and -6 have different signs, we find the difference between 2 and 6, giving us 4, and take the negative sign from the 6, since it is larger than the 2.

EXAMPLE 10 $-14 + 3 = -11$
The numbers -14 and 3 have different signs, so we find the difference between 14 and 3, which is 11, and take the negative sign for our answer, since $14 > 3$.

It is important that you learn how to add signed numbers exceedingly well, since much of the mathematics that follows is based on this skill. The following practice exercise (Example 11) will help you to master these fundamentals. Refer back as often as necessary to the rules for adding signed numbers.

EXAMPLE 11
(a) $-5 + 6 = 1$
(b) $-6 + (-5) = -11$
(c) $2 + (-6) = -4$
(d) $-8 + 4 = -4$
(e) $6 + 9 = 15$
(f) $-2 + (-9) = -11$
(g) $14 + (-2) = 12$
(h) $-7 + 7 = 0$
(i) $12 + (-12) = 0$

8.4 Exercises

▼ Add the following:

1. $6 + 8$
2. $4 + 9$
3. $3 + (-6)$
4. $4 + (-7)$
5. $7 + (-5)$
6. $6 + (-8)$
7. $13 + (-6)$
8. $15 + (-3)$
9. $-6 + 5$
10. $-8 + 4$
11. $-13 + 7$
12. $-11 + 6$
13. $-26 + 16$
14. $-28 + 12$
15. $-3 + (-5)$
16. $-6 + (-3)$
17. $-8 + (-9)$
18. $-7 + (-6)$
19. $18 + (-5)$
20. $16 + (-11)$
21. $-18 + (-12)$
22. $-16 + (-8)$
23. $19 + 12$
24. $11 + 16$
25. $-7 + 25$
26. $-8 + 19$
27. $-11 + (-11)$
28. $-15 + (-15)$
29. $-11 + 11$
30. $-15 + 15$
31. $\begin{array}{r} 18 \\ \underline{32} \end{array}$
32. $\begin{array}{r} 25 \\ \underline{26} \end{array}$
33. $\begin{array}{r} 29 \\ \underline{-14} \end{array}$

34. $\begin{array}{r} 57 \\ \underline{-25} \end{array}$

35. $\begin{array}{r} -36 \\ \underline{-85} \end{array}$

36. $\begin{array}{r} -64 \\ \underline{-39} \end{array}$

37. $\begin{array}{r} -52 \\ \underline{18} \end{array}$

38. $\begin{array}{r} -76 \\ \underline{24} \end{array}$

39. $\begin{array}{r} -11 \\ \underline{77} \end{array}$

40. $\begin{array}{r} -26 \\ \underline{94} \end{array}$

41. $\begin{array}{r} -193 \\ \underline{714} \end{array}$

42. $\begin{array}{r} -235 \\ \underline{512} \end{array}$

43. $-\frac{3}{8} + \left(-\frac{1}{8}\right)$

44. $\frac{1}{5} + \left(-\frac{2}{5}\right)$

45. $\frac{1}{7} + \left(-\frac{3}{7}\right)$

46. $-\frac{2}{9} + \left(-\frac{4}{9}\right)$

47. The distance between two numbers on the number line is 8. If one of the numbers is 5, find two possible answers for the other number.

48. If the distance between two numbers on the number line is 10 and one of the numbers is -4, find two possible answers for the other number.

49. What is the result if the sum of -8 and -5 is increased by 10?

50. If the sum of -8 and 6 is increased by 7, what is the result?

51. What number must be added to -7 to get -2?

52. What number must be added to 6 to get -4?

53. What number must be added to -8 to get 7?

54. What number must be added to 5 to get -3?

LIVING IN THE REAL WORLD

55. Janine started with a balance of \$300 in her checking account and then wrote checks for the following amounts: books, \$58.75; jacket, \$45.80; room rent, \$180.00; telephone deposit, \$25.00. What is her balance?

56. In two successive nights of playing poker, you lose \$2.58 the first night and win \$1.76 the second night. What are your total winnings or losses?

57. Starting with a balance of zero in his checking account, a financier writes checks for \$72.05, \$31.16, \$24.17, and \$12.12. The next day he rushes to the bank and makes a deposit of \$125.00 to cover them. What is the status of his account?

58. On Monday your stock in Hexagon Oil Co. goes up $2\frac{1}{3}$ points. On Tuesday it falls $6\frac{2}{3}$ points. What is the net amount of the change for the two days and in what direction?

59. Death Valley is the lowest point in North America at 282 feet below sea level. Mount McKinley in Alaska is 20,320 feet above sea level, the highest point in North America. What is the difference in altitude between these two places?

60. Place one of the signed numbers -5, -4, -3, -2, -1, 0, 1, 2, and 3 in each circle to make the sum of each side of the triangle equal to -3. (There is more than one correct answer.)

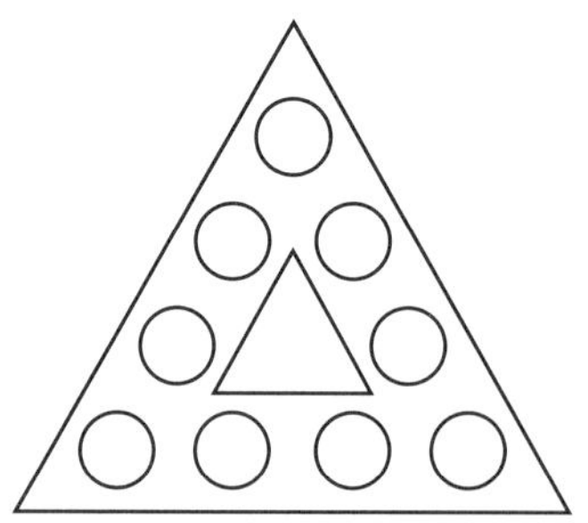

Complete the following magic squares. The sum of each column, row, and diagonal must all equal the same number.

61.

−1		
4		
3		5

62.

$-1\frac{1}{2}$		
	−1	
	0	$-\frac{1}{2}$

MENTAL MATHEMATICS

Add the following pairs of numbers mentally, without using a pencil or paper

63. $3 + 8$

64. $-5 + 9$

65. $-4 + 10$

66. $6 + (-3)$

67. $5 + 6$

68. $7 + (-7)$

69. $-5 + (-5)$

70. $-8 + (-7)$

71. $-11 + 4$

72. $-14 + 9$

73. $6 + (-13)$

74. $11 + (-3)$

75. $12 + 6$

76. $14 + (-7)$

77. $-32 + 32$

78. $-18 + 18$

79. $13 + (-7)$

80. $16 + (-16)$

81. $-10 + 5$

82. $-4 + (-8)$

83. $-6 + (-7)$

84. $6 + (-3)$

85. $18 + (-12)$

86. $-4 + 5$

87. $-8 + 2$

88. $-11 + 7$

89. $-8 + (-7)$

90. $8 + (-7)$

91. $6 + 9$

92. $-7 + 1$

93. $14 + (-12)$

94. $3 + (-11)$

95. $-7 + (-11)$

96. $-14 + 14$

97. $6 + (-6)$

98. $13 + (-14)$

99. $-6 + (-10)$

100. $-1 + (-15)$

101. $-7 + (-3)$

102. State the rule for adding signed numbers in your own words.

8.5 Subtracting Signed Numbers

The Negative of a Number

The concept of the **negative** of a number suggests the opposite of that number. The opposite of moving 3 steps to the right on the number line would be moving 3 steps to the left. Since $+3$ represents a move of 3 steps to the right, the negative of $+3$ would mean a move of 3 steps to the left, or -3. So the negative of $+3$ is -3.

Following this same thinking, since -3 represents a move of 3 units to the left, the opposite or negative of -3 would be a move of 3 units to the right. So the negative of -3 is $+3$, and we write this as $-(-3) = +3 = 3$.

EXAMPLE 1 **(a)** The negative of 7 is -7.

(b) The negative of -4 is $-(-4) = 4$.

(c) The negative of -6 is 6.

The rule for subtraction of signed numbers is stated in terms of the rule for adding signed numbers, as follows:

> **To subtract one signed number from another,** add its negative.

EXAMPLE 2 $7 - 9 = 7 + (-9)$ To subtract 9, add its negative, -9.

$= -2$

EXAMPLE 3 $-4 - (-3) = -4 + 3$ To subtract -3, add its negative, $+3$.

$= -1$

EXAMPLE 4 $6 - (-1) = 6 + 1$ To subtract -1, add its negative, 1.

$= 7$

EXAMPLE 5 $-8 - 4 = -8 + (-4)$ To subtract 4, add its negative, -4.

$= -12$

EXAMPLE 6 $-4 - (+5) = -4 + (-5) = -9$

EXAMPLE 7 $6 - (+7) = 6 + (-7) = -1$

EXAMPLE 8 $7 - (+3) = 7 + (-3) = +4$

EXAMPLE 9

$$\begin{array}{rr} & -4 \\ - & \underline{+3} \end{array} = \begin{array}{rr} & -4 \\ + & \underline{-3} \\ & -7 \end{array}$$

EXAMPLE 10

$$\begin{array}{rr} & +9 \\ - & \underline{+16} \end{array} = \begin{array}{rr} & +9 \\ + & \underline{-16} \\ & -7 \end{array}$$

EXAMPLE 11

$$\begin{array}{rr} & -214 \\ - & \underline{+123} \end{array} = \begin{array}{rr} & -214 \\ + & \underline{-123} \\ & -337 \end{array}$$

EXAMPLE 12

$$\begin{array}{r} -394 \\ - \underline{-236} \end{array} = \begin{array}{r} -394 \\ + \underline{+236} \\ -158 \end{array}$$

EXAMPLE 13

$$\begin{array}{r} +325 \\ - \underline{+112} \end{array} = \begin{array}{r} +325 \\ + \underline{-112} \\ +213 \end{array}$$

Subtraction of Integers and Number Line • • Now that we have learned to subtract integers, it would be nice to gain some intuitive understanding of the process. This is perhaps best done by using the number line once again.

When we subtract two integers—in other words, find the difference between two integers—what we are really doing is finding the *distance* from the second integer to the first integer on the number line. To illustrate our point, let's take a look at some examples.

EXAMPLE 14

$(+6) - (+2) = +4$, because it represents a move of 4 units from the second integer to the first integer (from $+2$ to $+6$). Since we are moving to the right, the 4 is positive and our answer is $+4$.

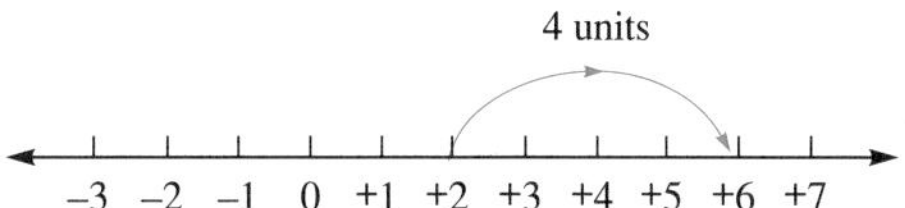

When we use the rule for subtracting integers, we get the same result:

$$(+6) - (+2) = (+6) + (-2) = +4.$$

EXAMPLE 15

$(-7) - (-4) = -3$, because when we move from -4 to -7 on the number line, we move a distance of 3 units to the left, or -3.

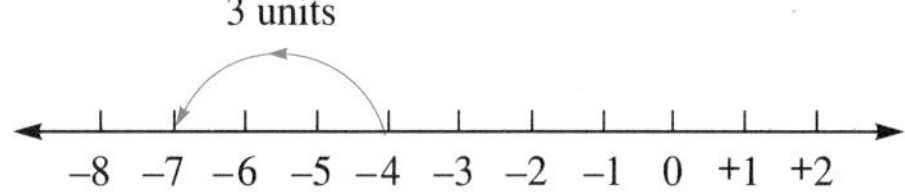

Again, when we use the rule for subtraction of integers, we get the same result:

$$(-7) - (-4) = (-7) + (+4) = -3$$

EXAMPLE 16

$(-2) - (+5) = -7$, because the move from $+5$ to -2 is a distance of 7 units to the left.

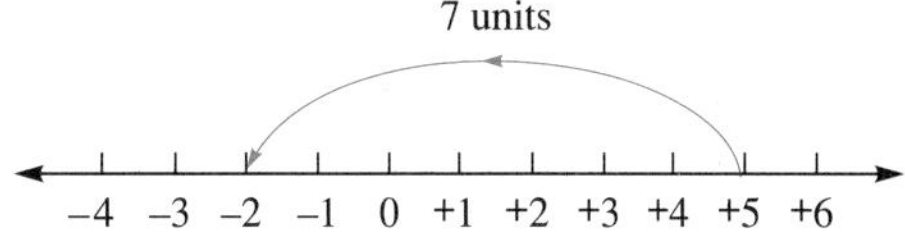

The rule for subtracting integers gives us

$$(-2) - (+5) = (-2) + (-5) = -7$$

8.5 Exercises

Subtract:

1. $6 - 8$
2. $9 - 12$
3. $4 - 9$
4. $5 - 7$
5. $16 - 7$
6. $14 - 8$
7. $-6 - 5$
8. $-8 - 4$
9. $-11 - 8$
10. $-14 - 6$
11. $-18 - 12$
12. $-16 - 9$
13. $7 - 7$
14. $5 - 5$
15. $6 - (-4)$
16. $8 - (-5)$
17. $8 - (-9)$
18. $4 - (-8)$
19. $11 - (-5)$
20. $14 - (-3)$
21. $-6 - (-2)$
22. $-8 - (-5)$
23. $-4 - (-8)$
24. $-6 - (-9)$
25. $-11 - (-12)$
26. $-14 - (-11)$
27. $-7\frac{3}{8} - 1\frac{1}{8}$
28. $-4\frac{2}{9} - 3\frac{1}{9}$
29. $-\frac{2}{7} - \left(-\frac{3}{7}\right)$
30. $\frac{2}{8} - \left(-\frac{5}{8}\right)$
31. $-8 - (-8)$
32. $-14 - (-14)$
33. $\begin{array}{r} +17 \\ -\ \underline{-16} \end{array}$
34. $\begin{array}{r} -26 \\ -\ \underline{+21} \end{array}$
35. $\begin{array}{r} +83 \\ -\ \underline{-12} \end{array}$
36. $\begin{array}{r} +226 \\ -\ \underline{-728} \end{array}$
37. $\begin{array}{r} -526 \\ -\ \underline{+218} \end{array}$
38. $\begin{array}{r} -72.85 \\ -\ \underline{+36.94} \end{array}$
39. $\begin{array}{r} +7\frac{5}{8} \\ -\ \underline{-2\frac{3}{4}} \end{array}$
40. $\begin{array}{r} -6\frac{2}{3} \\ -\ \underline{-5\frac{1}{5}} \end{array}$

For each of the following, sketch the problem on a number line and also solve by use of the rule for subtraction of integers.

41. $(-3) - (-5)$
42. $(+2) - (+5)$
43. $(-6) - (+5)$
44. $(+6) - (-6)$
45. $(+6) - (+6)$

46. Subtract -8 from 5.

47. Subtract -6 from -8.

48. What number must be subtracted from 3 to obtain 9?

49. What number must be subtracted from 5 to obtain -3?

50. What number must be subtracted from -2 to obtain -7?

51. What number must be subtracted from -5 to obtain -1?

LIVING IN THE REAL WORLD

52. At 8 P.M., Alex's temperature was 102.1° F. At midnight, it was 99.4° F. How much did his temperature drop?

53. At 3 A.M., the temperature in Billings, Montana, was $-16°$ F. By noon it had risen to 22° F. What was the rise in temperature?

54. Elise has a balance of \$38.40 in her checking account. Find her new balance if she writes a check for \$50.00.

55. Jim's checking account is overdrawn by \$31.03. What will his new balance be if he makes a deposit of \$100.00?

56. Sue Myers overdrew her checking account by \$14.12. How much money should she deposit so that her new balance will be \$75.00?

MENTAL MATHEMATICS

▼ In Exercises 57–101, do the subtractions mentally, without the aid of pencil or paper.

57. $8 - 3$ 58. $9 - 6$

59. $5 - 9$ 60. $6 - 7$

61. $13 - 5$ 62. $6 - 13$

63. $-8 - 3$ 64. $-4 - 5$

65. $-2 - 3$ 66. $6 - 6$

67. $11 - 11$ 68. $6 - (-3)$

69. $4 - (-2)$ 70. $5 - (-6)$

71. $7 - (-4)$ 72. $-3 - (-5)$

73. $-4 - (-7)$ 74. $-8 - (-5)$

75. $-8 - (-8)$ 76. $-2 - (-2)$

77. $-5 - (-13)$ 78. $0 - 6$

79. $0 - (-5)$ 80. $0 - 12$

81. $8 - 14$ 82. $9 - 11$

83. $-6 - 14$ 84. $-5 - 10$

85. $11 - 8$ 86. $15 - 7$

87. $4 - (-5)$ 88. $6 - (-7)$

89. $3 - (-9)$ 90. $-4 - (-7)$

91. $-7 - (-5)$ 92. $-3 - (-1)$

93. $6 - (-11)$ 94. $14 - 17$

95. $18 - 12$ 96. $-4 - 0$

97. $-7 - 0$ 98. $12 - (-12)$

99. $8 - (-8)$ 100. $-4 - (-4)$

101. $-9 - (-9)$

102. In your own words, state the rule for subtracting signed numbers.

8.6 Adding and Subtracting More than Two Signed Numbers

EXAMPLE 1

$4 - 6 - 5 = 4 + (-6) + (-5)$ Change all subtractions to additions.

$= -2 + (-5)$ Add from left to right.

$= -7$

EXAMPLE 2

$6 - (-3) - 7 = 6 + 3 + (-7)$ Change all subtractions to additions.

$= 9 + (-7)$ Add from left to right.

$= 2$

EXAMPLE 3 We frequently encounter problems in algebra that look like this:

$$6 + 4 - 7 + 3 - (-8) - 4 - 2$$

SOLUTION The easiest way to solve a long problem of this type is to change all subtractions to additions, add all of the positive values, add all of the negative values, and then combine these two results.

$6 + 4 - 7 + 3 - (-8) - 4 - 2$
$= 6 + 4 + (-7) + 3 + 8 + (-4) + (-2)$ Change all subtractions to additions.

$6 + 4 + 3 + 8 = 21$ Add the positive values.

$(-7) + (-4) + (-2) = -13$ Add the negative values.

$21 + (-13) = 8$ Combine the results from the preceding steps.

EXAMPLE 4 A man on a diet keeps a careful daily record of his weight change. Here is an account:

- On Monday he loses 2 pounds.
- On Tuesday he loses 4 pounds.
- On Wednesday he gains 1 pound.
- On Thursday he loses 2 pounds.
- On Friday he gains 2 pounds.
- On Saturday he loses 2 pounds.
- On Sunday he gained 1 pound.

(a) How much weight does he gain?

(b) How much weight does he lose?

(c) What is his *net* gain or loss for the week?

(d) If he weighed 230 pounds when he started, how much does he weigh at the end of the week?

SOLUTION **(a)** $1 + 2 + 1 = 4$ pounds (gained)

(b) $(-2) + (-4) + (-2) + (-2) = -10$ pounds (lost)

(c) $4 + (-10) = -6$ pounds (net loss)

(d) $230 + (-6) = 224$ pounds at the end of the week

As you go through the exercises, you may notice shortcuts that simplify your work. Go ahead and use any shortcuts you find, providing they always give you correct answers.

8.6 Exercises

Combine:

1. $-6 - 3 + 4 - 5 + 1$
2. $-6 + 3 - 8 - 4 + 7$
3. $-4 - 6 - 8 - 3 + 1 - 9$
4. $8 - 6 + 8 - 6 + 8 - 6$
5. $-3 - (-8) + 2 - 7 - 4$
6. $-2 - (-4) - (-5) - 5$
7. $-6 + 5 - 7 - (-1) - 4$
8. $-10 - 1 - (-9) - (-2)$
9. $-4 - 6 - 7 - 5 + 11 + 3$
10. $-8 - 2 - 1 - 4 + 6 + 8$
11. $3 - (-5) + 4 - 5 + 6$
12. $-3 - (-4) + 6 - 5 + 3$
13. $-1 + 1 - 5 + 5 - 4 + 4$
14. $8 - 8 + 6 - 6 + 7 - 7$
15. $13 - (-2) - 4 - 6 + 1 - (-2)$
16. $11 - 5 + 4 - 6 - (-2) - (-1)$
17. $4 - (-6) - (-2) - (-5) + 8$
18. $-2 - (-2) - (-6) + 5 - (-7)$
19. $0 - 5 + 8 - (-2) - 7$
20. $0 - (-4) + 3 - 12 - (-6)$
21. $-2.03 + 4.71 + 5.32 - (-1.1) - 2.05$

22. $3.14 - (-1.41) - 2.24 + 1.73 - 6.09$

23. $-\frac{2}{5} + \frac{3}{10} - \frac{1}{2} - \frac{1}{5} + \frac{7}{10}$

24. $-\frac{3}{8} + \frac{1}{4} - \frac{5}{8} - \frac{1}{2} + \frac{3}{4}$

LIVING IN THE REAL WORLD

25. The Big Bang Uranium Co. stock shows the following gains and losses for the week:

- Monday: $+\frac{1}{8}$
- Tuesday: $+\frac{3}{8}$
- Wednesday: $-\frac{1}{2}$
- Thursday: $-2\frac{1}{4}$
- Friday: $+1$

a. What is the net gain or loss for the week?

b. If the stock starts the week at a value of $64\frac{1}{4}$, what is its value at the end of the week?

26. Valerie starts with \$143.50 in her checking account. She makes deposits and writes checks in the following amounts:

7/5 check	\$18.75
7/5 deposit	\$62.44
7/7 check	\$128.70
7/8 check	\$12.50
7/9 deposit	\$46.60

What is Valerie's balance?

27. Julio's checking account has \$141.10 at the beginning of the month. During the month he writes the following checks: rent, \$310.00; groceries, \$62.58 and \$41.15; telephone, \$31.08; college bookstore, \$20.00 and \$12.50; utility company, \$24.50. He also makes deposits of \$350.00, \$100.00, and \$41.25. What is his balance at the end of the month?

28. You write checks for \$36.10, \$12.45, \$19.75, and \$220.00 during January. You also make a deposit of \$150.00. If your beginning balance was \$112.43, what is your balance at the end of the month?

8.7 Multiplying Signed Numbers

Recall from our earlier work with multiplication that the answer to any multiplication problem is called the **product,** and the numbers being multiplied are called **factors.**

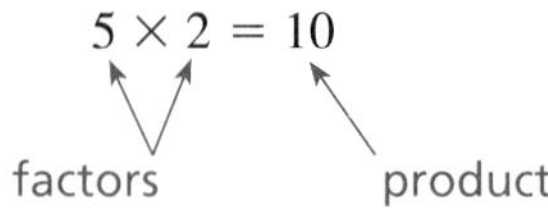

There are several different symbols used to indicate multiplication:

1. $5 \times 2 = 10$ "×" means to multiply.

2. $5 \cdot 2 = 10$ "·" means to multiply.

3. $(5)(2) = 10$ Parentheses written right next to each other mean multiply.

4. $5(2) = 10$ | **5.** $(5)2 = 10$ — Frequently only one set of parentheses is used to mean multiplication.

6. xy — This means x times y. Writing two letters next to each other indicates multiplication.

It is helpful to remember that in arithmetic, multiplication is defined as repeated addition. For example, $5 \times 2 =$ five 2s $= 2 + 2 + 2 + 2 + 2 = 10$. We can apply this same principle to multiplication of signed numbers.

EXAMPLE 1

$$5 \times (-2) = \text{five } (-2)\text{s} = (-2) + (-2) + (-2) + (-2) + (-2) = -10$$

EXAMPLE 2

$$(-3) \times 2 = 2 \times (-3) = \text{two } (-3)\text{s} = (-3) + (-3) = -6$$

Examples 1 and 2 show that when two numbers with different signs are multiplied, their product is negative.

EXAMPLE 3

$$(-1)(3) = (3)(-1) = \text{three } (-1)\text{s} = (-1) + (-1) + (-1) = -3$$

EXAMPLE 4

$$(-1)(5) = -5 = \text{the negative of } 5$$

We can see from Examples 3 and 4 that multiplying by -1 gives us the negative of that number. We will use this concept now to show that the product of two negative numbers is positive.

EXAMPLE 5

$(-2)(-5) = (-1)(2)(-5)$ — Since $-2 = (-1)(2)$

$= (-1)(-10)$ — Since $(2)(-5) = -10$

$=$ negative of -10 — Because -1 times a number gives the negative of that number

$= 10$ — Since the negative of a number is found by changing its sign

EXAMPLE 6

$$(-3)(-4) = (-1)(3)(-4) = (-1)(-12) = 12$$

Examples 5 and 6 show us that the product of two negative numbers is a positive number.

These rules for multiplying signed numbers are summarized as follows:

To multiply two signed numbers:

1. Multiply their absolute values.
2. The product is positive if their signs are alike. The product is negative if their signs are different.

EXAMPLE 7 Multiply: $(-4)(6)$.

SOLUTION

$$(-4)(6) = -24$$

The product of the absolute values is $4 \times 6 = 24$.
The sign is negative because the signs of -4 and 6 are different.

EXAMPLE 8 Multiply: $(-3)(-6)$.

SOLUTION

$$(-3)(-6) = +18$$

$3 \times 6 = 18$
The sign is positive because the signs of -3 and -6 are alike.

EXAMPLE 9 Multiply: $(-3)(5)$.

SOLUTION

$$(-3)(5) = -15$$

$3 \times 5 = 15$
The product is negative because the signs are different.

EXAMPLE 10 Multiply: $\left(-3\frac{1}{3}\right)\left(4\frac{1}{2}\right)$.

SOLUTION

$$\left(-3\tfrac{1}{3}\right)\left(4\tfrac{1}{2}\right) = \left(-\frac{10}{3}\right)\left(\frac{9}{2}\right)$$

$$= -\left(\frac{\overset{5}{\cancel{10}}}{\underset{1}{\cancel{3}}} \cdot \frac{\overset{3}{\cancel{9}}}{\underset{1}{\cancel{2}}}\right)$$

$$= -\frac{15}{1} = -15$$

EXAMPLE 11 Multiply: $(-1.6)(-3.5)$.

SOLUTION

$$(-1.6)(-3.4) = +(1.6 \times 3.4)$$

$$= 5.44$$

EXAMPLE 12

(a) $(-7)(-4) = 28$

(b) $(-3)(7) = -21$

(c) $(9)(8) = 72$

(d) $(3.6)(-0.002) = -0.0072$

(e) $\left(-5\frac{1}{4}\right)\left(-2\frac{2}{7}\right) = \left(-\frac{21}{4}\right)\left(-\frac{16}{7}\right)$

$$= +\left(\frac{\overset{3}{\cancel{21}}}{\underset{1}{\cancel{4}}} \cdot \frac{\overset{4}{\cancel{16}}}{\underset{1}{\cancel{7}}}\right) = 12$$

Multiplying More than Two Signed Numbers

Let's extend the notion that we have just learned. Consider a problem in which there are more than two signed numbers to be multiplied. One way to solve the problem is to multiply the first two signed numbers together, multiply this result by the third signed number, multiply that result by the fourth signed number, and so on. To illustrate:

$$(-2)(+1)(-3)(-2)(+4)$$

$$= (-2)(-3)(-2)(+4) \qquad \text{Multiplying } (-2)(+1) = -2$$

$$= (+6)(-2)(+4) \qquad \text{Multiplying } (-2)(-3) = +6$$

$$= (-12)(+4) \qquad \text{Multiplying } (+6)(-2) = -12$$

$$= (-48) \qquad \text{Multiplying } (-12)(+4) = -48$$

There is an easier method for solving this type of problem. Each pair of negative signed numbers, when multiplied together, yields a positive answer. Therefore, if there is an even number of negative signs in the problem, the product will be positive. If there is an odd number of negative signs, each pair of negative factors yields a positive answer, but there will be one negative factor remaining, which will make the final product negative. The numerical part of the product will always be the product of the absolute values of all the factors in the problem.

To multiply more than two signed numbers:

1. Multiply the absolute values of the factors.
2. If there is an **even** number of negative signs, the product is **positive.** If there is an **odd** number of negative signs, the product is **negative.**

EXAMPLE 13 Multiply: $(-2)(+1)(-3)(-2)(+4)$.

SOLUTION

$$(2)(1)(3)(2)(4) = 48 \qquad \text{Multiply the absolute values.}$$

Since there are three negative factors, an *odd* number, our answer is *negative.* Therefore

$$(-2)(+1)(-3)(-2)(+4) = -48$$

EXAMPLE 14 Multiply: $(-2)(-2)(-2)(-2)(-2)$.

SOLUTION

$$(2)(2)(2)(2)(2) = 32 \qquad \text{Multiply the absolute values.}$$

Five negative signs is an odd number, so our answer is negative.

$$(-2)(-2)(-2)(-2)(-2) = -32$$

EXAMPLE 15 Multiply: $(-1)(2)(3)(-1)(4)$.

SOLUTION

$$(1)(2)(3)(1)(4) = 24.$$

Two negative factors, an even number, tells us our answer is positive.

$$(-1)(2)(3)(-1)(4) = +24 = 24$$

8.7 Exercises

Multiply:

1. $(-4)(-6)$
2. $3(-8)$
3. $(-7)(4)$
4. $(-10)(-10)$
5. $-8(2)$
6. $(-9)8$
7. $(6)(11)$
8. $(-12)(-8)$
9. $(10)(-6)$
10. $(-13)(-4)$
11. $(-11)(-12)$
12. $(-14)(31)$
13. $33(-16)$
14. $(-30)5$
15. $(-14)(7)$
16. $(-7)(-20)$
17. $(-1.2)(3.4)$
18. $\left(-\frac{2}{3}\right)\left(\frac{3}{8}\right)$
19. $(4.5)(-12)$
20. $(-400)(-700)$
21. $\left(-3\frac{1}{5}\right)\left(1\frac{1}{4}\right)$
22. $\left(-4\frac{1}{2}\right)\left(-1\frac{1}{3}\right)$
23. $(-8.6)(4.5)$
24. $(17.2)(-0.046)$
25. $(-12)(+4)(-3)(-1)$
26. $(+3)(+1)(+2)(+5)$
27. $(-1)(-2)(-3)(-4)(-5)$
28. $(3)(-2)(1)(1)(1)(2)$
29. $(-1)(1)(-1)(-1)(1)$
30. $(-3)(0)(6)(-2)(-2)(-1)$
31. $(-2.1)(1.4)(2)(-3)$
32. $\left(-\frac{2}{3}\right)\left(-\frac{3}{4}\right)\left(-\frac{1}{8}\right)\left(+\frac{4}{5}\right)$
33. $(-3)(-3)(-3)(-3)$
34. $(-3)(-3)(-3)(-3)(-3)$
35. $\left(-1\frac{1}{3}\right)\left(-\frac{2}{3}\right)\left(-1\frac{1}{2}\right)\left(-\frac{3}{4}\right)$
36. $\left(-\frac{3}{4}\right)\left(-\frac{5}{8}\right)(4)\left(\frac{1}{5}\right)\left(-\frac{8}{15}\right)\left(1\frac{2}{3}\right)$
37. Find the product of -8 and -7.
38. Find the product of -9 and 4.
39. What number must be multiplied by -6 to obtain -24?

40. What number must be multiplied by −5 to obtain 35?

41. What number must be multiplied by 12 to obtain −36?

42. What number must be multiplied by −7 to obtain −63?

LIVING IN THE REAL WORLD

43. If an investor buys 800 shares of stock at $72 per share and then sells them at a loss of $14 per share, find the total amount of the loss and indicate it as a signed number.

44. Sara Barbagallo buys 450 shares of stock for $36 per share. She sells them two years later for $29 per share. How much is her loss per share? What is her total loss? Indicate your results as signed numbers.

MENTAL MATHEMATICS

▼ Multiply the following mentally, without using a pencil or paper.

45. (6)(8)
46. (9)(6)
47. (−3)(7)
48. (−5)(6)
49. (−9)(3)
50. (−2)(−6)
51. (−3)(−8)
52. (6)(−8)
53. (10)(−7)
54. (9)(−8)
55. (0)(−4)
56. (−8)(0)
57. (−6)(−6)
58. (−5)(−9)
59. (−20)(5)
60. (−30)(6)
61. (−50)(−5)
62. (−25)(6)
63. (−3)(4)(−2)
64. (−2)(−5)(−6)
65. (4)(−1)(−2)
66. (−3)(−2)(5)
67. (−2)(−3)(−5)(−1)
68. (3)(−1)(−2)(−5)
69. (−7)(0)(−3)(4)
70. (−6)(−3)(0)(−4)
71. (−1)(−1)(−1)(−1)(−5)
72. (−3)(−2)(−1)(−2)(−2)

73. When multiplying signed numbers, why does the product of an even number of negative numbers always produce a positive result? Why will the product of an odd number of negative factors always be negative?

8.8 Dividing Signed Numbers

Division can be shown in three different ways:

$15 \div 3 = 5$ (15: dividend; 3: divisor; 5: quotient)

$3\overline{)15}$ with quotient 5 (3: divisor; 15: dividend; 5: quotient)

$\frac{15}{3} = 5$ (15: dividend; 3: divisor; 5: quotient)

In all cases, the number you divide by is called the **divisor,** the number being divided is called the **dividend,** and the result is called the **quotient.**

To check a division problem, we multiply the divisor by the quotient to obtain the dividend.

$$15 \div 3 = 5 \qquad \text{because} \qquad 3 \times 5 = 15$$

Applying this principle to signed numbers gives us rules for dividing signed numbers:

Since $(3)(5) = 15$, $\quad 15 \div 3 = 5$
Since $(-3)(5) = -15$, $\quad -15 \div -3 = 5$
Since $(3)(-5) = -15$, $\quad -15 \div 3 = -5$
Since $(-3)(-5) = 15$, $\quad 15 \div -3 = -5$

These examples show that division of signed numbers follows the same rule of signs as multiplication of signed numbers. These rules are stated as follows:

To divide one signed number by another:

1. Divide their absolute values.
2. The quotient is positive if their signs are alike. The quotient is negative if their signs are different.

EXAMPLE 1 Divide: $-42 \div 6$

SOLUTION $-42 \div 6 = -7$ $\quad 42 \div 6 = 7$, and the answer is negative because the signs are different. ▮▮

EXAMPLE 2 Divide: $\frac{-72}{-9}$.

SOLUTION $\frac{-72}{-9} = 8$ $\quad \frac{72}{9} = 8$, and the quotient is positive because the signs are alike. ▮▮

EXAMPLE 3 Divide: $-4\overline{)24}$.

SOLUTION $-4\overline{)24}$ with quotient -6 $\quad 4\overline{)24}$ with quotient 6, and the quotient is negative because the signs are different. ▮▮

EXAMPLE 4 **(a)** $(-28) \div (-4) = 7$ $\quad$ Signs alike

(b) $\frac{-36}{9} = -4$ $\quad$ Signs different

(c) $\frac{49}{7} = 7$ $\quad$ Signs alike

(d) $-8\overline{)-64}$ with quotient 8 — Signs alike

(e) $18 \div (-3) = -6$ — Signs different

EXAMPLE 5 Divide: $-\frac{9}{16} \div \frac{3}{8}$.

SOLUTION

$$-\frac{9}{16} \div \frac{3}{8} = -\frac{\overset{3}{\cancel{9}}}{\underset{2}{\cancel{16}}} \times \frac{\overset{1}{\cancel{8}}}{\underset{1}{\cancel{3}}}$$ Invert the divisor and multiply.

$$= -\frac{3}{2}$$ The quotient is negative because the signs are different.

EXAMPLE 6 Divide: $2.4\overline{)-1.44}$

SOLUTION

$$\begin{array}{r} -.6 \\ 2_{\wedge}4.\overline{)-1_{\wedge}4.4} \\ 1\ 4\ 4 \\ \hline 0 \end{array}$$ The quotient is negative because the signs are different.

EXAMPLE 7 On his diet, Bruce lost a total of 21 pounds in 12 weeks. What was his average weight loss per week?

SOLUTION If we think of losing 21 pounds as -21 pounds, we have the following:

$$\frac{-21 \text{ lb}}{12 \text{ weeks}} = -1\frac{9}{12}\ \frac{\text{lb}}{\text{week}}$$ Reduce the fraction.

$$= -1\frac{3}{4}\ \frac{\text{lb}}{\text{week}}$$ Average weight loss

Division Involving Zero

Previously we showed that division by zero is not allowed. Recall that

$$\frac{0}{3} = 0$$ Because $(0)(3) = 0$

$$\frac{0}{-3} = 0$$ Because $(0)(-3) = 0$

Now consider the equation $\frac{3}{0} = ?$ What number can replace the question mark so that $? \times 0 = 3$? Since any number multiplied by zero gives zero, there is no possible replacement for the question mark. We conclude then that $\frac{3}{0}$ has no meaning.

Next consider the equation $\frac{0}{0} = ?$ What number can replace the question mark so that $? \times 0 = 0$. Since *any* number times zero is zero, we can replace the question mark by any number we like. This gives us $\frac{0}{0} = 5, -6, 14$, or any other number. We say that $\frac{0}{0}$ is **indeterminant,** or not uniquely defined. This situation is certainly not allowable, so we conclude that division by zero in all cases is not permitted, and we will say that it is **undefined.**

Division by zero is undefined.
Zero may never by used as a divisor.

EXAMPLE 8 **(a)** $\frac{-74}{0}$ is undefined.

(b) $\frac{0}{-4} = 0$ Because $(0)(-4) = 0$

(c) $\frac{0}{0}$ is undefined.

8.8 Exercises

Divide:

1. $\frac{-12}{-4}$
2. $\frac{-21}{-7}$
3. $\frac{-12}{-12}$
4. $\frac{-16}{-16}$
5. $\frac{-56}{8}$
6. $\frac{-32}{4}$
7. $\frac{63}{-7}$
8. $\frac{28}{-7}$
9. $\frac{-3}{18}$
10. $\frac{-4}{20}$
11. $\frac{0}{-8}$
12. $\frac{0}{-6}$
13. $\frac{-7}{0}$
14. $\frac{-4}{0}$
15. $-28 \div (-7)$
16. $54 \div (-9)$
17. $-16 \div (-4)$
18. $-24 \div (-8)$
19. $18 \div (-3)$
20. $28 \div (-7)$
21. $56 \div 7$
22. $81 \div 9$
23. $-36 \div 4$
24. $-27 \div 9$
25. $42 \div 0$
26. $-17 \div 0$
27. $0 \div (-5)$
28. $0 \div 14$
29. $-\frac{3}{4} \div \left(-\frac{1}{2}\right)$
30. $-\frac{3}{8} \div \left(-\frac{5}{6}\right)$
31. $-4\frac{3}{8} \div 1\frac{3}{4}$
32. $-7\frac{1}{7} \div 3\frac{1}{3}$
33. $8.4 \div (-21)$
34. $9.6 \div (-16)$
35. $41.8 \div 0$
36. $-54.6 \div 0$
37. Find the quotient of -42 and -7.
38. Find the quotient of 56 and -8.
39. Find the quotient of -18 and -6.
40. Find the quotient of -24 and 3.
41. What number must be divided by -3 to obtain -7?
42. What number must be divided by -4 to obtain 9?

43. What number must you divide -35 by to obtain 7?

44. What number must you divide 63 by to obtain -9?

45. What number must you divide by 8 to obtain -8?

LIVING IN THE REAL WORLD

46. A man lost \$24 in one day at the racetrack. If he lost the same amount in each of four races, how much did he lose per race?

47. Eight youngsters playing baseball break a neighbor's window and must share equally the expense of replacing the glass. If the repair costs \$28.80, how much will it cost each youngster?

48. An investor sells 250 shares of the Heavy Balloon Corporation stock at a total loss of \$437.50. What is his loss per share?

49. JoAnn Penton paid \$219.60 for cable TV for one year. How much did she pay per month?

MENTAL MATHEMATICS

▼ Do the following division problems mentally, without using a pencil or paper.

50. $\frac{24}{6}$

51. $\frac{63}{7}$

52. $\frac{-28}{7}$

53. $\frac{-48}{8}$

54. $\frac{64}{-8}$

55. $\frac{72}{-9}$

56. $\frac{-36}{-6}$

57. $\frac{-54}{-9}$

58. $\frac{-36}{-4}$

59. $\frac{-45}{-9}$

60. $\frac{0}{-6}$

61. $\frac{0}{-7}$

62. $\frac{-14}{0}$

63. $\frac{-18}{0}$

64. $\frac{56}{-7}$

65. $\frac{21}{-3}$

66. $\frac{-16}{-16}$

67. $\frac{0}{0}$

68. $\frac{-23}{-23}$

69. $\frac{26}{-13}$

70. $\frac{-38}{-19}$

71. $\frac{-18}{2}$

72. $\frac{-24}{-6}$

73. $\frac{-22}{11}$

74. $\frac{26}{13}$

75. $\frac{15}{-3}$

76. $\frac{0}{18}$

77. $\frac{55}{-11}$

78. $\frac{-60}{-12}$

79. $\frac{-80}{-20}$

8.9 Properties of Signed Numbers

Previously we discussed properties of whole numbers. We will now investigate these same properties and how they apply to signed numbers.

Whenever you reverse the order of two signed numbers in an addition problem, the sum does not change.

EXAMPLE 1

(a) $6 + 3 = 9$
$3 + 6 = 9$

(b) $(-4) + 5 = 1$
$5 + (-4) = 1$

(c) $(-6) + (-5) = -11$
$(-5) + (-6) = -11$

This property is true in general and is called the **commutative property of addition.**

The commutative property of addition:

If a and b represent any signed numbers, then $a + b = b + a$.

Similarly, if the order of the factors in a multiplication problem is reversed, the product remains the same.

EXAMPLE 2

(a) $6 \cdot 3 = 18$
$3 \cdot 6 = 18$

(b) $(-4)(5) = -20$
$(5)(-4) = -20$

(c) $(-6)(-5) = 30$
$(-5)(-6) = 30$

This property is called the **commutative property of multiplication** and is stated formally as follows:

The commutative property of multiplication:

If a and b represent any signed numbers, then $a \cdot b = b \cdot a$.

Note that in subtraction and division of signed numbers, reversing the order *does* make a difference. You get a different result.

EXAMPLE 3

(a) $7 - 5 = 2$
$5 - 7 = -2$ } Not the same

(b) $20 \div 4 = 5$
$4 \div 20 = \frac{4}{20} = \frac{1}{5}$ } Not the same

Therefore, neither subtraction nor division of signed numbers is commutative.

When we add three numbers, we have a choice of which two to add first. Does it make any difference? We use parentheses to indicate which two numbers we are adding first.

EXAMPLE 4 Add: $3 + 2 + 6$.

(a) $3 + 2 + 6 = (3 + 2) + 6$ Add 3 + 2 first.

$= 5 + 6$

$= 11$

(b) $3 + 2 + 6 = 3 + (2 + 6)$ Add 2 + 6 first.

$= 3 + 8$

$= 11$

Example 4 illustrates that when adding three signed numbers, it makes no difference which two we add first. This property is always true and is called the **associative property of addition.**

The associative property of addition:

If a, b, and c represent any signed numbers, then

$$(a + b) + c = a + (b + c)$$

An example suggests that a similar property holds for multiplication of three signed numbers.

EXAMPLE 5 Multiply: $-3 \cdot 4 \cdot 2$.

(a) $-3 \cdot 4 \cdot 2 = (-3 \cdot 4) \cdot 2$ Multiply inside the parentheses first.

$= -12 \cdot 2$ $-3 \cdot 4 = -12$

$= -24$

(b) $-3 \cdot 4 \cdot 2 = -3 \cdot (4 \cdot 2)$ Multiply inside the parentheses first.

$= -3 \cdot 8$ $4 \cdot 2 = 8$

$= -24$

As you can see, we obtained the same result no matter which two numbers we multiplied first.

The associative property of multiplication:

If a, b, and c represent any signed numbers, then $(a \cdot b) \cdot c = a \cdot (b \cdot c)$.

The commutative and associative properties are quite intuitive, but we state them here because we will use them frequently as we continue on in algebra. As is the case with the commutative property, neither subtraction nor division is associative. This is illustrated in the following example.

EXAMPLE 6 **(a)** $(6 - 3) - 7 = 3 - 7 = -4$
$6 - (3 - 7) = 6 - (-4) = 6 + 4 = 10$ } Results differ.

(b) $(20 \div -4) \div 2 = -5 \div 2 = -\frac{5}{2}$
$20 \div (-4 \div 2) = 20 \div (-2) = -10$ } Results differ. ▮▮

Study the next example carefully to discover still another important property of signed numbers.

EXAMPLE 7 Evaluate (a) $4(2 + 7)$; (b) $4 \cdot 2 + 4 \cdot 7$.

SOLUTION **(a)** $4(2 + 7) = 4 \cdot 9$ Add 2 + 7 inside the parentheses first.
$= 36$ Multiply.

(b) $4 \cdot 2 + 4 \cdot 7 = 8 + 28$ Multiply first.
$= 36$ Add.

Since both results are the same, we see that $4(2 + 7) = 4 \cdot 2 + 4 \cdot 7$. ▮▮

This property is true in general and is called the **distributive property of multiplication over addition.**

The distributive property of multiplication over addition:

If a, b, and c are signed numbers, then $a \cdot (b + c) = a \cdot b + a \cdot c$.

EXAMPLE 8 Verify that the distributive property holds in each case.

(a) $3(2 + 6) = 3 \cdot 2 + 3 \cdot 6$

(b) $-7(4 + 2) = (-7)(4) + (-7)(2)$

(c) $12(\frac{1}{4} + \frac{1}{6}) = 12 \cdot \frac{1}{4} + 12 \cdot \frac{1}{6}$

SOLUTION

Left Side	Right Side
(a) $3(2+6) = 3(8) = 24$	$3 \cdot 2 + 3 \cdot 6 = 6 + 18 = 24$
(b) $-7(4+2) = -7(6) = -42$	$(-7)(4) + (-7)(2) = (-28) + (-14) = -42$
(c) $12\left(\frac{1}{4}+\frac{1}{6}\right) = 12\left(\frac{3}{12}+\frac{2}{12}\right) = \overset{1}{\cancel{12}} \cdot \frac{5}{\underset{1}{\cancel{12}}} = 5$	$12 \cdot \frac{1}{4} + 12 \cdot \frac{1}{6} = \overset{3}{\cancel{12}} \cdot \frac{1}{\underset{1}{\cancel{4}}} + \overset{2}{\cancel{12}} \cdot \frac{1}{\underset{1}{\cancel{6}}} = 3 + 2 = 5$

8.9 Exercises

In Exercises 1–14, answer true or false and give a reason for your answer.

1. Changing the order of addition does not change the result.
2. Changing the order of division does not change the result.
3. Changing the grouping in a subtraction problem does not change the difference.
4. Changing the order of multiplication does change the product.
5. Changing the grouping in an addition problem has no effect on the sum.
6. $6 - 4 = 4 - 6$
7. $r(st) = (rs)t$
8. $\frac{a}{b} = \frac{b}{a}$
9. $(e \div f) \div g = e \div (f \div g)$
10. $(x + 3) + 4 = x + (3 + 4)$
11. $x - 15 = 15 - x$
12. $(7 - 5) - 2 = 7 - (5 - 2)$
13. $6 + m = m + 6$
14. $m \cdot 6 = 6m$

In Exercises 15–20, verify the distributive property.

15. $6(3 + 5) = 6 \cdot 3 + 6 \cdot 5$
16. $-4(-5 + 7) = (-4)(-5) + (-4)(7)$
17. $-8(4 + 7) = (-8)(4) + (-8)(7)$
18. $\frac{1}{2}(7 + 5) = \frac{1}{2} \cdot 7 + \frac{1}{2} \cdot 5$

19. $4.3(1.2 + 0.3) = (4.3)(1.2) + (4.3)(0.3)$

20. $6.01(-2.1 + 0.05) = (6.01)(-2.1) + (6.01)(0.05)$

Using the given property, complete each statement.

21. The commutative property of multiplication: $(6)(-7) =$

22. The commutative property of addition: $6 + 5 =$

23. The associative property of addition: $(2 + 5) + 7 =$

24. The distributive property: $6(4 + 5) =$

25. Tell why subtraction is not commutative. Use an example.

26. Explain the meaning of the commutative property of multiplication.

8.10 Powers and Roots of Signed Numbers

Powers of Signed Numbers

In a previous section we discussed powers of whole numbers. We now expand that work to include powers of signed numbers.

When the same factor appears many times, a shorthand notation is usually used. For example:

$$2^6 \quad \text{means} \quad \underbrace{2 \cdot 2 \cdot 2 \cdot 2 \cdot 2 \cdot 2}_{\text{6 times}} = 64$$

In the expression 2^6, the number 2 is called the **base,** 6 is called the **exponent,** and 2^6 is read "2 to the sixth power."

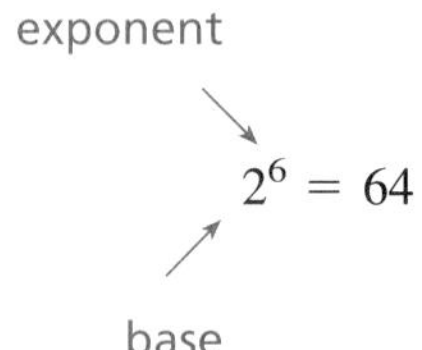

EXAMPLE 1

(a) $2^1 = 2$ 2 to the first power is 2.

(b) $2^2 = 2 \cdot 2 = 4$ 2 to the second power, or "2 squared," is 4.

(c) $2^3 = 2 \cdot 2 \cdot 2 = 8$ 2 to the third power, or "2 cubed," is 8.

(d) $2^4 = 2 \cdot 2 \cdot 2 \cdot 2 = 16$ 2 to the fourth power is 16.

EXAMPLE 2 Evaluate each of the following:

(a) $3^2 = 3 \cdot 3 = 9$

(b) $(-3)^2 = (-3)(-3) = 9$

(c) $(-3)^3 = (-3)(-3)(-3) = -27$

(d) $5^1 = 5$

(e) $1^5 = 1 \cdot 1 \cdot 1 \cdot 1 \cdot 1 = 1$

(f) $(-1)^5 = (-1)(-1)(-1)(-1)(-1) = -1$

(g) $\left(\frac{1}{2}\right)^3 = \left(\frac{1}{2}\right)\left(\frac{1}{2}\right)\left(\frac{1}{2}\right) = \frac{1}{8}$

(h) $(10)^3 = (10)(10)(10) = 1000$

(i) $(0.1)^3 = (0.1)(0.1)(0.1) = 0.001$

(j) $\left(-\frac{3}{4}\right)^2 = \left(-\frac{3}{4}\right)\left(-\frac{3}{4}\right) = \frac{9}{16}$

(k) $\left(-\frac{4}{5}\right)^3 = \left(-\frac{4}{5}\right)\left(-\frac{4}{5}\right)\left(-\frac{4}{5}\right) = -\frac{64}{125}$

Roots of Signed Numbers

Subtraction is called the *inverse operation* of addition, and division is considered the inverse operation of multiplication. There is also an inverse operation of raising a number to a power: this is called *finding the root of a number.*

The inverse of squaring is called **finding the square root.** For example, to find the square root of 9, which is written $\sqrt{9}$, we must find a number that, when squared, gives us 9.

$$\text{Since } 3^2 = 9 \qquad \text{we have} \qquad \sqrt{9} = 3$$

In a similar fashion we have

$$\sqrt{16} = 4 \qquad \text{because} \qquad 4^2 = 16$$
$$\sqrt{4} = 2 \qquad \text{because} \qquad 2^2 = 4$$
$$\sqrt{49} = 7 \qquad \text{because} \qquad 7^2 = 49$$

and so on

EXAMPLE 3 Find the following square roots:

(a) $\sqrt{36} = 6$ because $6^2 = 36$

(b) $\sqrt{1} = 1$ because $1^2 = 1$

(c) $\sqrt{0} = 0$ because $0^2 = 0$

(d) $\sqrt{121} = 11$ because $11^2 = 121$

It may have occurred to you that not only does $3^2 = 9$ but also $(-3)^2 = 9$. However, whenever we use the symbol $\sqrt{\ }$ it is understood to mean the *positive* square root, which is called the **principal square root.**

$\sqrt{9} = 3$ 3 is the principal square root of 9

$-\sqrt{9} = -3$ because $\sqrt{9} = 3$, whereas $-\sqrt{9} = -3$.

We may be asked to find roots other than square roots. For example, $\sqrt[3]{}$, $\sqrt[4]{}$, $\sqrt[5]{}$ mean the third root (or cube root), fourth root, fifth root, and so on.

The symbol $\sqrt{}$ means positive square root only.

EXAMPLE 4 Find the following roots:

(a) $\sqrt[3]{8} = 2$ Read "the cube root of 8 is 2" (because $2^3 = 8$).

(b) $\sqrt[4]{81} = 3$ Read "the fourth root of 81 is 3" (because $3^4 = 81$).

(c) $\sqrt[5]{32} = 2$ Read "the fifth root of 32 is 2" (because $2^5 = 32$).

(d) $\sqrt[3]{-8} = -2$ Read "the cube root of −8 is −2" (because $(-2) = -8$).

(e) $\sqrt[5]{-1} = -1$

(f) $-\sqrt[3]{8} = -2$

(g) $-\sqrt[3]{-8} = -(-2) = 2$ ❙❙

8.10 Exercises

Evaluate the following:

1. 2^4
2. 3^2
3. $(-2)^2$
4. $(-2)^3$
5. 3^4
6. 5^4
7. $(-2)^6$
8. $(-3)^4$
9. 2^8
10. 10^2
11. 10^3
12. 10^4
13. 1^4
14. $(-1)^4$
15. $(-10)^3$
16. $(\frac{3}{4})^2$
17. $(\frac{1}{2})^2$
18. $(0.4)^2$
19. $(0.03)^2$
20. $(-0.1)^5$
21. $(-\frac{2}{3})^3$
22. $(-\frac{3}{4})^3$
23. $(-1)^{50}$
24. $(-1)^{51}$

25. From Exercises 10, 11, and 12, what general rule can be established for powers of 10?

Find the roots:

26. $\sqrt{25}$
27. $\sqrt{81}$
28. $-\sqrt{36}$
29. $\sqrt{144}$
30. $\sqrt{49}$
31. $\sqrt[3]{27}$
32. $\sqrt[3]{-27}$
33. $-\sqrt[3]{27}$
34. $-\sqrt[3]{-27}$
35. $\sqrt[3]{125}$
36. $\sqrt[6]{1}$
37. $\sqrt[7]{-1}$
38. $\sqrt{400}$
39. $\sqrt[3]{64}$

40. $\sqrt[3]{216}$

41. $\sqrt[3]{-1000}$

42. $\sqrt[6]{64}$

43. $\sqrt[3]{343}$

44. Compare $\frac{1}{2}$ with $\left(\frac{1}{2}\right)^3$. Which is greater?

45. Compare $\frac{1}{3}$ with $\left(\frac{1}{3}\right)^2$. Which is greater?

46. Compare $\frac{1}{3}$ with $\left(\frac{1}{3}\right)^4$. Which is greater?

CRITICAL THINKING

47. If you raise a negative number to an even power, what sign will the answer have?

48. If you raise a negative number to an odd power, what sign will the answer have?

49. If you raise a fraction to a positive whole-number power, which is larger, the result or the original fraction?

50. What two numbers can be raised to the fourth power to obtain 16?

MENTAL MATHEMATICS

▼ Evaluate the following powers and roots mentally, without the aid of pencil or paper.

51. 5^2

52. 7^2

53. 9^2

54. 6^2

55. $(-2)^2$

56. $(-5)^2$

57. 2^3

58. 5^3

59. $(-2)^3$

60. $(-3)^3$

61. 10^3

62. $(-10)^3$

63. $\sqrt{36}$

64. $\sqrt{64}$

65. $\sqrt{81}$

66. $\sqrt{100}$

67. $\sqrt{16}$

68. $\sqrt{49}$

69. $-\sqrt{25}$

70. $-\sqrt{1}$

71. $-\sqrt{9}$

72. $\sqrt{0}$

73. $\sqrt[3]{27}$

74. $\sqrt[3]{-27}$

75. $\left(\frac{1}{2}\right)^2$

76. $\left(\frac{2}{3}\right)^2$

77. $\left(\frac{1}{8}\right)^2$

78. $\left(\frac{3}{4}\right)^2$

79. $\left(\frac{1}{2}\right)^3$

80. $\left(\frac{2}{3}\right)^3$

8.11 Order of Operations

If we are given an expression containing more than one operation, we must be careful to perform the operations in the proper order. The following convention is used:

Order of operations:

1. Evaluate the expression inside any parentheses ( ), brackets [ ], or braces { }.
2. Powers and roots are performed.
3. Multiplications and divisions are done as they appear, from left to right.
4. Additions and subtractions are done as they appear, from left to right.

EXAMPLE 1 Evaluate: $27 \div (12 - 3) \cdot 2 + \sqrt{36}$.

SOLUTION

$27 \div (12 - 3) \cdot 2 + \sqrt{36}$	Evaluate inside the parentheses.
$= 27 \div 9 \cdot 2 + \sqrt{36}$	Take the square root.
$= 27 \div 9 \cdot 2 + 6$	Divide and multiply from
$= 3 \cdot 2 + 6$	left to right.
$= 6 + 6$	Add.
$= 12$	

EXAMPLE 2 Evaluate: $\sqrt{49} - 6 + 3^2$.

SOLUTION

$\sqrt{49} - 6 + 3^2$	Do the root and the power.
$= 7 - 6 + 9$	Subtract and add
$= 1 + 9$	from left to right.
$= 10$	

EXAMPLE 3 Evaluate: $\sqrt{16} - 4(\sqrt{25} - 3) \div 4$.

SOLUTION

$\sqrt{16} - 4\left(\sqrt{25} - 3\right) \div 4$	Evaluate the expression inside
$= \sqrt{16} - 4(5 - 3) \div 4$	the parentheses.
$= \sqrt{16} - 4(2) \div 4$	Take the square root.
$= 4 - 4(2) \div 4$	Do multiplication and
$= 4 - 8 \div 4$	division from left to right.
$= 4 - 2$	Subtract.
$= 2$	

EXAMPLE 4 Evaluate: $3 \cdot 3^3 - 4\sqrt{64} + 3 \cdot 5$.

SOLUTION

$3 \cdot 3^3 - 4\sqrt{64} + 3 \cdot 5$	Do the powers and roots.
$= \underbrace{3 \cdot 27} - \underbrace{4 \cdot 8} + \underbrace{3 \cdot 5}$	Do the multiplications.
$= \quad 81 \quad - \quad 32 \quad + \quad 15$	Subtract.
$= 49 + 15$	Add.
$= 64$	

EXAMPLE 5 Evaluate: $9 - (3 \cdot 5^2 + 1) - \sqrt[3]{8}$.

SOLUTION

$9 - (3 \cdot 5^2 + 1) - \sqrt[3]{8}$	Square the 5 inside the parentheses.
$= 9 - (3 \cdot 25 + 1) - \sqrt[3]{8}$	Multiply inside the parentheses.
$= 9 - (75 + 1) - \sqrt[3]{8}$	Add inside the parentheses.
$= 9 - 76 - \sqrt[3]{8}$	Take the cube root.
$= 9 - 76 - 2$	Subtract from left to right.
$= -67 - 2$	
$= -69$	

8.11 Exercises

Evaluate using the correct order of operations.

1. $6 \div 2 + 4$
2. $9 \div 3 + 6$
3. $6 \div (2 + 4)$
4. $9 \div (3 + 6)$
5. $7 + 3 \cdot 4$
6. $4 - 3^2 + 6 \cdot 5$
7. $\sqrt{36} - 4^2 + 5 - 2$
8. $4 - \left(\sqrt[3]{8} + 1\right) \cdot 3$
9. $3\sqrt{9} - 5 + 2^4 \div 8$
10. $3 + 44 \div 11 - 6$
11. $(5^2)\sqrt{4} \div 10$
12. $(6 - 14) - \left(\sqrt{36} - 5\right)$
13. $3^3 - 14 \div 2 + 3 \cdot 6$
14. $(3^3 - 14) \div 2 + 3 \cdot 6$

15. $(-2)^3 + \sqrt[3]{-8} \div 2$

16. $-\sqrt{81} \div 9 + 4^2 - (2 + 3^2)$

17. $-\sqrt{16} \cdot 4 - 8 \cdot 0$

18. $(3 \cdot 4^2 - 8) \div (-10)$

19. $(2\sqrt{121} + 3) \cdot 4$

20. $-\sqrt{196} \cdot 0 - 1$

21. $6 - 3^4 + (\sqrt{49} + 9) \div 4$

22. $\sqrt[3]{125} - 2(4^2 - 6 \div 3)$

8.12 Grouping Symbols

In the preceding section we introduced these grouping symbols:

( )	parentheses
[ ]	brackets
{ }	braces

These symbols provide a means for changing the normal order of operations. For example, in the expression $6 - 4 \cdot 5$, using the correct order of operations, we multiply first and then subtract:

$$6 - 4 \cdot 5 = 6 - 20 = -14$$

If it were our intention to subtract first and then to multiply, we would use parentheses to indicate this:

$$(6 - 4) \cdot 5 = 2 \cdot 5 = 10$$

Grouping symbols can also be used in more complicated expressions. When grouping symbols occur within other grouping symbols, the procedure is to evaluate the innermost grouping first. The key phrase to remember is "inside out."

EXAMPLE 1 Evaluate: $14 - [7 - (9 - 4)]$.

SOLUTION

$14 - [7 - (9 - 4)]$	Evaluate the inner parentheses.
$= 14 - [7 - 5]$	Evaluate the brackets.
$= 14 - 2$	Subtract.
$= 12$	

EXAMPLE 2 Evaluate: $7 + 2\{3 - [11 - (3 + 4)]\}$.

SOLUTION

$7 + 2\,\{3 - [11 - (3 + 4)]\}$	Evaluate (3 + 4).
$= 7 + 2\,\{3 - [11 - 7]\}$	Evaluate [11 − 7].
$= 7 + 2\,\{3 - 4\}$	Evaluate {3 − 4}.
$= 7 + 2(-1)$	Multiply.
$= 7 - 2$	Subtract.
$= 5$	

EXAMPLE 3 Evaluate: $6 - [3 - (4 - 8)]$.

SOLUTION

$6 - [3 - (4 - 8)]$	Evaluate (4 − 8).
$= 6 - [3 - (-4)]$	Replace −(−4) by +4
$= 6 - [3 + 4]$	Evaluate [3 + 4].
$= 6 - 7$	Subtract.
$= -1$	

EXAMPLE 4 Compare (a) $3 - 6 \cdot 2$ and (b) $(3 - 6) \cdot 2$.

SOLUTION

(a) $3 - 6 \cdot 2 = 3 - 12$ Multiply first.

$= -9$

(b) $(3 - 6) \cdot 2 = -3 \cdot 2$ Evaluate the parentheses first.

$= -6$

As you can see, the results are quite different.

8.12 Exercises

▼ Simplify:

1. $6(2 - 7)$
2. $-4(3 - 12)$
3. $(7 - 5) + 11$
4. $7 - (5 + 11)$
5. $-4(2 - 8) + 5$
6. $-8(7 - 8) - 4$
7. $19 - [7 + (4 - 7)]$
8. $6 + [2 - (5 - 8)]$
9. $[6 + (2 - 19)] - 12$
10. $3 + 5[4 - (6 - 2)]$
11. $3[4 - 2(3 - 8)]$
12. $7 - \{4 + [2 + (3 - 4)]\}$

13. $\{3 - [6 + (8 - 9)]\} - 5$

14. $2 + [5 - (\sqrt{16} + 3)]$

15. $23 - \{6 - [4 - 3(7 - 5)]\}$

16. $7 - [4(3)^2 - (7 + 4)]$

17. $32 - 3\{4 - 7[6 - (5 - 8)]\}$

18. $17 + 4\{5 - 6[4 - 3(7 - 5)]\}$

CRITICAL THINKING

Compare the results in parts (a) and (b) in each of Exercises 19–34.

19. **a.** $9 - 6 \cdot 2$ **b.** $(9 - 6) \cdot 2$

20. **a.** $3 - 4 \cdot 8$ **b.** $(3 - 4) \cdot 8$

21. **a.** $8 - 5 + 9$ **b.** $8 - (5 + 9)$

22. **a.** $3 - 11 + 2$ **b.** $3 - (11 + 2)$

23. **a.** $3 + 4^2$ **b.** $(3 + 4)^2$

24. **a.** $7 - 3^2$ **b.** $(7 - 3)^2$

25. **a.** $24 \div 4 + 2$ **b.** $24 \div (4 + 2)$

26. **a.** $18 - 6 \div 2$ **b.** $18 - (6 \div 2)$

27. **a.** $7 - 3 \cdot 4$ **b.** $7 - (3 \cdot 4)$

28. **a.** $-3 \cdot 4 + 10 \div 2$ **b.** $-3(4 + 10 \div 2)$

29. **a.** -8^2 **b.** $(-8)^2$

30. **a.** -6^2 **b.** $(-6)^2$

31. **a.** $4^2 + 5^2$ **b.** $(4 + 5)^2$

32. **a.** $7^2 - 4^2$ **b.** $(7 - 4)^2$

33. **a.** $2^2 \cdot 4^2$ **b.** $(2 \cdot 4)^2$

34. **a.** $6^2 \div 3^2$ **b.** $(6 \div 3)^2$

SUMMARY—CHAPTER 8

The numbers in brackets refer to the section in the text in which each concept is presented.

[8.1] **Integers** or **signed numbers** are the positive and negative counting numbers.

EXAMPLES

[8.1]

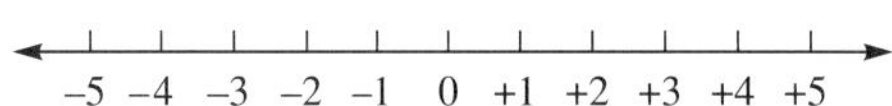

[8.2] The **absolute value** of a signed number a is the number of units that it is from zero on the number line. It is written $|a|$.

[8.2] $|-6| = 6$
$|+4| = 4$

[8.3] The symbol $>$ is read "is greater than." The symbol $<$ is read "is less than."

[8.3] $6 > 4$
$-6 < 5$

[8.4] **To add two signed numbers:**

1. If both numbers have the same sign, add the absolute values and keep the common sign for the answer.
2. If the numbers have different signs, find the difference in their absolute values and take the sign of the number with the larger absolute value.

[8.4] $6 + 4 = 10$
$-6 + (-4) = -10$
$-6 + 4 = -2$
$6 + (-4) = 2$

[8.5] **To subtract one signed number from another,** add its negative

[8.5] $5 - 9 = 5 + (-9) = -4$
$-4 - (-6) = -4 + 6 = 2$

[8.6] **To add or subtract more than two signed numbers:**

1. Change all subtractions to additions by adding the negatives.
2. Add all the positive values.
3. Add all the negative values.
4. Add the results from Steps 2 and 3.

[8.6] $-4 + 5 - (-7) - 3 + 4 - 1$
$= -4 + 5 + 7 + (-3) + 4 + (-1)$

Add the positive values:
$5 + 7 + 4 = 16$

Add the negative values:
$-4 + (-3) + (-1) = -8$

Combine the two preceding results:
$16 + (-8) = 8$

[8.7] **To multiply two signed numbers:**

1. Multiply their absolute values.
2. The product is positive if their signs are alike. The product is negative if their signs are different.

[8.7] $(-7)(-4) = 28$
$(7)(4) = 28$
$(-7)(4) = -28$
$(7)(-4) = -28$

[8.7] **To multiply more than two signed numbers:**

1. Multiply the absolute values of the factors.
2. If there are an even number of negative signs, the product is positive. If there are an odd number of negative signs, the product is negative.

[8.7] $(-5)(-1)(3)(1)(2)(-2) = -60$
3 negative signs is an odd number.

$(-2)(-3)(2)(-1)(2)(-3) = 72$
4 negative signs is an even number.

EXAMPLES

[8.8] **To divide one signed number by another:**

1. Divide their absolute values.
2. The quotient is positive if their signs are alike. The quotient is negative if their signs are different.

[8.8] $\frac{-36}{-9} = 4$

$\frac{14}{-7} = -2$

$\frac{-18}{6} = -3$

$\frac{0}{-6} = 0$

[8.8] Division by zero is undefined.

[8.8] $\frac{-3}{0}$ is undefined.

$\frac{0}{0}$ is undefined

[8.9] Properties of signed numbers:
The commutative property of addition: If *a* and *b* represent any signed numbers, then $a + b = b + a$.

[8.9] $-4 + 7 = 7 + (-4) = 3$

The commutative property of multiplication: If *a* and *b* represent any signed numbers, then $a \cdot b = b \cdot a$.

$(-4)(6) = (6)(-4) = -24$

The associative property of addition: If *a, b,* and *c* are any signed numbers, then $(a + b) + c = a + (b + c)$.

$(-6 + 3) + 2 = -6 + (3 + 2)$
$-3 + 2 = -6 + 5$
$-1 = -1$

The associative property of multiplication: If *a, b,* and *c* are any signed numbers, then $(a \cdot b) \cdot c = a \cdot (b \cdot c)$.

$(-6 \cdot 3) \cdot 2 = -6 \cdot (3 \cdot 2)$
$-18 \cdot 2 = -6 \cdot 6$
$-36 = -36$

The distributive property of multiplication over addition: If *a, b,* and *c* are any signed numbers, then $a \cdot (b + c) = a \cdot b + a \cdot c$.

$-2(3 + 5) = (-2)(3) + (-2)(5)$
$-2 \cdot 8 = -6 + (-10)$
$-16 = -16$

[8.10] **Powers of numbers** indicate repeated multiplication.

[8.10] $\underbrace{a \cdot a \cdot a \cdot a \cdot a}_{\text{5 factors}} = a^5$

exponent or power → 5; base → a

[8.10] Finding the root of a number is the inverse operation of finding the power.
The symbol $\sqrt{\ }$ means the positive square root, called the **principle square root.**

[8.10] $\sqrt{16} = 4$
$-\sqrt{16} = -4$
$\sqrt[3]{64} = 4$

[8.11] **Order of operations:**

1. Evaluate the expression inside any parentheses, ( ), brackets, [ ], or braces, { }.
2. Powers and roots are performed.
3. Multiplications and divisions are done as they appear, from left to right.
4. Additions and subtractions are done as they appear, from left to right.

[8.11] $\sqrt{9} + 2(6 + 4) \div (-5)$
$= \sqrt{9} + 2 \cdot 10 \div (-5)$
$= 3 + 2 \cdot 10 \div (-5)$
$= 3 + 20 \div (-5)$
$= 3 + (-4)$
$= -1$

[8.12] When grouping symbols occur within other grouping symbols, evaluate the innermost grouping first. Remember: "inside out."

EXAMPLES

[8.12]
$$\begin{aligned} &13 - [4 - 2(3 - 6)] \\ &= 13 - [4 - 2(-3)] \\ &= 13 - [4 + 6] \\ &= 13 - 10 \\ &= 3 \end{aligned}$$

REVIEW EXERCISES—CHAPTER 8

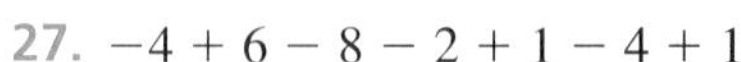

The numbers in brackets refer to the section in the text in which similar problems are presented.

[8.1] Locate each signed number on the number line.

1. -4 2. $+3$ 3. -1

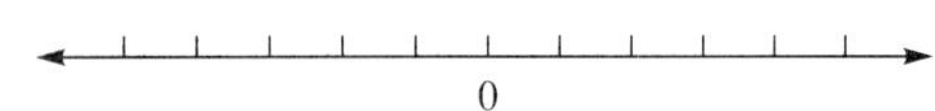

[8.1] Represent each quantity by a signed number.

4. The temperature is six degrees below zero.

5. The stock market goes up eight points.

6. You get a raise of \$30 per week.

[8.2] Find the absolute value of each of the following signed numbers.

7. -4 8. $+5$

9. -12 10. 0

[8.3] Insert > or < between each pair of numbers to make a true statement.

11. $-3 \quad 5$ 12. $14 \quad -14$

13. $3 \quad -6$ 14. $-2 \quad -12$

[8.4—8.8] Calculate as indicated.

15. $4 - 8$ 16. $(-6)(-3)$

17. $(-8) \div (-4)$ 18. $3 - (-3)$

19. $-6 - (-5)$ 20. $-6 - 8$

21. $-21 \div 7$ 22. $(4.6)(-1.12)$

23. $0 - (-9)$ 24. $12.8 - 15.6$

25. $-8 \div 0$ 26. $-3\frac{5}{8} + 2\frac{3}{8}$

27. $-4 + 6 - 8 - 2 + 1 - 4 + 1$

28. $7 - (-5) + 2 - (-8) + 1 - 2 - 4$

29. $3.81 - (-1.19) - 2.08 - 4.02 - 0.07 + 1.11$

30. $(-1)(4)(-1)(-1)(2)(1)(-2)(-2)$

31. $(-2)(-3)(1)(-1)(-2)(-2)(1)(-3)$

[8.9]

32. Does changing the grouping in a subtraction problem change the result?

[8.9]

33. Does $(a \div b) \div c = a \div (b \div c)$?

[8.9]

34. Does changing the order in a multiplication problem affect the result?

[8.10] Evaluate the following powers and roots.

35. 3^3 36. 5^2

37. $(-3)^4$ 38. 10^3

39. 10^6 40. $(-1)^{30}$

41. $\sqrt{36}$ 42. $\sqrt{49}$

43. $\sqrt[3]{8}$ 44. $-\sqrt{100}$

45. $\sqrt[3]{1000}$ 46. $\sqrt[3]{-1000}$

[8.11] Evaluate, using the correct order of operations:

47. $4 - 3 \cdot 5$

48. $6 - 27 \div 3 \cdot 3 + 1$

49. $-\sqrt{25} \cdot 3 + 6 + 3$

50. $3 - (\sqrt[3]{8} \div 2) - 2 \cdot 1$

[8.12] Evaluate each of the following:

51. $13 - [5 - (6 - 10)]$

52. $11 + 2(-6 - 1)$

53. $[4 - (2 + 3)] - 8$

54. $12 - \{4 - [3 + 2(1 - 4)]\}$

[8.6]

55. Bill Fehn has $31.44 in his checking account. He makes a deposit of $148.55 and then writes checks for $12.85, $19.52, and $103.75. What is his new balance?

[8.8]

56. Kim Zarro pays $1,246.20 per year for car insurance. How much is this per month?

CRITICAL THINKING EXERCISES—CHAPTER 8

Say whether each of the following is true or false and tell why.

1. $-6 > -15$

2. $-14 > 12$

3. $15 \div 0 = 15$

4. $(24 \div 6) \div 2 = 24 \div (6 \div 2)$

5. $-\sqrt[3]{-8} = 2$

There is an error in each of the following problems. Can you find it?

6. $(-6)(-3) = -9$

7. $-6 - 3 = 18$

8. $10 - (-10) = 0$

9. $0 \div (-4)$ is undefined.

10. $-6 - (-3) = (-6)(3) = -18$

11. $\dfrac{-8}{-8} = 0$

12. $-4 \div -12 = -3$

13. $18 \div 9 = 9 \div 18$

14. $-15 > -6$

15. $|-6| + |+5| = |-1|$

16. $5 + 2 \cdot 7 = 49$

17. $18 \div 4 + 5 = 2$

18. $-8^2 = 64$

19. $\sqrt{-49} = -7$

20. $\sqrt{16} = -4$ since $(-4)^2 = 16$

21. $(2 + 5)^2 = 2^2 + 5^2 = 4 + 25 = 29$

22. $\frac{0}{0} = 1$ because any number divided by itself is 1.

NAME ▮▮▮▮▮ CLASS

ACHIEVEMENT TEST—CHAPTER 8

▼ Draw a number line and locate the following points.

1. -4 2. 0 3. $+5$

▼ Represent each quantity by a signed number.

4. You overdraw your checking account by $24. 4. ______

5. You receive a bonus of $50. 5. ______

▼ Find the absolute value of each of the following numbers.

6. 0 6. ______

7. -8 7. ______

8. $+13$ 8. ______

▼ Insert $>$ or $<$ between each pair of numbers to make a true statement.

9. $-8 \quad -4$ 9. ______

10. $6 \quad -5$ 10. ______

11. $-4 \quad 0$ 11. ______

▼ Evaluate each of the following:

12. $4 - 5$ 12. ______

13. $-4 - (-5)$ 13. ______

14. $-4 + 5$ 14. ______

15. $(-4)(-5)$ 15. ______

16. $(-16) \div (-4)$ 16. ______

17. $-7 \div 0$ 17. ______

18. $0 \div (-7)$ 18. ______

19. $-3 + 4 - 8 - 2 + 4 + 1 - 5$ 19. ______

20. $-2 + 1 - (-3) - 5 - 2 - (-1)$ 20. ______

21. $(-2)(-1)(2)(-1)(-3)(-2)(2)$ 21. ______

22. $4.19 - 7.35$ 22. ______

23. $-2\frac{1}{5} + 7\frac{3}{5}$ 23. ______

24. 4^3 24. ______

25. $(-2)^4$ 25. ______

26. $-\sqrt{64}$ 26. ______

27. 10^5 27. ______

28. $\sqrt[3]{27}$ 28. ______

Evaluate using the correct order of operations:

29. $3 + 4 \cdot 7$ 29. ______

30. $3 - 2^2 + 5 \cdot 2$ 30. ______

31. $2 - 3[3 - (2 + 3)]$ 31. ______

32. $4 - \{2[4 - (3 - 5)]\}$ 32. ______

33. Frank Baca bought ten shares of I.B.M. stock at a price of $94.00 per share. He sold it eight months later for $106.25 per share. How much was his profit? 33. ______

34. Jane Hansen paid for her furniture in six equal monthly installments. If her furniture cost $1008.00, how much was each payment? 34. ______

35. Christy's checking account is overdrawn by $21.24, so the bank charges her an additional fee of $10.00. What will her new balance be if she makes a deposit of $60.00? 35. ______

CHAPTER 9

Algebra

INTRODUCTION

People often see little reason for algebra in their everyday lives. They see only letters and numbers being moved around according to some rules and procedures, but this process has little value to most.

In this chapter we will begin to see some genuine real-world applications of algebra to business, science, personal finance, sports, and other areas. When you finish Chapter 9, hopefully you will have a better idea of how algebra touches on everyone's lives and how we can best use it to our advantage.

CHAPTER 9—NUMBER KNOWLEDGE

Ann Marie Morgan paid $3600 for a used Ford Escort. She paid $600 down and financed $3000 for 4 years. The interest on the loan amounted to $820. After she had paid on the loan for two years she decided to pay it off using her income tax refund, thinking that she could save half of the finance charge, or $410. The bank gave her back $209.18, only about half of the $410 she expected. How come?

When she asked for an explanation, she was told that most lending institutions refund interest on loans paid off early by using a rather complicated-looking formula called the "rule of 78." The formula is

$$u = \frac{f \cdot k(k + 1)}{n(n + 1)}$$

As it turns out, she might have been better off putting her money into a savings account than paying off her loan early.

In this chapter we will learn how to use this formula as well as several others, to find everything from areas to bowling handicaps.

9.1 Formulas and the Substitution Principle

In almost every field of study, relationships between quantities are expressed using formulas. A formula is a brief, easy-to-use rule for finding the value of an unknown quantity when the values of other quantities are known. Formulas are convenient, and they are easier to remember than the English-language statement of the rule, as illustrated by the following examples.

EXAMPLE 1 The area of a triangle is equal to one-half the product of its base and its height. This rather complicated statement translates into a much shorter algebraic formula:

$$A = \frac{1}{2}bh$$

The letters in the formula are usually chosen as the first letter in the word they represent. Here A = area, b = base, and h = height.

EXAMPLE 2 The area of a circle is equal to π (pi) times the square of the radius of the circle, where π is a constant approximately equal to 3.14. The formula that represents this statement is

$$A = \pi r^2$$

where A = area, $\pi \approx 3.14$, and r = radius.

EXAMPLE 3 The amount of interest on a loan is given by the product of the principal, the yearly rate of interest, and the time in years. The corresponding formula is

$$I = prt$$

where I = interest, p = principle, r = rate, and t = time.

In algebraic expressions, both **variables** and **constants** are used. Variables are quantities that can assume different values, while constants are quantities that do not change.

In the formula for the area of a circle,

$$A = \pi r^2$$

both A and r are variables because their values change from one problem to the next. However, π is a constant equal to approximately 3.14, and it never changes.

Evaluating Formulas

Formulas are evaluated by substituting numbers for the letters.

To evaluate a formula:

1. Write the formula from the given English statement (if necessary).
2. Replace each letter by its numerical value.
3. Do all arithmetic operations, using the correct order of operations.

EXAMPLE 4 The area (A) of a triangle is given by the formula

$$A = \frac{1}{2}bh$$

where b = base and h = height. Find A if $b = 16$ and $h = 6$.

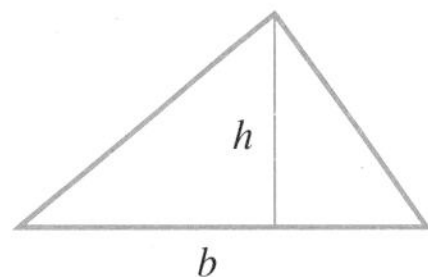

SOLUTION

$A = \frac{1}{2}bh$ — $b = 16$, $h = 6$

$= \frac{1}{2}(16)(6)$ — Substitute 16 for b and 6 for h.

$= \left(\frac{1}{\cancel{2}_1}\right)(\cancel{16}^8)(6)$ — Multiply.

$= 48$ square units — (Area is always given in square units.)

EXAMPLE 5 Celsius (C) and Fahrenheit (F) temperatures are related by the formula

$$C = \frac{5}{9}(F - 32)$$

If the boiling point of water is 212° Fahrenheit, what is it in Celsius?

SOLUTION

$$C = \frac{5}{9}(F - 32)$$

$$= \frac{5}{9}(212 - 32)$$ Substitute 212 for F.

$$= \frac{5}{9}(180)$$ Work in parentheses first.

$$= \frac{5}{\underset{1}{9}}(\overset{20}{\cancel{180}})$$ Cancel and multiply.

$$= 100$$

Therefore 212° F = 100° C. Be sure to use correct units. ▮▮

EXAMPLE 6 The total resistance R of two resistances a and b connected in parallel is given by the product of a and b divided by their sum. Find the total resistance R if $a = 100$ ohms and $b = 400$ ohms. (An ohm is a unit of electrical resistance.)

SOLUTION

$$R = \frac{a \cdot b}{a + b}$$ Write the formula from the English statement.

$$= \frac{100 \cdot 400}{100 + 400}$$ Substitute $a = 100$ and $b = 400$.

$$= \frac{40{,}000}{500}$$ Calculate the product and the sum.

$$= 80 \text{ ohms}$$ Divide. ▮▮

EXAMPLE 7 A bowler's handicap is calculated using the formula $H = 0.8(200 - A)$, where H is the handicap and A is the bowler's three-game average.

(a) What is your handicap if your three-game average for the night is 135?

(b) Your final score is equal to your actual score added to your handicap. Find your final score if you bowl a game of 142.

SOLUTION **(a)** $H = 0.8(200 - A)$

$$= 0.8(200 - 135)$$ Substitute 135 for A.

$$= 0.8(65)$$ Subtract inside the parentheses.

$$= 52$$ Multiply

(b) Final score = actual score + handicap

$= 142 + 52$ Substitute.

$= 194$ Add.

EXAMPLE 8 You may have heard a weather reporter refer to a number called the *THI* (the temperature–humidity index). The *THI* tells you how comfortable or uncomfortable you will be under any specific combination of temperature and humidity. The formula is given as

$$THI = t - (0.55)(1 - h)(t - 58)$$

where t = temperature (in degrees F) and h = relative humidity (expressed as a decimal). Meteorologists tell us the following:

At a *THI* of 70 or below, most people are comfortable.

At a *THI* of 75, half are comfortable and half are uncomfortable.

At a *THI* of 79 or above, almost everyone will be uncomfortable

Find the *THI* when the temperature is 85° F and the relative humidity is 0.7

SOLUTION

$THI = t - (0.55)(1 - h)(t - 58)$

$= 85 - (0.55)(1 - 0.7)(85 - 58)$ Substitute 85 for t and 0.7 for h.

$= 85 - (0.55)(0.3)(27)$ Work inside the parentheses.

$= 85 - 4.455$ Multiply.

$= 80.545$ Subtract.

Since the *THI* is greater than 79, most people would be uncomfortable.

9.1 Exercises

In Exercises 1–10, evaluate each formula using the values of the letters as given. Use a value of $\pi = 3.14$. Some of these formulas should be familiar to you.

1. $A = lw$ (area of a rectangle); $l = 12$ ft (length), $w = 9$ ft (width)

2. $P = 2l + 2w$ (perimeter of a rectangle); $l = 12$ ft, $w = 9$ ft

3. $I = prt$ (interest); $p = \$5000$ (principle); $r = 0.15$ (rate), $t = 2$ (time)

4. $P = S - C - H$ (net profit); $S = \$40$ (selling price), $C = \$20$ (cost), $H = \$5$ (overhead)

5. $d = rt$ (distance); $r = 55$ mph (rate), $t = 3.2$ hr (time)

6. $C = \frac{5}{9}(F - 32)$ (temperature conversion); $F = 68°$

7. $A = \frac{1}{2}h(b_1 + b_2)$ (area of a trapezoid); $h = 6$ (height), $b_1 = 7$ (one base), $b_2 = 9$ (the other base)

8. $A = \pi r^2$ (area of a circle); $r = 4$ m (radius)

9. $C = \pi d$ (circumference of a circle); $d = 4.6$ cm (diameter)

10. $A = \frac{1}{2}bh$ (area of a triangle); $b = 24$ cm (base), $h = 5$ cm (height)

▼ In Exercises 11–20, evaluate each formula using the values of the letters as given. Use $\pi = 3.14$ as an approximate value.

11. $I = \dfrac{E}{R}$, where $E = 114$, $R = 6$

12. $S = \frac{1}{2}gt^2$, where $g = 32$, $t = 10$

13. $S = \frac{1}{2}gt^2$, where $g = 32$, $t = 20$

14. $R = \dfrac{ab}{a + b}$, where $a = 10$, $b = 20$

15. $R = \dfrac{ab}{a + b}$, where $a = 20$, $b = 80$

16. $V = \frac{4}{3}\pi r^3$, where $r = 3$

17. $P = i^2r$, where $i = 3$, $r = 100$

18. $d = \sqrt{x^2 + y^2}$, where $x = 3$, $y = 4$

19. $V = lwh$, where $l = 2.1$, $w = 4.6$, $h = 7$

20. $V = \pi r^2h$, where $r = 5$, $h = 12$

LIVING IN THE REAL WORLD

21. How much carpet should you purchase to cover a floor that measures 5 yd by $3\frac{1}{3}$ yd? Use the formula $A = lw$.

22. What is the area of a pizza with a radius of 8 in.? Use the formula $A = \pi r^2$ and a value of 3.14 for π.

23. Maria Lyons's refrigerator measures 2.5 ft wide by 4 ft high by 2 ft deep. The volume, which is given by $V = lwh$, is measured in cubic feet. How many cubic feet does Maria's refrigerator have?

24. The top end of Mr. Goodman's house is in the shape of a triangle that measures 24 ft wide with a height of 8 ft. How much vinyl siding must he purchase to cover this area? Use the formula for the area of a triangle, $A = \frac{1}{2}bh$.

25. Helen Malinowski wants to put new baseboard around the perimeter of her living room. How much baseboard should she buy if the room measures 22 ft by 16 ft? Use the formula for the perimeter of a rectangle, $P = 2l + 2w$.

26. Calculate the temperature–humidity index using the formula $THI = t - (0.55)(1 - h)(t - 58)$ when:

a. $t = 68°$, $h = 0.4$;

b. $t = 90°$, $h = 0.22$;

c. $t = 78°$, $h = 0.95$.

27. When a consumer pays off a loan before it is due, part of the interest is refunded using a formula called the *rule of 78,* given by

$$u = \frac{f \cdot k(k + 1)}{n(n + 1)}$$

where f is the finance charge, k is the number of months left in the loan period, and n is the total number of months in the loan period. How much interest (u) is refunded when:

a. A 36-payment loan ($n = 36$) is paid off 12 months early ($k = 12$) and the total finance charge is \$180 ($f = \$180$)?

b. The total finance charge is \$158 on a 12-payment loan that is paid off in 10 months? (There are two payments left, so $k = 2$.)

c. A 48-month loan is paid off in two years and the total finance charge is \$710? The borrower paid off the loan in half the time—is half of the total finance charge refunded?

28. The number of stars in each of the triangles below is given by the formula $T = \frac{1}{2}n(n + 1)$, where n equals the number of rows in the triangle.

a. Verify that the formula works for each of the given triangles.

b. How many stars are there in a triangle with 10 rows?

c. How many stars are there in a triangle with 100 rows?

29. From your own experiences, write about a situation that requires you to use a formula to solve some practical problem.

9.2 Evaluating Algebraic Expressions

Algebraic expressions contain numbers (called constants) and letters (called variables) that represent numbers. In the algebraic expression $-6x^2y + 5z$, -6 and 5 are constants and x, y, and z are variables.

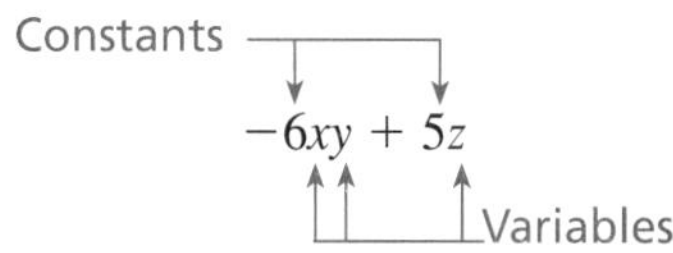

We evaluate algebraic expressions by substituting numerical values for these variables in the same way that we did when we evaluated formulas in the previous section.

To evaluate algebraic expressions:

1. Replace each variable (letter) by its given numerical value.
2. Perform all the arithmetic operations, using the correct order of operations.

EXAMPLE 1 Evaluate $3x - 5$ if $x = 4$.

SOLUTION

$3x - 5 = 3(4) - 5$	Replace x by 4.
$= 12 - 5$	Multiply before you subtract.
$= 7$	Subtract.

EXAMPLE 2 Find the value of $-2x + 3y$ when $x = -5$ and $y = 4$.

SOLUTION

$-2x + 3y = -2(-5) + 3(4)$	Replace x by -5 and y by 4.
$= 10 + 12$	Multiply.
$= 22$	Add.

EXAMPLE 3 Evaluate $4x^2 - 3y$ if $x = 2$ and $y = 4$.

SOLUTION

$4x^2 - 3y = 4 \cdot 2^2 - 3 \cdot 4$	Replace x by 2 and y by 4.
$= 4 \cdot 4 - 3 \cdot 4$	Perform powers first.
$= 16 - 12$	Multiply.
$= 4$	Subtract.

EXAMPLE 4 Find the value of $-3xy + x^2$ when $x = -2$ and $y = -5$.

SOLUTION

$-3xy + x^2 = -3(-2)(-5) + (-2)^2$	Replace x by -2 and y by -5.
$= -3(-2)(-5) + 4$	Do powers first.
$= -30 + 4$	Multiply.
$= -26$	Add.

STOP Be careful when evaluating expressions like -4^2 and $(-4)^2$:

$-4^2 = -16$	The exponent 2 applies only to the 4.
$(-4)^2 = 16$	The exponent applies to -4.

EXAMPLE 5 Evaluate the following expressions:

(a) x^2 when $x = -3$;

(b) $-x^2$ when $x = -3$;

(c) $(-x)^2$ when $x = -3$.

SOLUTION **(a)** $x^2 = (-3)^2 = 9$

(b) $-x^2 = -(-3)^2 = -9$

(c) $(-x)^2 = [-(-3)]^2 = 3^2 = 9$

9.2 Exercises

Find the value of each expression when $x = 3$, $y = 7$, and $z = -2$.

1. $6x$
2. $3y$
3. $-8x$
4. $-2z$
5. $6 - 2y$
6. $-3x + 5$
7. $x + 3y$
8. $2x - 3y$
9. $x - y$
10. $x + y + z$
11. $2x - 3y + 2z$
12. $4x + 7y - z$
13. x^2
14. $-x^2$
15. $(-x)^2$
16. $2x^2 + 3y$
17. $-5x^2 - 7y$
18. $2xz$
19. xyz
20. $-6z^2 - 8y$
21. $4z^2 + 6y$
22. $x - y^2 + z$
23. $x^2 + y^2$
24. $xy^2 - z^2$
25. $2(x + z)$
26. $2(x - y)$
27. $3x - 2(y - z)$
28. $4y + 2(z + 5)$
29. $3x + 2y - z$
30. $4x + 4y - z$

9.3 Adding and Subtracting Like Terms

The addition signs (+) and subtraction signs (−) separate an algebraic expression into parts called **terms.**

EXPRESSION	TERMS
$3x^2y + 4xy$	$3x^2y,\ 4xy$
$5x^2 + 2x - 1$	$5x^2,\ 2x,\ -1$
$-17xy + 2y^3 + 5$	$-17xy,\ 2y^3,\ 5$

The number in a term is called the **numerical coefficient,** or just the **coefficient,** and the variables are called the **literal parts.**

$$\underbrace{-5}_{\text{Coefficient}}\ \underbrace{x^3y}_{\text{Literal part}}$$

Two terms are called **like terms** if they differ *only* in their coefficient. That is, like terms must have exactly the same literal parts (the same variables with the same exponents).

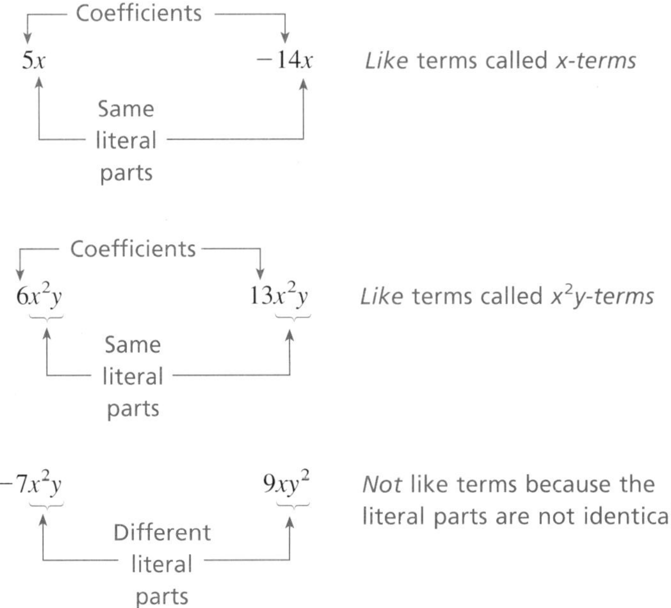

Combining Like Terms

The following examples will illustrate how the distributive property allows us to add or subtract like terms.

EXAMPLE 1 Combine $6x + 7x$.

SOLUTION

$$6x + 7x = (6 + 7)x \quad \text{Use the distributive property.}$$
$$= 13x \quad \text{Add } 6 + 7.$$

Adding, in both arithmetic and algebra, is really *counting,* and we could say that *6 things* + 7 *things* = *13 things,* where *things* refer to similar objects or quantities. This thinking leads us to the same conclusion:

$$6x + 7x = 13x$$

EXAMPLE 2 Combine $7xy^2 + 9xy^2 + 2xy^2$.

$$7xy^2 + 9xy^2 + 2xy^2 = (7 + 9 + 2)xy^2 \quad \text{Use the distributive property.}$$
$$= 18xy^2 \quad \text{Add inside the parentheses.}$$

EXAMPLE 3

(a) $14x - 3x = 11x$

(b) $-12x^2y - 11x^2y = -23x^2y$

(c) $4y^2 + 8y^2 + 7y^2 = 19y^2$

(d) $x^2y + 4x^2y = 5x^2y$ (Note that the coefficient of x^2y is 1 even though it is not usually written.)

(e) $4xyz - xyz = 3xyz$

Our rule then is as follows:

To add or subtract like terms:

1. Add or subtract the numerical coefficients.
2. Keep the same literal part.

EXAMPLE 4 Combine $6xy + 7xy^2 + 5xy + 2xy^2$.

SOLUTION Since addition is both commutative and associative, we can change both the order and the grouping and rewrite the problem:

$$6xy + 7xy^2 + 5xy + 2xy^2 = 6xy + 5xy + 7xy^2 + 2xy^2$$
$$= (6xy + 5xy) + (7xy^2 + 2xy^2)$$
$$= 11xy + 9xy^2$$

Notice that $11xy$ and $9xy^2$ are *not like terms* and cannot be combined further.

Usually a problem is not rewritten and the answer is obtained directly:

Combine these like terms

$$3x^3 + 7xy + 5x^3 - 9xy = 8x^3 - 2xy$$

Combine these like terms

EXAMPLE 5

(a) $-2x^2y + 13xy^2 + 5x^2y - 2xy^2 = 3x^2y + 11xy^2$

(b) $xyz + 4xy^4 - 3xyz - 7xy^4 = -2xyz - 3xy^4$

(c) $-2xy + 7y^3 + 3xy - 5y^3 + 3x^4 = xy + 2y^3 + 3x^4$

(d) $7ab^3 - 4abc - 7ab^3 + 2 = -4abc + 2$

(e) $-2x^2y + 5xy^2 + 3xy - 3$: There are no like terms in this expression, so it cannot be simplified.

9.3 Exercises

In Exercises 1–10, identify the terms, numerical coefficients, and literal parts of each expression.

1. $2x + 3y$
2. $5x^2y + 3xyz$
3. $-6x^3y + 2x^2y^3$
4. $2x^3 + 3x - 1$
5. $7x^3y - 14xy^3 + 2z - 5$
6. $r^2 + 2s - 3t + 7$
7. $-7xy^4 + 2x^2 - 3xyz$
8. $-4a^2 + 2b^2c + 3b$
9. $-11a^3 - 14b^3 + 17ac^4$
10. $-21x^2 + 6y^2 + 2xz - 3yz + 7$

In Exercises 11–40, combine where possible.

11. $7x - 4x$
12. $3x + 7x$
13. $10a - 5a$
14. $7x^2 + 2x^2$
15. $6x + 2x + 9x$
16. $3a - 5a + 7a$
17. $2x + x$
18. $6a - a$
19. $3a^2b + 2a^2b - a^2b$
20. $6y + y - (-5y)$
21. $x^2y^2 - 3x^2y^2$
22. $6xyz - 9xyz + 3xyz$
23. $3x^3 - 5x^3 - 7x^3 + x^3$
24. $-16x^2y + 2x^2y - 3x + 5x$
25. $7x^3 + 2x + 3x^3 + 5x$
26. $3a + 2b - 5a + 7b$
27. $7mm - (-3mn) + 4mn$
28. $3a^2 + 5b^2 - 7c^2$
29. $3ab - 7b + 5ab - b + 3$
30. $8w + 5w - 6$
31. $-11cd - 5c^2d + 4cd^2$
32. $-4a - 5b + 7a - 6b + 7$
33. $-7 + 4x^2y - 8 + 5x^2y + 6$
34. $-14rs^2 + 2r^2s^3 - 3r^2s^3 + 2rs^2$

35. $3m + 6 - 4 + 5n - (-n) + 2m$

36. $x^3 - 3x^2 + 2x - x^3 + 3x^2 - 2x$

37. $5a - 6b + 4c + 3a - 5b - 7c$

38. $4c - 3d^2 - 7b - 5c + 3d^2 - 5b$

39. $6x^2y + 5xy^2 - 3xy - 5xyz$

40. $-4x^2y + 7xy - 8x^2y - 6 - 7xy + 5$

9.4 The Meaning of Equations and Their Solutions

Equations are perhaps the most useful and important concept you will learn in algebra. The main reason for studying algebra is to learn to solve problems in the real world, and many of these problems are solved using equations.

An equation contains two algebraic expressions separated by an **equal sign** (=). The expression on the left side of the equal sign is called the *left side* or *left member,* and the expression on the right side of the equal sign is called the *right side* or *right member* of the equation.

Equal sign
↓

$$7x - 3 \quad = \quad 5x - 9$$

↑ Left side or left member

↑ Right side or right member

The symbol "=" is an important one and means that the left and right sides of an equation are different names for the same thing. This also means that $7x - 3 = 5x - 9$ and $5x - 9 = 7x - 3$ are the same.

Solution to an Equation

A **solution** to an equation is a value that, when substituted for the variable, makes the equation a true statement.

For example, if we substitute 12 for x in the equation $x + 5 = 17$, we get

$$x + 5 = 17$$

$$12 + 5 = 17 \quad \text{Substitute 12 for } x.$$

$$17 = 17$$

which is a true statement, and we say that 12 is the *solution* to the equation $x + 5 = 17$, or that $x = 12$ *satisfies* the equation.

Therefore the first question to be answered is, how do we recognize the solution to an equation if someone gives it to us? This process is called *checking the solution to the equation.*

To check a solution to an equation:

1. Substitute the proposed solution for the variable wherever it appears in the equation.
2. Evaluate each side of the equation.
3. If the left and right sides of the equation are equal, then the proposed solution *satisfies* the equation.
   If the left and right sides are not the same, then the proposed solution *does not satisfy* the equation.

EXAMPLE 1 Is $y = 5$ a solution to the equation $3y - 5 = 2y$?

SOLUTION

$3y - 5 = 2y$	
$3 \cdot \boxed{5} - 5 \stackrel{?}{=} 2 \cdot \boxed{5}$	Substitute 5 for *y* wherever it appears.
$15 - 5 \stackrel{?}{=} 10$	Evaluate each side of the equation.
$10 = 10$	Left side = right side.

The number 5 is a solution. ❙❙

EXAMPLE 2 Is $x = 7$ a solution to $3x + 2 = 4x - 6$?

SOLUTION

$3x + 2 = 4x - 6$	
$3 \cdot \boxed{7} + 2 \stackrel{?}{=} 4 \cdot \boxed{7} - 6$	Substitute 7 for *x*.
$21 + 2 \stackrel{?}{=} 28 - 6$	Evaluate each side of the equation.
$23 \neq 22$	$\neq$ means *is not equal to.*

Therefore the number 7 is not a solution, since the two sides are not equal. ❙❙

❙❙

EXAMPLE 3

Equation	Solution	Reason
$y + 4 = 7$	$y = 3$	$\boxed{3} + 4 = 7$
$13 = 18 - x$	$x = 5$	$13 = 18 - \boxed{5}$
$6 - x = 8$	$x = -2$	$6 - \boxed{(-2)} = 8$
$3(z + 2) = 18$	$z = 4$	$3(\boxed{4} + 2) = 18$

9.4 Exercises

1. Determine whether $x = 4$ is a solution to each of the following equations.

   a. $3x - 4 = 8$
   b. $x - 2 = 3$
   c. $x + 2 = 3x - 6$
   d. $6 - x = 3 + x$
   e. $2x - 10 = 2 - x$

2. Does $y = -4$ satisfy each of the following equations?

   a. $-8 = 2y$
   b. $3 + 5y = 17$
   c. $-3y + 3 = 15$
   d. $5 = y + 1$
   e. $-2y - 5 = y + 7$

3. Is $z = -6$ a solution to each of the following equations?

   a. $4z = -28$
   b. $\frac{12}{z} = -2$
   c. $-7 = -1 + z$
   d. $-2(z + 1) = 10$
   e. $3(z + 6) = 2z + 12$

9.5 Solving Equations—The Addition Principle

Now that we have learned what a solution to an equation looks like, the next step is to discover how to find one.

In the previous section, it was stated that the left and right sides of an equation are different ways of representing the same quantity. In this way, we can think of an equation as a balance, with the left member on one side and the right member on the other. The equation $x - 5 = -1$ can then be represented by the following diagram.

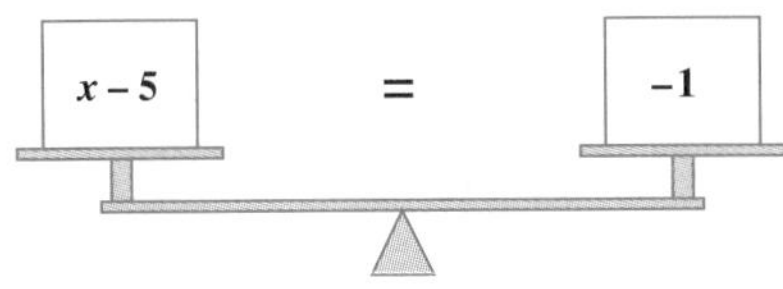

If we add the same quantity, say 7, to both sides of the balance, it should stay in equilibrium, since both sides still carry the same amount.

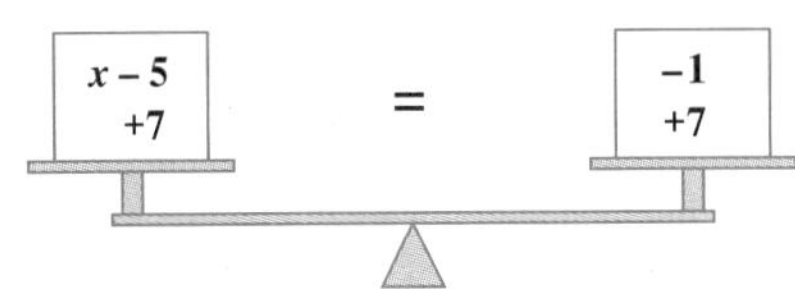

Algebraically, we are adding 7 to both sides of the original equation:

$$\begin{array}{r} x - 5 = -1 \\ \underline{+7} \quad \underline{+7} \\ x + 2 = \quad 6 \end{array}$$

If you check each of the equations $x - 5 = -1$ and $x + 2 = 6$, you will find that $x = 4$ is a solution to both.

$$\begin{array}{rcl} x - 5 = -1 & \quad & x + 2 = 6 \\ \boxed{4} - 5 = -1 & & \boxed{4} + 2 = 6 \\ -1 = -1 & & 6 = 6 \end{array}$$

Therefore the number 4 is a solution to both equations.

Two equations that have the same solution are called **equivalent equations.** It is apparent from the illustration that if we add the same number to both sides of any equation, we will obtain a new equation that is **equivalent** to the original one, that is, they both have the same solution.

In a similar way, we can show that if we subtract the same quantity from both sides of an equation, the result is an equivalent equation. Actually, this is the same as adding a negative quantity to both sides of the equation. This principle is basic to the solution of equations and is stated as follows:

The addition principle:

The same number may be added to (or subtracted from) each side of any equation and the solution will remain the same.

Let's now apply this principle to finding the solutions to equations. Our purpose is always the same, to isolate the variable all by itself on one side of the equation. This involves adding or subtracting any numbers that are found on the same side of the equation as the variable. Always do the opposite operation—if a number is added, subtract it (i.e., add its negative). Remember, whatever we do to one side of the equation we must do to the other side of the equation to keep it in balance.

EXAMPLE 1 Solve $x + 6 = 14$.

SOLUTION

$x + 6 = 14$	Isolate x on the left side.
$x + 6 + (-6) = 14 + (-6)$	Add −6 to remove the +6 from the left side.
$x + 0 = 8$	
$x = 8$	The solution is 8.

We can check the solution by putting $x = 8$ in the original equation:

Check $\boxed{8} + 6 = 14$ ✓

EXAMPLE 2 Solve $x - 7 = -3$.

SOLUTION

$x - 7 = -3$	Isolate x by eliminating -7.
$x - 7 + 7 = -3 + 7$	Add 7 to both sides.
$x = 4$	The solution is 4.

Check $4 - 7 = -3$ ✓

EXAMPLE 3 Solve $-9 = y + 3$.

SOLUTION

$-9 = y + 3$	Isolate y by eliminating 3.
$-9 + (-3) = y + 3 + (-3)$	Add -3 to both sides.
$-12 = y$	The solution is -12.

Note that $-12 = y$ and $y = -12$ say the same thing.

Check $-9 = -12 + 3$ ✓

EXAMPLE 4 Solve $3.1 + y = 2.4$.

SOLUTION

$3.1 + y = 2.4$	Isolate y.
$3.1 + (-3.1) + y = 2.4 + (-3.1)$	Add -3.1 to both sides.
$y = -0.7$	The solution is -0.7.

Check $3.1 + (-0.7) = 2.4$ ✓

EXAMPLE 5 Solve $8 + x = -5$.

SOLUTION

$8 + x = -5$	Eliminate the 8 on the left side.
$8 + (-8) + x = -5 + (-8)$	Add -8 to both sides.
$x = -13$	The solution is -13.

Check $8 + (-13) = -5$ ✓

STOP In many of the exercises that follow, you will be able to see the solutions immediately. However, you should work through each problem carefully to learn the methods involved. These are the same methods that we will use to solve more difficult equations for which the solutions are not at all obvious.

9.5 Exercises

▼ Solve and check the following equations:

1. $x - 4 = 7$
2. $y + 5 = 14$
3. $y - 6 = 5$
4. $-6 + x = 4$
5. $5 = x + 7$
6. $-6 = h - 7$
7. $7 + h = 17$
8. $14 = x - 11$
9. $-13 = -12 + x$
10. $x - 16 = -9$
11. $2 + x = -7$
12. $-6 = x - 4$
13. $4 + y = -11$
14. $12 + h = -14$
15. $-16 = -14 + m$
16. $-71 = s - 26$
17. $3.75 + x = 8$
18. $6.4 + x = -3.6$
19. $x + \frac{3}{4} = 7\frac{3}{4}$
20. $x - 1\frac{2}{3} = 3\frac{2}{3}$
21. $4.62 = x - 3.91$
22. $y + 2\frac{1}{5} = 7\frac{3}{5}$
23. $x - 4.1 = -3.8$
24. $x - 1.75 = 8.5$
25. $4.75 + y = -2\frac{1}{4}$

9.6 Solving Equations—The Multiplication Principle

If we return to our idea of the balance that represents an equation, like $2x = 18$, we have the following diagram:

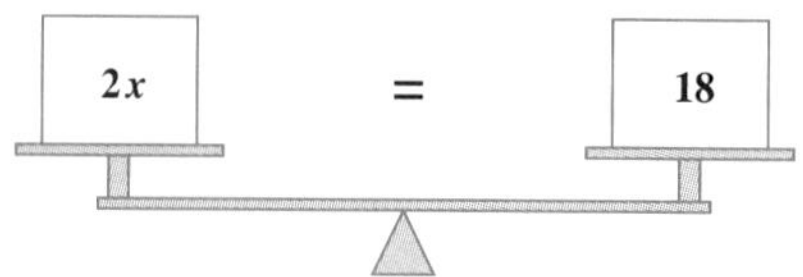

If we decided to multiply each side of our equation by 3, we would have three times as much on each side of the balance and it would remain in equilibrium. Algebraically, it would look like this:

$$2x = 18$$
$$3 \cdot 2x = 3 \cdot 18$$
$$6x = 54$$

If we substitute $x = 9$ in each of our equations, $2x = 18$ and $6x = 54$, we will find it is a solution for each of them:

$$\begin{aligned} 2x &= 18 \qquad & 6x &= 54 \\ 2 \cdot 9 &= 18 \qquad & 6 \cdot 9 &= 54 \\ 18 &= 18 \qquad & 54 &= 54 \end{aligned}$$

Therefore 9 is a solution to both equations.

Reciprocal

Dividing by a number such as 3 is the same as multiplying by its reciprocal, $\frac{1}{3}$, where the **reciprocal** of a number is 1 divided by the number. So, for example,

$$\frac{1}{5} \text{ is the reciprocal of } 5$$

The reciprocal of $\frac{1}{5}$ is

$$\frac{1}{\frac{1}{5}} = \frac{\frac{1}{1}}{\frac{1}{5}} = \frac{1}{1} \div \frac{1}{5} = \frac{1}{1} \cdot \frac{5}{1} = 5$$

So $\frac{1}{5}$ and 5 are reciprocals of each other.

The reciprocal of $\frac{2}{3}$ is

$$\frac{1}{\frac{2}{3}} = \frac{\frac{1}{1}}{\frac{2}{3}} = \frac{1}{1} \div \frac{2}{3} = \frac{1}{1} \cdot \frac{3}{2} = \frac{3}{2}$$

This is true in general:

> The reciprocal of $\frac{a}{b}$ is $\frac{b}{a}$ $\quad (a, b \neq 0)$.
>
> a and $\frac{1}{a}$ are reciprocals of each other $\quad (a \neq 0)$.

It seems reasonable that we can multiply or divide both sides of an equation by the *same* number and produce an equivalent equation. This is the second basic principle that we will use in solving equations.

The multiplication principle:

If both sides of an equation are multiplied by (or divided by) the *same* number (except zero), an equivalent equation results.

To apply this rule, keep in mind that we want to isolate the variable on one side of the equation. If the variable is multiplied by a number, we eliminate it by dividing both sides of the equation by that number. If the variable is divided by a number, we eliminate it by multiplying both sides of the equation by that number. Of course, we must remember that whatever we do to one side of the equation we have to do to the other side of the equation as well.

EXAMPLE 1 Solve $5y = 30$.

SOLUTION

$5y = 30$	Isolate y on the left side.
$\frac{\not{5}y}{\not{5}} = \frac{30}{5}$	Divide both sides by 5.
$1 \cdot y = 6$	The 5s cancel.
$y = 6$	$1 \cdot y = y$, which is what we want.

Check $5 \cdot 6 = 30$ ✓ The solution checks.

We can also obtain the same result by multiplying by the reciprocal of 5, which is $\frac{1}{5}$.

SOLUTION

$5y = 30$	
$\frac{1}{\not{5}} \cdot \not{5}y = \frac{1}{5} \cdot 30$	Multiply both sides by $\frac{1}{5}$.
$1 \cdot y = 6$	
$y = 6$	

Check $5 \cdot 6 = 30$ ✓ ▮▮

EXAMPLE 2 Solve $\frac{1}{4}x = 3$.

SOLUTION

$\frac{1}{4}x = 3$	Eliminate the $\frac{1}{4}$ on the left.
$4 \cdot \frac{1}{4}x = 4 \cdot 3$	Multiply by 4, the reciprocal of $\frac{1}{4}$.
$1 \cdot x = 12$	$1 \cdot x = x$
$x = 12$	

Check $\frac{1}{4} \cdot 12 = 3$ ✓ The solution checks.

EXAMPLE 3 Solve $\frac{3}{4}x = 18$.

SOLUTION

$\frac{3}{4}x = 18$	Eliminate the $\frac{3}{4}$.
$\frac{4}{3} \cdot \frac{3}{4}x = \frac{4}{3} \cdot \overset{6}{18}$	Multiply by $\frac{4}{3}$, the reciprocal of $\frac{3}{4}$.
$1 \cdot x = 24$	
$x = 24$	The solution is 24.

Check $\frac{3}{4} \cdot 24 = \frac{3}{\underset{1}{4}} \cdot \overset{6}{24} = \frac{18}{1} = 18$ ✓

EXAMPLE 4 Solve $32 = -8h$.

SOLUTION

$32 = -8h$	Eliminate the −8 on the right.
$\frac{32}{-8} = \frac{-8h}{-8}$	Divide both sides by −8.
$-4 = h$	The solution is −4.

Check $32 = (-8) \cdot (-4)$ ✓

EXAMPLE 5 Solve $\frac{x}{7} = -6$.

SOLUTION

$$\frac{x}{7} = -6 \quad \text{We need to eliminate the 7.}$$

$$7 \cdot \frac{x}{7} = 7 \cdot (-6) \quad \text{Multiply both sides by 7.}$$

$$x = -42 \quad \text{The solution is } -42.$$

Check $\frac{-42}{7} = -6$ ✓

EXAMPLE 6 Solve $-x = -14$.

SOLUTION Since $-x$ is the same as $(-1)x$, we divide both sides of the equation by -1:

$$-x = -14$$

$$-1 \cdot x = -14 \quad -x \text{ is the same as } -1 \cdot x.$$

$$\frac{-1 \cdot x}{-1} = \frac{-14}{-1} \quad \text{Divide by } -1 \text{ on both sides.}$$

$$x = 14$$

After you do a few problems of this type, you will probably notice a shortcut. If one side of the equation contains only a variable and a negative sign, to solve for the positive variable just change the sign on both sides of the equation.

EXAMPLE 7 **(a)** Solve $-x = 6$.

(b) Solve $-y = -8$.

SOLUTION **(a)** $x = -6$ Change the sign on both sides.

(b) $y = 8$

9.6 Exercises

▼ Solve each equation and check.

1. $6x = 26$
2. $3x = 21$
3. $5x = 35$
4. $9x = 36$
5. $48 = 6x$
6. $32 = 16x$
7. $-7y = 21$
8. $-8y = 32$
9. $-2y = -18$
10. $-3y = -12$
11. $-72 = -9y$
12. $-56 = -8y$

13. $\frac{1}{4}x = 8$

14. $\frac{1}{5}x = -7$

15. $\frac{2}{3}x = 12$

16. $\frac{2}{5}x = 14$

17. $\frac{3}{4}x = -18$

18. $\frac{3}{5}x = -24$

19. $\frac{x}{6} = 3$

20. $\frac{x}{9} = 4$

21. $-\frac{y}{8} = 7$

22. $-\frac{y}{7} = 4$

23. $-2 = -\frac{y}{3}$

24. $-7 = -\frac{x}{7}$

25. $39x = -13$

26. $48x = -24$

27. $-36 = -\frac{3}{4}x$

28. $-6 = -\frac{2}{7}y$

29. $2.4x = -14.4$

30. $4.2 = -3x$

31. $\frac{7}{8}x = -\frac{5}{16}$

32. $\frac{3}{16}x = -\frac{3}{2}$

9.7 Solving Equations by Combining Rules

Some equations involve the use of both the addition and the multiplication principles. Our objective is still the same: to try to isolate the variable on one side of the equation. To do this we first get all the terms containing the variable on one side of the equation and the plain numbers; or constants, on the other side of the equation.

To solve an equation using both the addition and the multiplication principles:

1. Put all the terms containing the variable on one side of the equation by using the addition principle.
2. Put all the constants on the *other* side of the equation by using the addition principle again.
3. Multiply both sides of the equation by the reciprocal of the coefficient of the variable.

EXAMPLE 1 Solve $5x + 1 = 3x + 7$.

SOLUTION

$5x + 1 = 3x + 7$	Get all x-terms on the left side.
$5x + (-3x) + 1 = 3x + (-3x) + 7$	Add $-3x$ to both sides.
$2x + 1 = 7$	All x-terms are now on the left side.
$2x + 1 + (-1) = 7 + (-1)$	Add -1 to both sides to get all the constants on the right side.
$2x = 6$	All constants are now on the right side.
$\frac{1}{\not{2}} \cdot \not{2}x = \frac{1}{2} \cdot 6$	Multiply both sides by $\frac{1}{2}$, the reciprocal of 2.
$x = 3$	The solution is 3.

Check

$$5x + 1 = 3x + 7$$
$$5 \cdot 3 + 1 \stackrel{?}{=} 3 \cdot 3 + 7$$
$$15 + 1 \stackrel{?}{=} 9 + 7$$
$$16 = 16 \checkmark$$

Did we have to put the x-terms on the *left* and the constants on the *right?* Could we have done it the other way around? Let's try it and see:

EXAMPLE 1 (alternate solution) Solve $5x + 1 = 3x + 7$.

SOLUTION #2

$5x + 1 = 3x + 7$	Get all x-terms on the right side.
$5x + (-5x) + 1 = 3x + (-5x) + 7$	Add $-5x$ to both sides.
$1 = -2x + 7$	All x-terms are on the right now.
$1 + (-7) = -2x + 7 + (-7)$	Add -7 to both sides.
$-6 = -2x$	Multiply both sides by $\frac{1}{-2}$ or,
$\frac{-6}{-2} = \frac{-\not{2}x}{-\not{2}}$	equivalently, divide by -2.
$3 = x$	The solution is the same.

EXAMPLE 2 Solve $3y - 2 = 13$.

SOLUTION

$$3y - 2 = 13$$ All y-terms are already on the left.

$$3y - 2 + 2 = 13 + 2$$ Eliminate the -2 on the left.

$$3y = 15$$ All constants are now on the right.

$$\frac{3y}{3} = \frac{15}{3}$$ Divide both sides by 3.

$$y = 5$$

Check

$$3 \cdot 5 - 2 \stackrel{?}{=} 13$$

$$15 - 2 = 13 \quad ✓$$

EXAMPLE 3 Solve $-22 - 3x = 5x$.

SOLUTION Since the only constant is already on the left side of the equation, there is some advantage to putting all of the variables on the right.

$$-22 - 3x = 5x$$

$$-22 - 3x + 3x = 5x + 3x$$ Add $3x$ to both sides.

$$-22 = 8x$$

$$\frac{-22}{8} = \frac{8x}{8}$$ Divide both sides by 8.

$$\frac{-22}{8} = -\frac{11}{4} = x$$ The solution is $-\frac{11}{4}$.

Check

$$-22 - 3 \cdot \left(-\frac{11}{4}\right) \stackrel{?}{=} 5 \cdot \left(-\frac{11}{4}\right)$$

$$-\frac{88}{4} + \frac{33}{4} \stackrel{?}{=} -\frac{55}{4}$$

$$-\frac{55}{4} = -\frac{55}{4} \quad ✓$$

EXAMPLE 4 Solve $5 + \frac{y}{6} = 8$.

SOLUTION

$$5 + \frac{y}{6} = 8$$

$$5 + (-5) + \frac{y}{6} = 8 + (-5) \quad \text{Add } -5 \text{ to both sides.}$$

$$\frac{y}{6} = 3$$

$$\not{6} \cdot \frac{y}{\not{6}} = 6 \cdot 3 \quad \text{Multiply both sides by 6.}$$

$$y = 18$$

Check

$$5 + \frac{18}{6} \stackrel{?}{=} 8$$

$$5 + 3 = 8 \quad ✓$$

Equations Containing Parentheses

When an equation contains parentheses, we must simplify it before we can solve it.

EXAMPLE 5 Solve $4(x + 3) = 8$.

SOLUTION Using the distributive rule, we multiply both x and 3 by 4 on the left side of the equation:

$$4(x + 3) = 8$$

$$4x + 12 = 8 \quad \text{Multiply parentheses by 4.}$$

$$4x + 12 + (-12) = 8 + (-12) \quad \text{Add } -12 \text{ to both sides.}$$

$$4x = -4$$

$$\frac{\not{4}x}{\not{4}} = \frac{-4}{4} \quad \text{Divide both sides by 4.}$$

$$x = -1$$

Check

$$4(x + 3) = 8$$

$$4(-1 + 3) \stackrel{?}{=} 8$$

$$4(2) \stackrel{?}{=} 8$$

$$8 = 8 \quad ✓$$

EXAMPLE 6 Solve $3x - 2(x + 4) = 6 - (x + 4)$.

SOLUTION

$3x - 2(x + 4) = 6 - (x + 4)$	
$3x - 2x - 8 = 6 - x - 4$	Remove the parentheses on both sides.
$x - 8 = -x + 2$	Combine like terms on both sides.
$x + x - 8 = -x + x + 2$	Add x to both sides.
$2x - 8 = 2$	
$2x - 8 + 8 = 2 + 8$	Add 8 to both sides.
$2x = 10$	
$\frac{\not{2}x}{\not{2}} = \frac{10}{2}$	Divide both sides by 2.
$x = 5$	5 is the solution.

Check

$$3x - 2(x + 4) = 6 - (x + 4)$$
$$3 \cdot 5 - 2(5 + 4) \stackrel{?}{=} 6 - (5 + 4)$$
$$15 - 2(9) \stackrel{?}{=} 6 - 9$$
$$15 - 18 \stackrel{?}{=} -3$$
$$-3 = -3 \checkmark$$

9.7 Exercises

▼ Solve each equation and check:

1. $4x + 3 = 11$

2. $3x + 6 = 15$

3. $6x - 2 = 10$

4. $5x - 3 = 7$

5. $2x - 8 = -6$

6. $7x - 3 = -17$

7. $7x = 35 + 2x$

8. $6x = 24 + 2x$

9. $23 = 8 + 5x$

10. $17 = 5 + 6x$

11. $-4x = -18 + 2x$

12. $-3x = 24 + 5x$

13. $4x - 6 = 3x + 2$

14. $6x - 3 = 2x + 5$

15. $6y + 9 = 2y - 7$

16. $3y + 1 = 6y - 26$

17. $2x + 7 = -4x - 7$

18. $5y + 6 = 8y - 3$

19. $-6x - 1 = -5x + 6$

20. $5 - 7x = 3x + 15$

21. $x = 5x - 8$

22. $y = 8y - 28$

23. $4 - x = -5x + 12$

24. $9 - x = -9x + 1$

25. $-8x + 2 = 3x - 4$

26. $5x + 6 = 3x + 5$

27. $3x - 5.2 = 4.1$

28. $5x + 1.7 = -3.3$

29. $4x + 5 = -3x + 5$

30. $6x - 8 = -4x - 8$

31. $\frac{x}{5} + 6 = 8$

32. $\frac{x}{3} - 5 = 2$

33. $\frac{1}{4}x - 3 = -5$

34. $\frac{1}{2}x + 3 = 7$

35. $\frac{2}{3}y + 4 = -4$

36. $\frac{3}{5}y + 6 = 12$

37. $3(x + 5) = -6$

38. $5(x - 2) = 20$

39. $7x = 2(x + 20)$

40. $12x = 2(x + 30)$

41. $3(x + 1) = 2x + 7$

42. $7(x - 2) = 6x - 5$

43. $y + 2(y + 4) = -1$

44. $3(x - 1) + 7 = 2x$

45. $3(x - 4) + 2(x + 6) = 20$

46. $3(x + 5) = 2(x - 1)$

47. $2(y - 5) = 2(3y - 5)$

48. $5(x + 5) = 3(2x - 1)$

49. $\frac{2}{3}(x - 6) = -4$

50. $\frac{3}{4}(x + 12) = -5$

9.8 Application Problems

Translating from English to Algebra

Problems in the real world are rarely stated in algebraic equations. The equations usually come well camouflaged within the English language. Most difficulties in solving application problems come not from solving the mathematics problems, but from improper translation of the English sentences into mathematical symbols and numbers.

We'll see if we can help you avoid such errors. Lets start by looking at some key words that tell us when we should add, subtract, multiply, or divide, and also which words mean "equal to."

Add	Subtract	Multiply	Divide	Equals
plus	minus	times	divide(d) by	equals
sum	difference	product	quotient	same as
add	subtract	multiply	shared	equivalent to
more than	less than	multiplied by		identical to
increased by	diminished by	of		is
	decreased by			are

Now let's consider some English phrases and translate them into mathematical expressions.

EXAMPLE 1 Write a mathematical expression that represents each English sentence.

(a) A number increased by 6 equals 9.

A number	increased by	6	equals	9
↓	↓	↓	↓	↓
x	$+$	6	$=$	9

or $$x + 6 = 9$$

(b) Three more than Fred's age is 22.

3	more than	Fred's age	is	22
↓	↓	↓	↓	↓
3	$+$	x	$=$	22

or $$3 + x = 22$$

(c) When the sum of x and 4 is divided by 3 the result is 6.

When the sum of x and 4	is divided by	3	the result is	6
↓	↓	↓	↓	↓
$(x + 4)$	$\div$	3	$=$	6

or $$(x + 4) \div 3 = 6$$

$$\frac{x + 4}{3} = 6$$

■

EXAMPLE 2 Write each sentence as an algebraic expression.

	ENGLISH	ALGEBRA
(a)	A number plus 3 is 14.	$x + 3 = 14$
(b)	The product of a number and 5 is 35.	$x \cdot 5 = 35$ or $5x = 35$
(c)	Six added to a number is 14.	$x + 6 = 14$
(d)	Twice the sum of a number and 6 is 18.	$2(x + 6) = 18$
(e)	The difference between some number and 4 is 10.	$x - 4 = 10$
(f)	Two-thirds of an unknown number is 8.	$\frac{2}{3}x = 8$
(g)	A certain number is 6 less than twice itself.	$x = 2x - 6$
(h)	A number is 10 more than 6 times itself.	$x = 6x + 10$
(i)	The quotient of some number and 6 is 8.	$\frac{x}{6} = 8$

■

Solving Application Problems

Now we combine our knowledge of solving equations with our ability to translate from English to mathematics. Solving application problems has traditionally been one of the most difficult parts of algebra for students. Following is a "battle-plan" that has helped many students learn to solve application problems. If you are willing to follow the steps of this strategy carefully through the examples and exercises, your efforts will be rewarded and, indeed, you *will* learn to solve application problems.

A strategy for solving application problems:

1. Read the problem several times, making note of any key words.
2. Write down what you're supposed to find and what is given.
3. Represent one of the unknown quantities by a variable. Express any other unknown quantities in terms of the *same* variable.
4. Find out which expressions are the same (equal) and write an equation. Sometimes a simple sketch helps determine which expressions are equal.
5. Solve the equation.
6. Check your solution in the *original word statement* to see whether or not it really meets the conditions of the problem. This will catch any error that might be made in writing the equation.

EXAMPLE 3 Three times a number increased by 7 is equal to 31. Find the number.

SOLUTION Let's let x represent the unknown number.

3	times	a number	increased by	7	is equal to	31
↓	↓	↓	↓	↓	↓	↓
3	$\cdot$	x	$+$	7	$=$	31

or

$$3x + 7 = 31$$

$$3x + 7 - 7 = 31 - 7$$ Subtract 7 from both sides.

$$3x = 24$$ Divide both sides by 8.

$$x = 8$$ The unknown number is 8.

Now let's put our solution, 8, back in the original word statement to see if it meets the condition of the problem.

Check

3	times	a number	increased by	7	is equal to	31
↓	↓	↓	↓	↓	↓	↓
3	$\cdot$	8	$+$	7	$\stackrel{?}{=}$	31

$$3 \cdot 8 + 7 \stackrel{?}{=} 31$$

$$24 + 7 \stackrel{?}{=} 31$$

$$31 = 31$$ ✓

EXAMPLE 4 Twice the difference of a number and 6 is 10. Find the number.

SOLUTION We let x represent the desired number.

Twice	the difference of a number and 6	is	10
↓	↓	↓	↓
$2\cdot$	$(x-6)$	$=$	10

$$2(x-6) = 10$$
$$2x - 12 = 10$$
$$2x = 22$$
$$x = 11$$ The desired number is 11.

Check $x = 11$ checks because twice the difference of 11 and 6, which is $2\cdot 5$, is indeed 10. ▮▮

EXAMPLE 5 The sum of two numbers is 84. If one number is one-half the other, find the numbers.

SOLUTION Let x = one of the numbers.

Then $\frac{1}{2}x$ = the second number.

The sum of two numbers			is	84
	↓		↓	↓
x	$+$	$\frac{1}{2}x$	$=$	84
↑		↑		
one number		second number		

$$x + \frac{1}{2}x = 84$$

$$\frac{3}{2}x = 84$$ Add x and $\frac{1}{2}x$ to get $\frac{3}{2}x$.

$$\frac{\not{2}}{\not{3}}\cdot\frac{\not{3}}{\not{2}}x = \frac{2}{\not{3}_1}\cdot \not{84}^{28}$$ Multiply by the reciprocal, $\frac{2}{3}$

$$x = 56$$ The first number

$$\frac{1}{2}x = \frac{1}{2}(56) = 28$$ The second number

So the numbers are 56 and 28.

Check 28 is one-half of 56 and $28 + 56 = 84$. ▮▮

EXAMPLE 6 In a math class of 30 students, there are 4 more males than females. How many of each sex are in the class?

SOLUTION Let $x =$ the number of females.
Then $x + 4 =$ the number of males (since there are 4 more males than females).

Number of females	plus	number of males	equals	total in the class
↓	↓	↓	↓	↓
x	$+$	$x + 4$	$=$	30

$$x + x + 4 = 30$$
$$2x + 4 = 30$$
$$2x = 26$$
$$x = 13 \quad \text{The number of females}$$
$$x + 4 = 17 \quad \text{The number of males}$$

Check There are 4 more males than females (17 is 4 greater than 13), and since $13 + 17 = 30$, the answers check.

9.8 Exercises

Use the proper steps to solve the following application exercises. Even if you can see the correct answer by inspection, take the time to write the equation that correctly describes the problem.

1. Six more than a number is 12. Find the number.

2. A number increased by 6 is 15. What is the number?

3. Seventeen is the same as the sum of 4 and some number. Find the number.

4. The letter x represents a number that is 5 less than 11. Find the value of x.

5. Some number divided by -4 is 5. What is the number?

6. Seven times an unknown number is 28. Find the number.

7. An unknown number is 5 greater than 8. Find the number.

8. The difference between an unknown number and 5 is equal to 12. What is the unknown number?

9. The product of a number and 7 is 42. What is the number?

10. Nine is 3 less than some number. What is the number?

11. Twice a number increased by 6 is 25. Find the number.

12. An unknown number is 6 less than twice itself. Find the unknown number.

13. Six times a number plus 4 times the same number is 30. Find the number.

14. Fifteen minus $\frac{5}{8}$ of a number equals 5. What is the number?

15. If you subtract a number from 8 and then add 6, the result is 10. Find the number.

16. Twice the sum of a number and 6 is 24. Find the number.

17. When 6 is added to two times a number, the result is 22. Find the number.

18. Four times a certain number when decreased by 5 is 27. What is the number?

19. When $\frac{1}{4}$ of a number is added to 4, the sum is 10. Find the number.

20. If 16 is subtracted from $\frac{2}{3}$ of a number, the result is 8. Find the number.

21. When a number is subtracted from 21, the result is twice the number. Find the number.

22. Three more than a number is the same as 2 subtracted from twice the number. What is the number?

23. When 6 is added to 4 times a number, the result is equal to 4 less than 6 times the same number. Find the number.

24. When 3 times an unknown number is subtracted from 16 and that result divided by 5, the quotient is equal to the unknown number. Find the unknown number.

25. One-third the sum of a number and 6 is equal to $\frac{3}{4}$ of the original number. Find that number.

26. Eight plus an unknown number gives the same result as twice the difference of the number and 2. What is the number?

27. One-third of a number subtracted from $\frac{1}{2}$ of the number is 6. Find the number.

28. When a number is decreased by 18, the result is $\frac{1}{2}$ of the number. What is the number?

29. The perimeter of a rectangle is 40. Find the length and width of the rectangle if the length is 4 more than the width.

30. A man is 3 times as old as his son. Find the age of each if the sum of their ages is 48.

31. A 78-inch board is cut into two pieces in such a way that one piece is twice as long as the other. How long is each piece?

32. The perimeter of a standard sheet of typing paper is 39 inches. If the length is $2\frac{1}{2}$ inches longer than the width, what are the dimensions of the paper?

33. Abraham Lincoln gave his famous Gettysburg Address in 1863. In his speech, he refers to the year 1776 as "four score and seven years ago." Represent "score" as an unknown in an equation and find out how many years a score is.

SUMMARY—CHAPTER 9

The numbers in brackets refer to the section in the text in which each concept is presented.

[9.1] A **formula** is a mathematical expression that describes a relationship between quantities in the real world.

EXAMPLE

[9.1] The area of a trapezoid is given by $A = \frac{1}{2}h(b_1 + b)_2$.

[9.1] **To evaluate a formula:**

1. Write the formula from the given English statement (if necessary).
2. Replace each letter by its numerical value.
3. Do all arithmetic operations, using the correct order of operations.

[9.1] Find A if $h = 8$, $b_1 = 6$, and $b_2 = 10$.

$$\begin{aligned} A &= \frac{1}{2}h(b_1 + b_2) \\ &= \frac{1}{2} \cdot 8(6 + 10) \\ &= 4 \cdot 16 \\ &= 64 \text{ sq. units} \end{aligned}$$

[9.2] A **constant** is a quantity that does not change its value. It is usually written as a number.

[9.2] **To evaluate an algebraic expression:**

1. Replace each variable (letter) by its given numerical value.
2. Perform all the arithmetic operations, using the correct order of operations.

[9.2] Find the value of $5(2x - y)$ if $x = -3$ and $y = 1$.

$$\begin{aligned} 5(2x - y) &= 5[2(-3) - 1] \\ &= 5[-6 - 1] \\ &= 5(-7) = -35 \end{aligned}$$

[9.3] The positive (+) and negative (−) signs in a mathematical expression separate it into **terms.**
The **numerical coefficient** is the number part of a term.
The **literal part** consists of the variables in a term and their exponents.
Like terms are terms that have the same literal parts.
Unlike terms are terms that have different literal parts.

[9.3] Given $4x^2y - 3xy + 7$:
The terms are $4x^2y$, $-3xy$, and 7.

The numerical coefficients are 4, −3, and 7.

The literal parts are x^2y and xy.

[9.3] **To add or subtract like terms:**

1. Add or subtract the numerical coefficients.
2. Keep the same literal part.

[9.3] $$\begin{aligned} &3xy^3 - 4x^2y - 5xy^3 + 7x^2y + 6 \\ &= -2xy^3 + 3x^2y + 6 \end{aligned}$$

[9.4] The **solution** to an equation is a value that, when substituted for the variable, makes the equation a true statement.

[9.4] Is 3 a solution to $2x + 3 = 5x - 6$?
Substitute 3 for x.

EXAMPLE

[9.4] **To check a solution to an equation:**

1. Substitute the proposed solution for the variable wherever it appears in the equation.
2. Evaluate each side of the equation.
3. If the left and right sides of the equation are equal, then the proposed solution satisfies the equation.
4. If the left and right sides of the equation are not equal, then the proposed solution does not satisfy the equation.

[9.4] $2 \cdot 3 + 3 \stackrel{?}{=} 5 \cdot 3 - 6$

$$6 + 3 \stackrel{?}{=} 15 - 6$$

$$9 = 9$$

Yes, 3 is a solution.

[9.5] **Addition principle:** The same number may be added to (or subtracted from) both sides of any equation and the solution will remain the same.

[9.5] Solve $x + 5 = -3$.

$$x + 5 + (-5) = -3 + (-5)$$

$$x = -8$$

[9.6] **Multiplication principle:** If both sides of an equation are multiplied by (or divided by) the same number (except zero), the new equation is equivalent to the original equation.

[9.6] Solve $\frac{2}{3}x = 12$.
Multiply both sides by $\frac{3}{2}$, the reciprocal of $\frac{2}{3}$.

[9.6] **Reciprocal.** The reciprocal of $\frac{a}{b}$ is $\frac{b}{a}$.

$$\frac{\cancel{3}}{\cancel{2}} \cdot \frac{\cancel{2}}{\cancel{3}} x = \frac{3}{\cancel{2}_1} \cdot \cancel{12}^{6}$$

$$x = 18$$

[9.7] **To solve an equation using both the addition and multiplication principles:**

1. Put all the terms containing the variable on one side of the equation by using the addition principle.
2. Put all the constants on the other side of the equation by again using the addition principle.
3. Multiply both sides of the equation by the reciprocal of the coefficient of the variable.

[9.7] Solve $8x + 1 = 5x - 11$.

$$8x + (-5x) + 1 = 5x + (-5x) - 11$$

$$3x + 1 = -11$$

$$3x + 1 + (-1) = -11 + (-1)$$

$$3x = -12$$

$$\frac{\cancel{3}x}{\cancel{3}} = \frac{-12}{3}$$

$$x = -4$$

[9.8] **A strategy for solving application problems:**

1. Read the problem several times, making note of any key words.
2. Write down what is given and what you're supposed to find.
3. Represent one of the unknown quantities by a variable, and express any other unknown quantities in terms of the same variable.
4. Find out which expressions are the same (equal) and write an equation. Sometimes a simple sketch helps determine which expressions are equal.
5. Solve the equation.
6. Check your solution in the original word statement to see if it really meets the conditions of the problem.

EXAMPLE

[9.8] The sum of three times some number and 6 is 33. Find the number.

Let x = the number.

$$3x + 6 = 33$$
$$3x = 27$$
$$x = 9$$

Check $3 \cdot 9 + 6 = 27 + 6$
$= 33$

REVIEW EXERCISES—CHAPTER 9

The numbers in brackets refer to the section in the text in which similar problems are presented.

[9.1] Evaluate each formula using the values of the letters as given. Use $\pi = 3.14$ as an approximate value.

1. $I = prt$, where $p = 6000$, $r = 0.17$, $t = 2$

2. $V = \pi r^2 h$, where $r = 3$, $h = 5$

3. $a = \frac{1}{2}h(b_1 + b_2)$, where $h = 7$, $b_1 = 4$, $b_2 = 10$

4. $THI = t - (0.55)(1 - h)(t - 58)$, where $t = 82$, $h = 0.7$

[9.2] Find the value of each expression when $x = 2$, $y = -6$, and $z = -1$.

5. $3x + 2y - 4z$

6. $7x - (x + 2z)$

7. $x - 2xy + yz$

8. $-2xy + 3xz - 4yz$

[9.3] Identify the terms, numerical coefficients, and literal parts of each expression.

9. $3a + 2b$

10. $-6x^2y + 2xy^2 - 5$

11. $-11m^2n + 2mn^2 - 1$

12. $21a^2b - 3ab^2 + 2ab - 5$

[9.3] Combine where possible:

13. $3a - 2b - 5a - 3b$

14. $3m - (-4n) + 2m - 3n$

15. $-3x^2y + 5xy^2 + 2xy - 7x^2y^2$

16. $3ef - (-5ef) - 2$

[9.4] In Exercises 17–22, determine whether $x = -3$ is a solution.

17. $-2x = 6$

18. $2x + 1 = 7$

19. $\frac{6}{x} = 2$

20. $3x - 1 = 4x + 4$

21. $2x + 4 = x - 1$

22. $4(x + 1) = -(5 - x)$

[9.5–9.7] Solve each equation and check:

23. $x - 4 = 7$

24. $-3x = -27$

25. $6 + x = 5$

26. $9x = -27$

27. $\frac{h}{3} = -4$

28. $\frac{2}{3}x = -12$

29. $7y + 5 = -6y + 4$

30. $6x - 4 = 7x + 5$

31. $2x + 6 = 6x - 18$

32. $7x - 6 = -4x - 35$

[9.7] Solve each equation and check:

33. $3(x - 2) = 12$

34. $3a - 5(a + 3) = 2 - (a + 4)$

35. $2y - 3(y - 1) = 2(y + 1)$

36. $5(y + 4) + 2 = 3(2 - y) - 4$

[9.8] Solve each application problem. Be sure to write an equation and check your solution in the original problem.

37. Six is the same as the sum of some number and 4. Find the number.

38. When 4 is subtracted from 3 times a number, the result is 17. Find the number.

39. An unknown number is 12 less than 3 times itself. What is the unknown number?

CRITICAL THINKING EXERCISES—CHAPTER 9

Tell whether each of the following statements is true or false and give reasons why.

1. To solve the equation $\frac{1}{3}x = 9$, we divide both sides of the equation by 3.

2. To solve the equation $6x = 6$, subtract 6 from both sides of the equation.

3. To solve the equation $24x = 6$, divide both sides of the equation by 6.

4. To solve the equation $32x = 5$, divide both sides of the equation by 32.

5. $6x + 7x = 13x^2$

There is an error in each of the following. Can you find it?

6. $4x = 24$
   $4x - 4 = 24 - 4$ Subtract 4 from both sides.
   $x = 20$

7. $x + 4 = 12$
   $x + 4 - 4 = 12 + 4$ Subtract 4 from the left and add 4 to the right to make up for it.
   $x = 16$

8. $3xy + 4x^2y = 7x^3y^2$

9. $12a^2 - 12 = a^2$

10. $3a^2 + 3b^2 = 3a^2b^2$

NAME ▮▮▮▮▮ CLASS

ACHIEVEMENT TEST—CHAPTER 9

Evaluate each formula using the values of the letters as given. Use $\pi = 3.14$ as an approximate value.

1. $V = \frac{4}{3}\pi r^3$, where $r = 2$ — 1. ______

2. $A = \frac{1}{2}bh$, where $b = 6$, $h = 9$ — 2. ______

Evaluate each expression when $a = -3$, $b = 4$, and $c = -2$.

3. $-3a + 4b - 5c$ — 3. ______

4. $-2a^2b - 3ac^2 + ab$ — 4. ______

Identify (a) the terms, (b) the numerical coefficients, and (c) the literal parts in the following expression:

5. $-2ab^3 + 6a^2b^2 + 2$ — 5. ______ ______ ______

Combine where possible:

6. $3x^2y - 2xy^2 + 5x^2y^2$ — 6. ______

7. $2a^2 - 3b^2 + c - 5a^2 - 4b^2 - c$ — 7. ______

8. $-3rst - (-5rst) - 2r + 1$ — 8. ______

Determine whether $y = 4$ is a solution to either of the following equations:

9. $2y + 3 = 5y - 7$ — 9. ______

10. $-2(y - 2) = 3(3 - y) - 1$ — 10. ______

▼ Solve each equation and check:

11. $x - 7 = 4$

12. $\frac{x}{3} = -2$

13. $7 - 2x = -3$

14. $3x - 4 = 5x + 10$

15. $3(x + 4) = 5(x - 2)$

16. $0.3x + 0.4 = 0.6x - 0.5$

11. ______

12. ______

13. ______

14. ______

15. ______

16. ______

▼ Solve each of the following:

17. When 6 times a number is added to 6, the sum is 48. Find the number.

18. Twice the sum of some number and 5 is 18. Find the number.

19. When 4 is subtracted from 3 times some number, the result is the same as twice the sum of the number and 5. What is the number?

20. The Freeport College basketball team beat Johnstown College by 12 points. If the total number of points scored in the game was 162, how many points did each team score?

17. ______

18. ______

19. ______

20. ______

NAME ▮▮▮▮▮ CLASS

CUMULATIVE REVIEW—CHAPTERS 7, 8, 9

1. Find the mean, median, and mode of the following set of data: 38, 41, 49, 37, 35, 35, 46, 43, 35, 42.

1. ______
______

2. The following line graph shows a person's cholesterol level at two-month intervals over a full year.

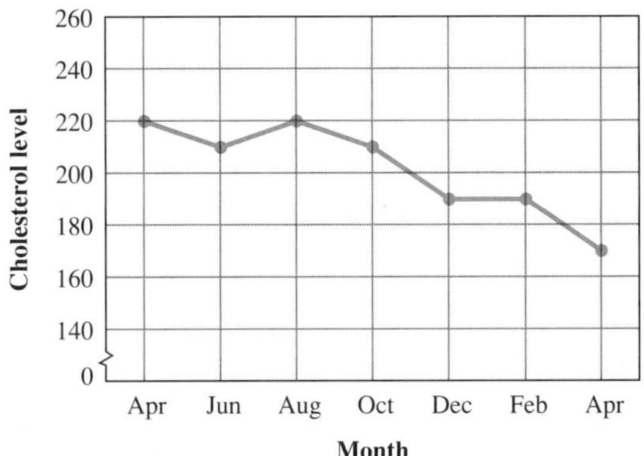

a. When was the cholesterol level at its highest?

2. (a) ______

b. When was it at its lowest?

(b) ______

c. How much did the cholesterol level drop from October to December?

(c) ______

3. Calculate the grade point average (GPA) for the following courses, assuming A = 4, B = 3, C = 2, D = 1, and F = 0.

3. ______

Course	Grade	Credits
French	A	3
Psychology	A	3
Biology	C	4
Typing	B	1
Chemistry	B	4

▼ Evaluate the expression in each of Exercises 4–14.

4. $(-4)(-5)$ 4. ______

5. $-4-(-5)$ 5. ______

6. $-4-5$ 6. ______

7. $-4+5$ 7. ______

8. $5-(-7)-9+6-4$ 8. ______

9. $\frac{-8}{0}$ 9. ______

10. $\sqrt[3]{64}$ 10. ______

11. 10^5 11. ______

12. $(-2)^3$ 12. ______

13. $19-21 \div 3 \cdot 2+4$ 13. ______

14. $14-\{3+[4-(3-1)]\}$ 14. ______

15. Given the formula $A=\frac{1}{2}bh$, find A when $b=13$ in. and $h=8$ in. 15. ______

16. Combine as much as possible: 16. ______

$$5x+7y-2x-10y+5$$

In Exercises 17–20, solve for x:

17. $x + 7 = -15$

18. $-3x = 36$

19. $6x - 1 = 4x + 7$

20. $4(x - 2) = 3x - 7$

21. The Chicago Bulls basketball team beat the Charlotte Hornets by 9 points. If the total number of points scored in the game was 199 points, what was the final score of the game?

17. ______

18. ______

19. ______

20. ______

21. ______

CHAPTER **10**

Geometry

INTRODUCTION

In this chapter we will investigate applications of mathematics to some of the more common geometric figures. The word **geometry** is derived from the Greek words *geo,* meaning earth, and *metron,* meaning measure. Geometry probably was developed in order to establish boundaries and to calculate the sizes of fields.

CHAPTER 10–NUMBER KNOWLEDGE

The following information is taken from the Pete's Pizza Palace menu.

MENU
PETE'S PIZZA PALACE

Pete's Pizza *It's The Best!*	Small 9″ (Serves 1–2)	Medium 13″ (Serves 2–4)	Large 15″ (Serves 4–6)	Medium Pair	Large Pair
Deluxe Cheese	$5.49	$7.49	$9.49	$11.99	$14.99
Each Topping (Pepperoni, sausage, meatballs, onions, olives, peppers, mushrooms)	1.00	1.25	1.50	2.00/pr.	2.50/pr.
Extra cheese	1.00	1.25	1.50	2.00/pr.	2.50/pr.

We have a choice of small (9″), medium (13″), or large (15″) pizzas, all at different prices. How are we to know which will give us the most pizza for our dollar? To keep things simple, let's answer the question only for deluxe cheese pizzas.

In the chart at the right, the unit price for each size has been calculated. This involves finding the area of a circular pizza, a technique we will investigate in this chapter.

The most expensive way to buy your pizza is to purchase the small size, which costs $0.0863 per square inch. The least expensive is the large pair at $0.0424 per square inch. The small size costs more than twice as much per square inch as the large pair.

Size	Total area	Price	$/in.^{2}
9″	63.6 in.^{2}	$ 5.49	$0.0863
13″	132.7 in.^{2}	7.49	$0.0564
15″	176.6 in.^{2}	9.49	0.0537
13″ pair	265.4 in.^{2}	11.99	0.0452
15″ pair	353.2 in.^{2}	14.99	0.0424

10.1 Names of Geometric Figures

A **polygon** is a closed geometric figure with sides that are all straight lines; see Figure 10.1

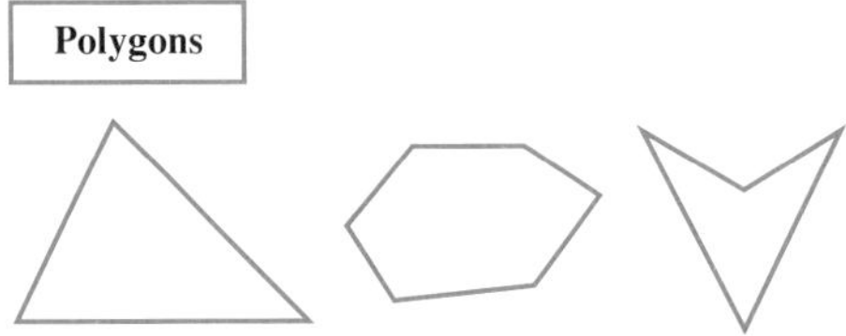

FIGURE 10.1. Polygons

A **triangle** is a polygon with three sides. Every triangle has three sides, three angles, and three vertices, as pictured in Figure 10.2.

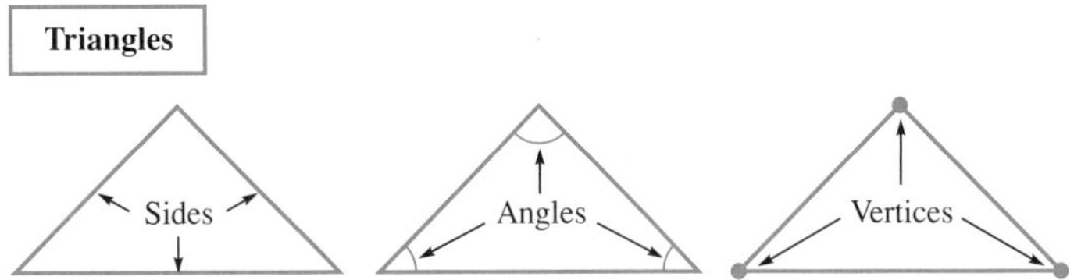

FIGURE 10.2. Triangles

An **equilateral triangle** has all three sides of equal length **(congruent sides).** An **equilateral triangle** also has all three angles of equal measure **(congruent angles);** see Figure 10.3.

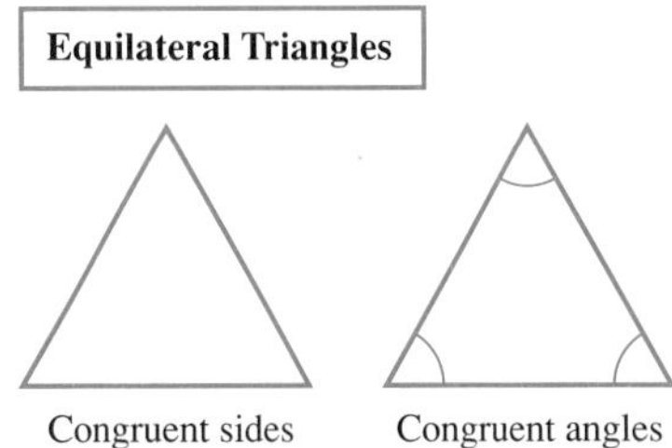

FIGURE 10.3. Equilateral triangles

A **right triangle** is a triangle with a **right angle** (a square corner measuring 90°) (Figure 10.4).

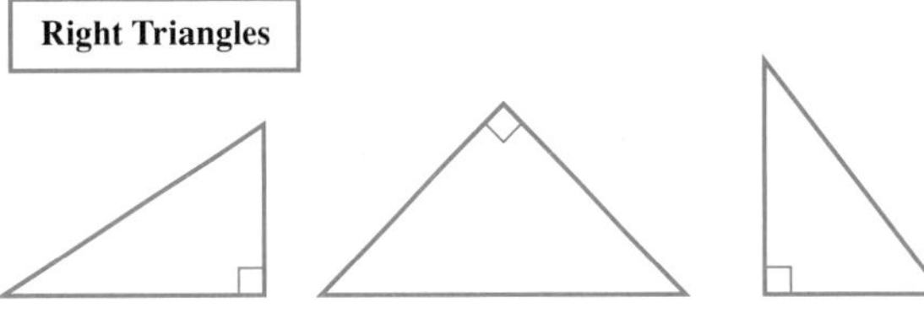

FIGURE 10.4. Right triangles

A **quadrilateral** is a polygon with four sides. Some examples are shown in Figure 10.5.

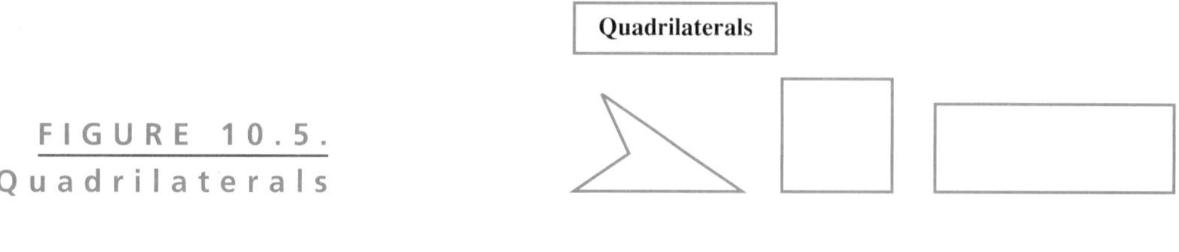

FIGURE 10.5. Quadrilaterals

A **rectangle** is a quadrilateral with four right angles (square corners), as shown in Figure 10.6.

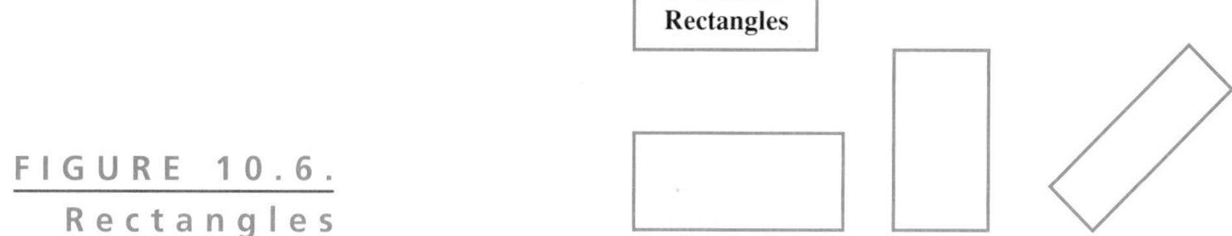

FIGURE 10.6. Rectangles

A **square** is a rectangle with all sides of equal length **(congruent sides);** see Figure 10.7.

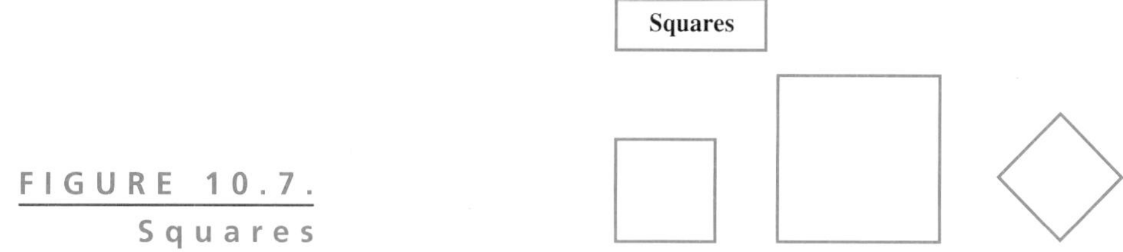

FIGURE 10.7. Squares

A **parallelogram** is a quadrilateral with two pairs of parallel sides. Three examples are pictured in Figure 10.8.

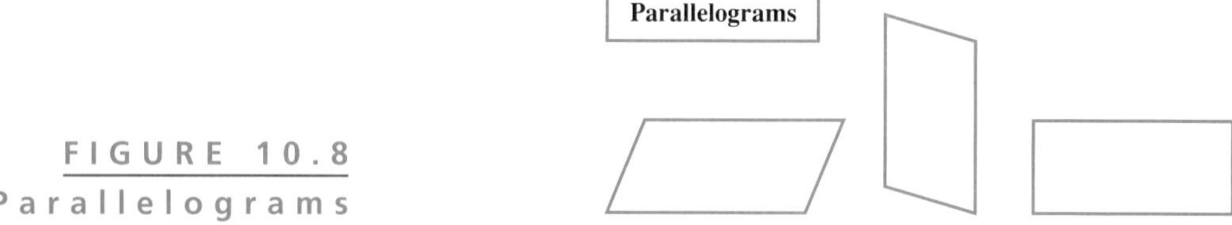

FIGURE 10.8 Parallelograms

A **trapezoid** is a quadrilateral with one pair of parallel sides; see Figure 10.9.

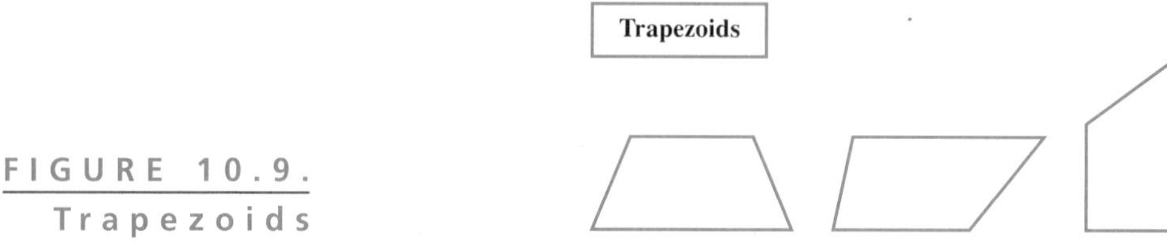

FIGURE 10.9. Trapezoids

Other common polygons are:

- a **pentagon** with five sides and five angles;
- a **hexagon** with six sides and six angles;
- an **octagon** with eight sides and eight angles.

A **circle** is a figure in which all points are the same distance from a fixed point, called the **center.** This distance is called the **radius.** (Figure 10.10)

FIGURE 10.10.
Circle

10.1 Exercises

Using the definitions of this section, match the geometric figure with the proper name.

- Trapezoid
- Parallelogram
- Octagon
- Circle
- Equilateral triangle
- Right triangle
- Rectangle
- Square
- Pentagon

1.

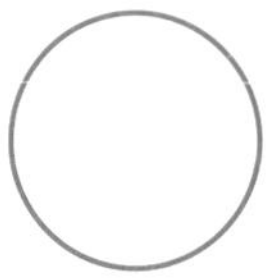

2.

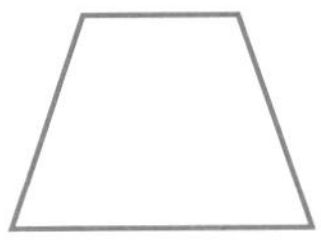

3.

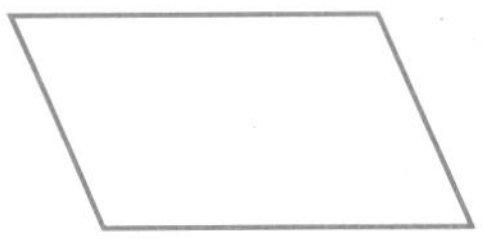

4.

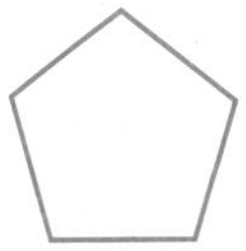

5.

6.

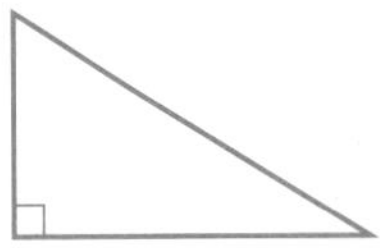

7.

9.

8.

10.2 Rectangles and Squares

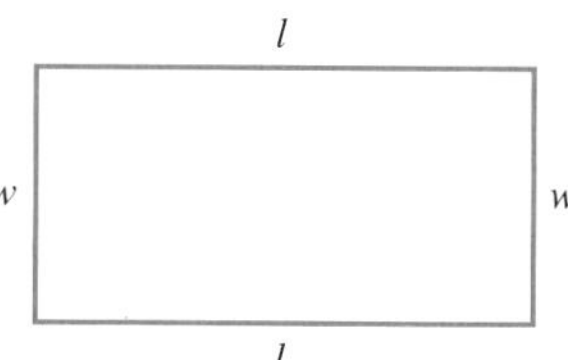

FIGURE 10.11.

The area A of the rectangle pictured in Figure 10.11 is the space inside the lines and is calculated by multiplying the length l times the width w. The result is given in square units, such as square inches or square centimeters, and is represented by the formula

$$\text{area} = A = l \times w$$

The perimeter P of a rectangle is the measure of the distance around the outside of the rectangle and is calculated by adding twice the length to twice the width. The formula is

$$\text{perimeter} = P = 2l + 2w$$

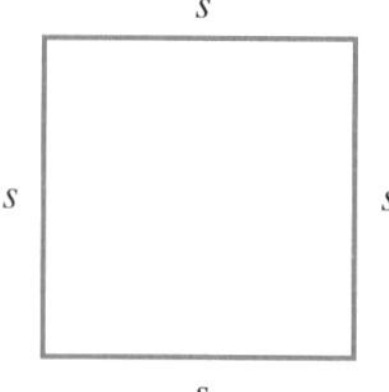

FIGURE 10.12.

A square (see Figure 10.12) is a special case of a rectangle in which the length and the width are the same. If we let s represent both the length and the width, we obtain the formulas for the area and perimeter of a square.

$$\text{area} = A = s \times s = s^2$$

$$\text{perimeter} = P = s + s + s + s = 4s$$

EXAMPLE 1 Find the area and perimeter of a rectangle that is 9 in. long and 6 in. wide.

SOLUTION

$$\text{area} = l \times w = 9 \text{ in.} \times 6 \text{ in.} = 54 \text{ in.}^2 \qquad (\text{in.} \times \text{in.} = \text{in.}^2)$$

$$\text{perimeter} = 2l + 2w = (2 \times 9 \text{ in.}) + (2 \times 6 \text{ in.}) = 18 \text{ in.} + 12 \text{ in.} = 30 \text{ in.}$$

Notice that the units for area are square inches (written in.^{2}) and the units for perimeter are inches.

EXAMPLE 2 Find the area and perimeter of a square that measures 6 ft on a side.

SOLUTION

$$\text{area} = s^2 = (6 \text{ ft})^2 = 36 \text{ ft}^2$$

$$\text{perimeter} = 4s = 4 \times 6 \text{ ft} = 24 \text{ ft}$$

EXAMPLE 3 How many square yards of carpeting are needed to cover a room that measures 4 yd by 3 yd?

SOLUTION

$$A = l \times w = 4 \text{ yd} \times 3 \text{ yd} = 12 \text{ yd}^2$$

EXAMPLE 4 How much will it cost to fence a yard that measures 75 ft by 130 ft if the fence costs \$6/ft?

SOLUTION

$$\text{perimeter} = 2l \times 2w = 2(130 \text{ ft}) + 2(75 \text{ ft})$$

$$= 260 \text{ ft} + 150 \text{ ft} = 410 \text{ ft}$$

$$\text{cost} = 410 \text{ ft} \times \$6/\text{ft} = \$2460$$

10.2 Exercises

In Exercises 1–6, find the area and the perimeter of each rectangle or square.

	LENGTH	WIDTH
1.	4 ft	3 ft
2.	18 cm	6 cm
3.	14 in.	14 in.
4.	4.35 m	7.14 m (express your answer to the nearest hundredth)
5.	4 yd	4 yd
6.	$17\frac{3}{4}$ ft	$11\frac{1}{2}$ ft

7. What is the cost of carpeting a room that measures 15 feet by 12 feet if the carpeting costs \$14 per square yard? Be careful of your units.

8. What is the length of a room that has an area of 144 square feet and is 9 feet wide?

9. Figure 10.13 shows the dimensions of a kitchen and dining room. How much will it cost to tile the floor if the tile is priced at \$1.49 a square foot? A baseboard is to be placed around the border of the room at a cost of \$1.10 per foot. How much will this cost?

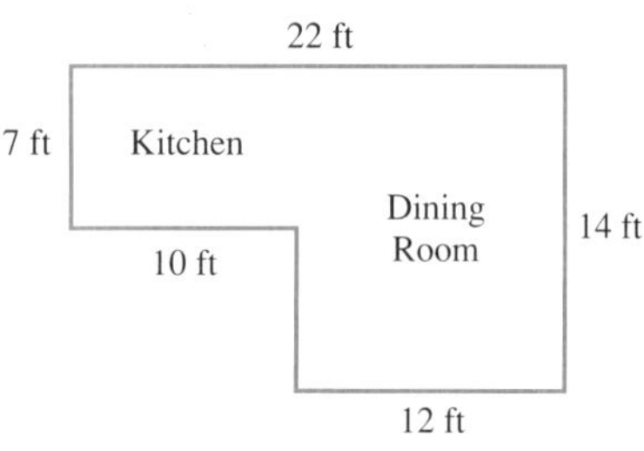

FIGURE 10.13.

10. A room measures 14 ft long by 12 ft wide, and the walls are $7\frac{1}{2}$ ft high. How many gallons of paint should you buy to paint the wall if one gallon of paint covers 275 sq. ft? How many gallons will be needed if two coats are required?

11. It is possible for the area and the perimeter of a rectangle to equal the same number. Find such a rectangle.

10.3 Triangles

Figure 10.14 shows two typical triangles. The **base *b*** of a triangle is the side on which it appears to rest. The **altitude** or **height *h*** of a triangle is the shortest distance from the base to the top of the triangle.

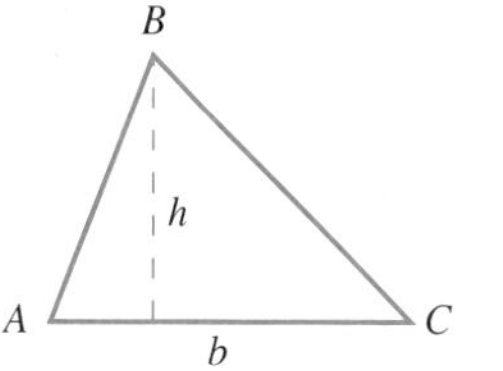

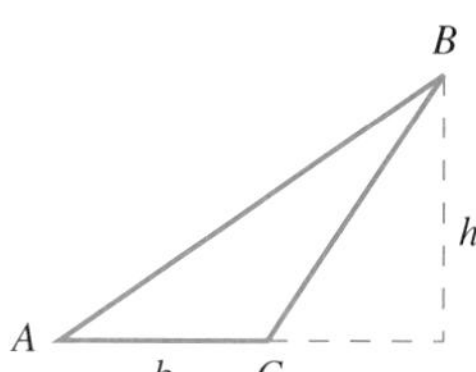

FIGURE 10.14.

The **area** A of a triangle is given by

$$A = \frac{1}{2} \text{ base} \times \text{height} = \frac{1}{2} bh$$

The **perimeter** P is found by adding the lengths of the three sides together:

$$P = s_1 + s_2 + s_3$$

See Figure 10.15.

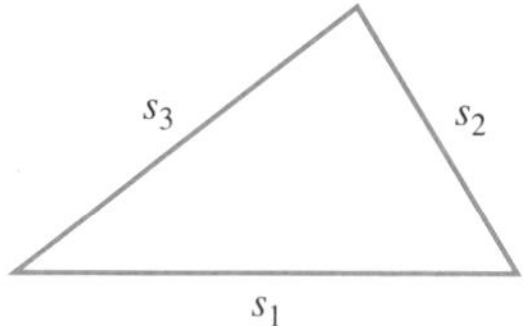

FIGURE 10.15.

EXAMPLE 1 Find the area and perimeter of the triangle in Figure 10.16.

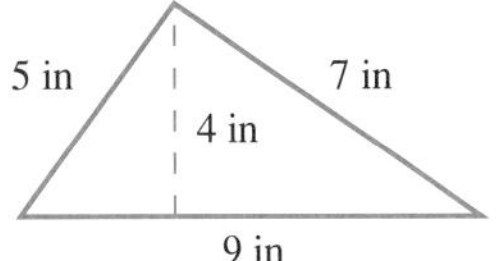

FIGURE 10.16.

SOLUTION

$$A = \frac{1}{2}bh = \left(\frac{1}{2}\right)(9 \text{ in.})(4 \text{ in.}) = 18 \text{ in.}^2$$

$$P = s_1 + s_2 + s_3 = 5 \text{ in.} + 7 \text{ in.} + 9 \text{ in.} = 21 \text{ in.}$$

EXAMPLE 2 Find the area and perimeter of the triangle in Figure 10.17.

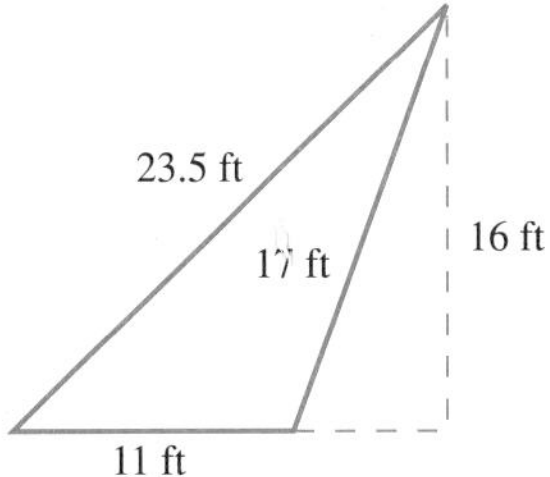

FIGURE 10.17.

SOLUTION

$$A = \frac{1}{2}bh = \left(\frac{1}{2}\right)(11 \text{ ft})(16 \text{ ft}) = 88 \text{ ft}^2$$

$$P = s_1 + s_2 + s_3 = 23.5 \text{ ft} + 11 \text{ ft} + 17 \text{ ft} = 51.5 \text{ ft}$$

EXAMPLE 3 Find the area and perimeter of the right triangle in Figure 10.18.

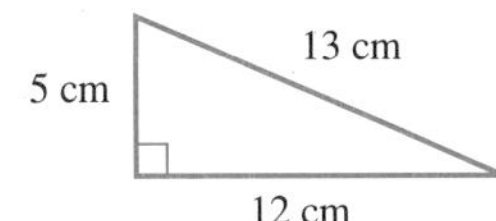

FIGURE 10.18.

SOLUTION

$$A = \frac{1}{2}bh = \left(\frac{1}{2}\right)(12 \text{ cm})(5 \text{ cm}) = 30 \text{ cm}^2$$

$$P = s_1 + s_2 + s_3 = 5 \text{ cm} + 12 \text{ cm} + 13 \text{ cm} = 30 \text{ cm}$$

10.3 Exercises

1. Find the area and perimeter of the triangle in Figure 10.19.

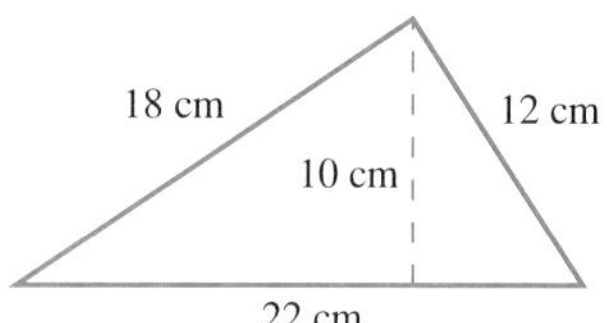

FIGURE 10.19.

2. Find the area and perimeter of the triangle in Figure 10.20.

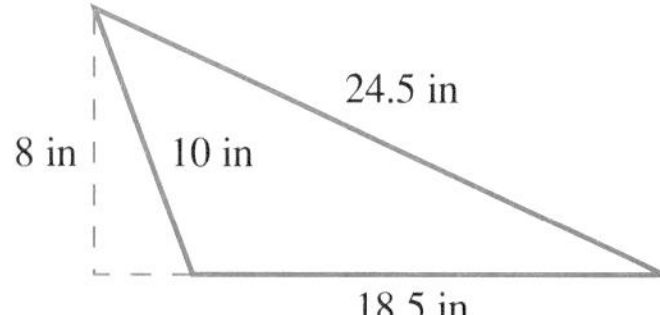

FIGURE 10.20.

3. Find the area and perimeter of the triangle in Figure 10.21.

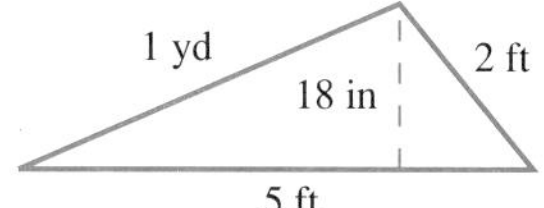

FIGURE 10.21.

4. Find the area and perimeter of the right triangle in Figure 10.22.

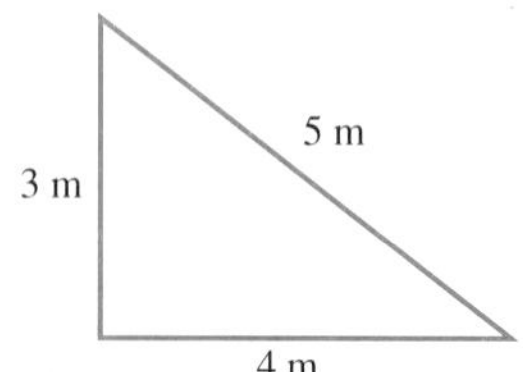

FIGURE 10.22.

5. Find the area and perimeter of the shape in Figure 10.23.

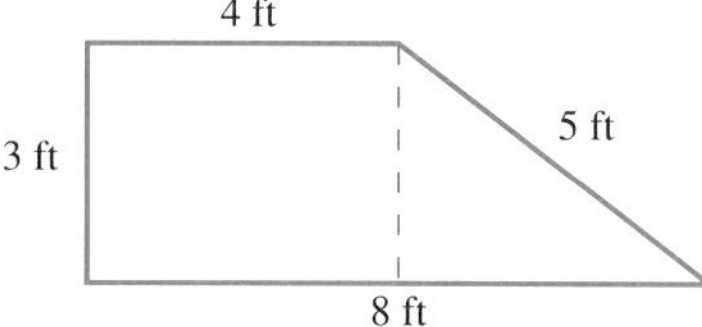

FIGURE 10.23.

6. How many square feet of siding will be required to cover the end of a house with the dimensions given in Figure 10.24?

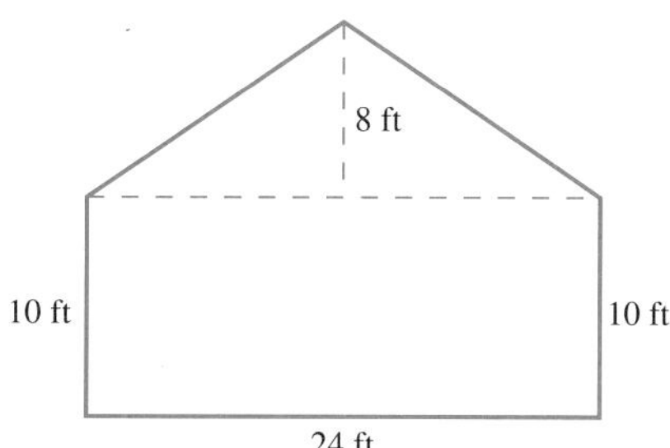

FIGURE 10.24.

7. What is the height of a triangle that has a base of 14 in. and an area of 49 sq. in.?

Find the area of the triangle with the given base and height.

	Base	Height	Area
8.	26 cm	26 cm	
9.	15 in.	9 in.	
10.	2.6 ft	3.4 ft	
11.	9.6 m	1.2 m	
12.	$3\frac{3}{4}$ in.	$5\frac{1}{2}$ in.	

10.4 Circles

A **circle** is shown in Figure 10.25:

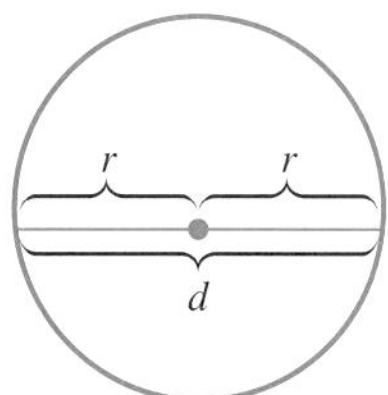

FIGURE 10.25.

It is defined to be a geometric figure on which every point is the same distance from the center. The distance from the center to any point on the circle is called the **radius,** *r*. The **diameter,** *d*, is the greatest line that can be drawn across the circle. It will always pass through the center and is equal to twice the radius.

The **circumference,** *C*, of the circle is the perimeter or length of the circle. In any circle, if we divide the circumference by the diameter we will always get the same result, a number represented by the Greek letter **pi,** written **π.** In any circle, no matter what the size,

$$\frac{\text{circumference}}{\text{diameter}} = \pi = 3.14159\ldots$$

The number π is used in calculating both the circumference and the area of a circle using the following formulas:

> **Area of a circle:** $A = \pi r^2$
> **Circumference of a circle:** $C = \pi d$ or $C = 2\pi r$

In using π for actual calculations, the value is rounded off to an appropriate number of places, depending on the degree of accuracy required. In the following examples, we will use a value for π of 3.14.

The **circumference** of a circle is the distance around the circle; see Figure 10.26.

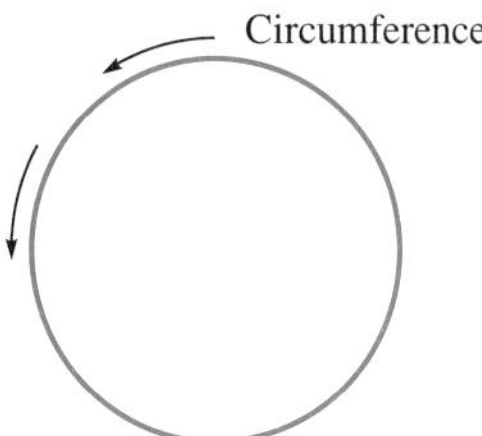

FIGURE 10.26.

The **area** of a circle is the portion of the plane enclosed by the circle; see Figure 10.27.

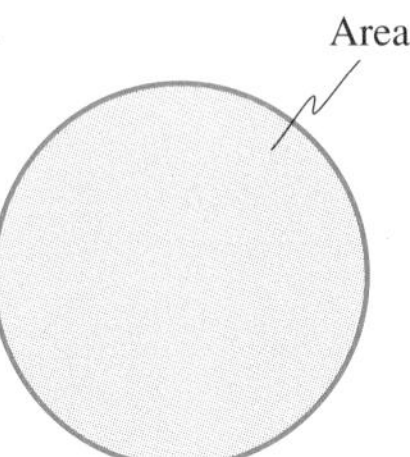

FIGURE 10.27.

EXAMPLE 1 Find the circumference and area of a circle that has a radius of 4 in.

SOLUTION

$$C = 2\pi r = (2)(3.14)(4 \text{ in.}) = 25.12 \text{ in.}$$
$$A = \pi r^2 = (3.14)(4 \text{ in.})^2 = (3.14)(16 \text{ in.}^2) = 50.24 \text{ in.}^2$$

EXAMPLE 2 Find the circumference and area of a circle having a diameter of 10 cm.

SOLUTION

$$C = \pi d = (3.14)(10 \text{ cm}) = 31.4 \text{ cm}$$

Now, since $r = \frac{1}{2}d = \left(\frac{1}{2}\right)(10 \text{ cm}) = 5 \text{ cm}$, we have

$$A = (3.14)(5 \text{ cm})^2$$
$$= (3.14)(25 \text{ cm}^2) = 78.5 \text{ cm}^2$$

EXAMPLE 3 In our problem at the beginning of this chapter involving the cost of pizzas, we needed to find the area of small (9-in.), medium (13-in.) and large (15-in.) pizzas. These numbers represent the diameters, so we need to find the radius of each pizza by taking half of these numbers, giving us 4.5 in., 6.5 in., and 7.5 in., respectively. We can now find the areas.

SOLUTION The formula for the area of a circle is

$$A = \pi r^2$$

- small (9″): $A = (3.14)(4.5 \text{ in.})^2 = (3.14)(20.25 \text{ in.}^2) = 63.6 \text{ in.}^2$
- medium (13″): $A = (3.14)(6.5 \text{ in.})^2 = (3.14)(42.25 \text{ in.}^2) = 132.7 \text{ in.}^2$
- large (15″): $A = (3.14)(7.5 \text{ in.})^2 = (3.14)(56.25 \text{ in.}^2) = 176.6 \text{ in.}^2$

10.4 Exercises

Complete the following table using an approximate value of $\pi = 3.14$ and round each answer to one decimal place. Be careful to use correct units for each answer.

	Radius	Diameter	Circumference	Area
1.		4 ft		
2.	6 in.			
3.	3 cm			
4.		7 m		
5.	1.4 in.			
6.		4.4 cm		
7.	$4\frac{1}{2}$ ft			

8. Compare the area of a 2-in.-radius circle with the area of a 1-in.-radius circle. Is the area twice as large?

9. Mike Worden plans to build a circular flower bed that has a diameter of 6 ft. A flat of flowers will cover 6 square feet. How many flats must be purchased?

10. The oddly shaped corner lot pictured in Figure 10.28 is a quarter circle with a radius of 100 ft. What is the area of the lot? What is its perimeter?

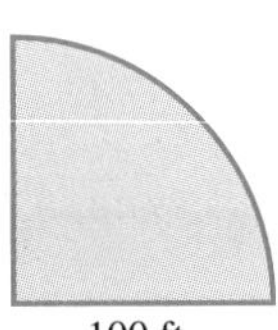

FIGURE 10.28.

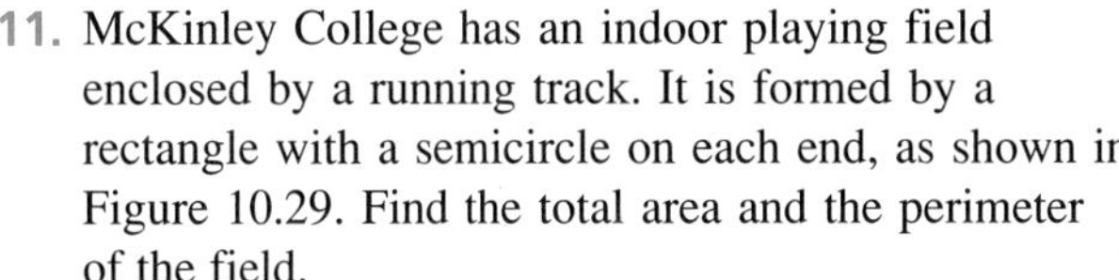

11. McKinley College has an indoor playing field enclosed by a running track. It is formed by a rectangle with a semicircle on each end, as shown in Figure 10.29. Find the total area and the perimeter of the field.

FIGURE 10.29.

10.5 Rectangular Solids

Figure 10.30 shows a **rectangular solid.** We are interested in finding the measure of the inside of such a solid, which is called its **volume.** The volume V of a rectangular solid is found by multiplying the length times the width times the height, and the result is given in cubic units.

$$\text{volume} = \text{length} \times \text{width} \times \text{height}$$

or

$$V = lwh$$

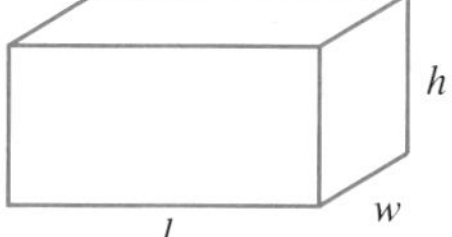

FIGURE 10.30.

A cube, shown in Figure 10.31; is a special case of a rectangular solid in which the length, the width, and the height are all the same size. We represent these equal lengths by the letter s (for side) and the volume will be given by

$$V = s^3$$

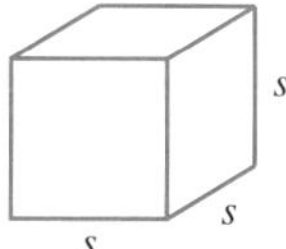

FIGURE 10.31.

EXAMPLE 1 Find the volume of a rectangular solid that measures 6 in. long, 2 in. wide, and 3 in. high.

SOLUTION

$$V = lwh = (6 \text{ in.})(2 \text{ in.})(3 \text{ in.}) = 36 \text{ in.}^3$$

EXAMPLE 2 Find the volume of a cube that measures 8 cm on each edge.

SOLUTION

$$V = s^3 = (8 \text{ cm})^3 = (8 \text{ cm})(8 \text{ cm})(8 \text{ cm}) = 512 \text{ cm}^3$$

EXAMPLE 3 The interior measurements of a refrigerator are 2.8 ft wide by 2.5 ft deep by 3 ft high. What is the capacity of the refrigerator in cubic feet?

SOLUTION

$$V = lwh = (2.8 \text{ ft})(2.5 \text{ ft})(3 \text{ ft}) = 21 \text{ ft}^3$$

10.5 Exercises

Find the following:

1. Find the volume of a box that measures 12 in. long by 6 in. wide by 6 in. high.

2. Find the volume of a rectangular solid that is 26 cm high, 14 cm wide, and 20 cm long.

3. How much earth came from a rectangular hole that measures 6 ft deep, 3 ft wide, and 6 ft long?

4. What is the volume of a cube that measures 2 m on an edge?

5. What is the length of the edge of a cube that has a volume of 64 cu. in.?

6. A child's sandbox has a square base that measures 4 ft by 4 ft, and it is 1 ft high. How many cubic feet of sand are required to fill the sandbox half full?

7. How many cubic feet of air are contained in a room that measures 10 ft by 14 ft and has an 8-ft ceiling?

8. How many cubic yards of cement are needed for a sidewalk that is 3 ft wide, 4 in. thick, and 30 ft long?

9. What is the capacity in cubic feet of a bin 7 ft 4 in. long, 4 ft 6 in. wide, and 3 ft 9 in. deep?

10. Find the volume and surface area of a closed box that measures 9 in. by 8 in. by 5 in.

10.6 Volumes of Other Solids

Spheres

A **sphere** is a three-dimensional figure that is shaped like a ball; it is shown in Figure 10.32. The **radius** r is the distance from the center of the sphere to any point on the surface.

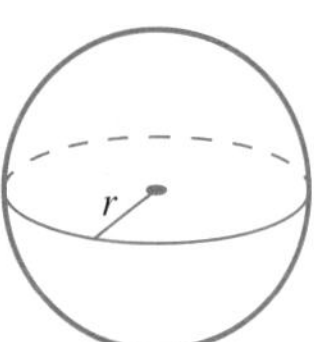

FIGURE 10.32.

The **volume** V of a sphere is given by the formula

$$V = \frac{4}{3}\pi r^3$$

EXAMPLE 1 Find the volume of a sphere that has a radius of 2 ft. Use $\pi = 3.14$ as an approximation and round your answer to one decimal place.

SOLUTION

$$V = \frac{4}{3}\pi r^3 = \frac{4}{3}(3.14)(2 \text{ ft})^3$$

$$= \frac{4}{3}(3.14)(8 \text{ ft}^3) = 33.5 \text{ ft}^3$$

EXAMPLE 2 Find the volume of a 20-in. sphere.

SOLUTION "A 20-in. sphere" means that it has a *diameter* of 20 in. Therefore, its radius is one-half of that, or $r = 10$ in.

$$V = \frac{4}{3}\pi r^3 = \frac{4}{3}(3.14)(10 \text{ in.})^3$$

$$= \frac{4}{3}(3.14)(1000 \text{ in.}^3) = 4186.7 \text{ in.}^3$$

Cylinders

A **cylinder,** also called a **right circular cylinder,** is shown in Figure 10.33. It is basically the shape of a tin can. The base has **radius** r and the side has **height** h.

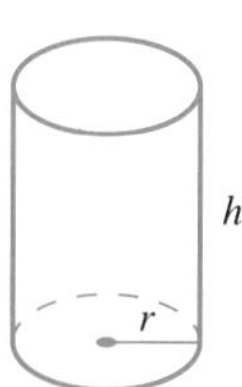

FIGURE 10.33.

The **volume** of a cylinder is given by the formula

$$V = \pi r^2 h$$

The volume is actually the area of the circular base (πr^2) multiplied by the height (h).

EXAMPLE 3 Find the volume of the cylinder shown in Figure 10.34.

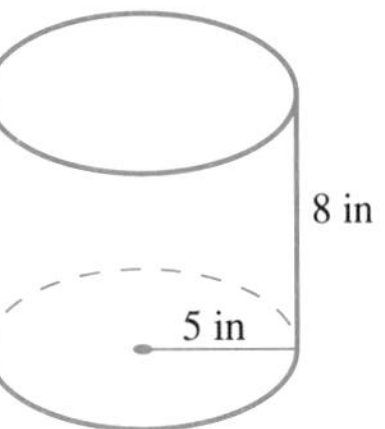

FIGURE 10.34.

SOLUTION

$$V = \pi r^2 h = (3.14)(5\text{ in.})^2(8\text{ in.})$$
$$= (3.14)(5\text{ in.})(5\text{ in.})(8\text{ in.}) = 628\text{ in.}^3$$

EXAMPLE 4 What is the volume of a tin can that is 6 in. high and measures 4 in. across the base? (See Figure 10.35.)

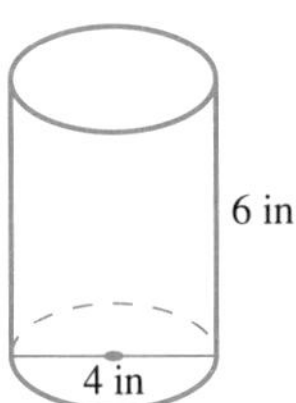

FIGURE 10.35.

SOLUTION The diameter of the circular base is 4 in. and therefore the radius is 2 in. So

$$V = \pi r^2 h = (3.14)(2\text{ in.})^2(6\text{ in.})$$
$$= (3.14)(2\text{ in.})(2\text{ in.})(6\text{ in.}) = 75.36\text{ in.}^3$$

10.6 Exercises

Complete the following table by finding the volumes of the given spheres. Use $\pi \approx 3.14$ and round answers to one decimal place.

	Radius	Diameter	Volume
1.	3 in.		
2.	6 in.		
3.		10 cm	
4.	3.2 ft		
5.		7 ft	
6.		1 m	

7. Find the volume of a 6-ft diameter spherical weather balloon.

8. **a.** Find the volume of a sphere with radius 1 ft.

b. Find the volume of a sphere with radius 2 ft.

c. What happens to the volume of a sphere when the radius is doubled?

9. A **hemisphere** is half of a sphere. Find the volume of the hemisphere in Figure 10.36.

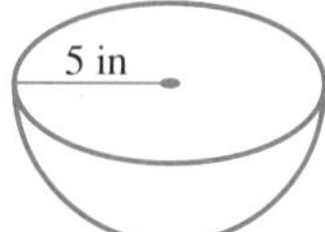

FIGURE 10.36.

▼ Complete the following table by finding the volumes of the given cylinders. Use $\pi \approx 3.14$ and round your answers to one decimal place.

	Radius	Diameter	Height	Volume
10.	3 in.		8 in.	
11.		4 ft	6 ft	
12.	2.4 cm		6.8 cm	
13.		7 in.	9 in.	
14.	$1\frac{3}{4}$ m		$2\frac{1}{2}$ m	
15.		$6\frac{2}{3}$ m	5 m	

16. Find the volume of a paint can with radius 4 in. and height 12 in.

17. Find the volume of the cylinder in Figure 10.37.

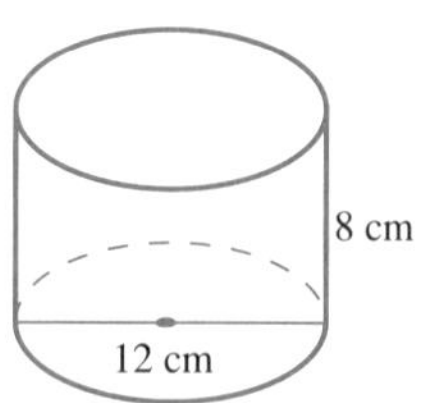

FIGURE 10.37.

18. Find the volume of a cylindrically shaped hot water heater that is 4 ft high and has a base radius of 1.2 ft.

19. Figure 10.38 shows a cylinder with a hemispherical cap on top. What is the total volume?

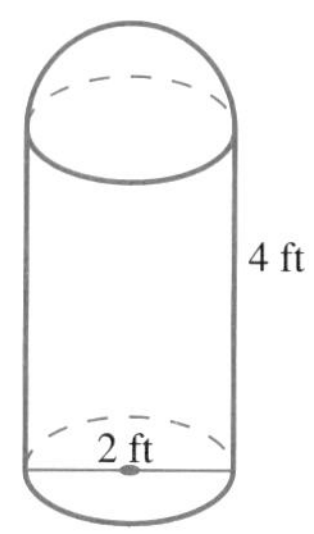

FIGURE 10.38.

20. Figure 10.39 shows a cylinder with a hemisphere on each end. What is its total volume?

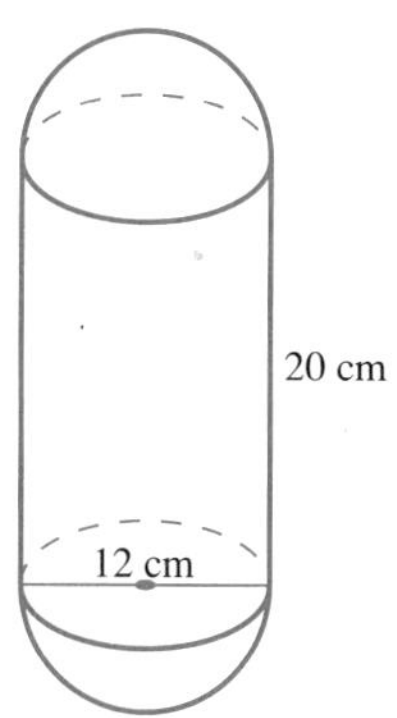

FIGURE 10.39.

SUMMARY—CHAPTER 10

The numbers in brackets refer to the section in the text in which each concept is presented.

EXAMPLES

[10.1] The definition of all of the geometric figures in the first section should be reviewed.

[10.2] **Rectangle**

- area: $A = lw$
- perimeter: $P = 2l + 2w$

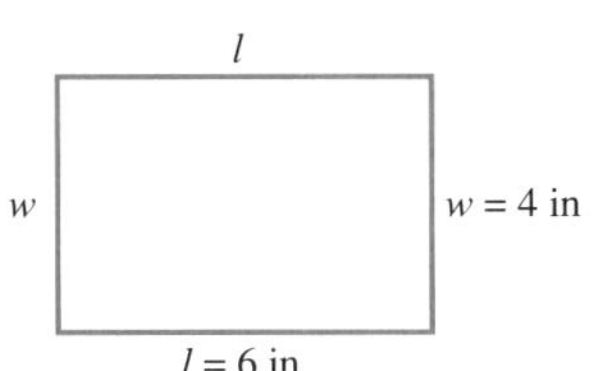

[10.2] $A = (6 \text{ in.})(4 \text{ in.}) = 24 \text{ in.}^2$
$P = 2(6 \text{ in.}) + 2(4 \text{ in.})$
$= 12 \text{ in.} + 8 \text{ in.}$
$= 20 \text{ in.}$

[10.2] **Square**

- area: $A = s^2$
- perimeter: $P = 4s$

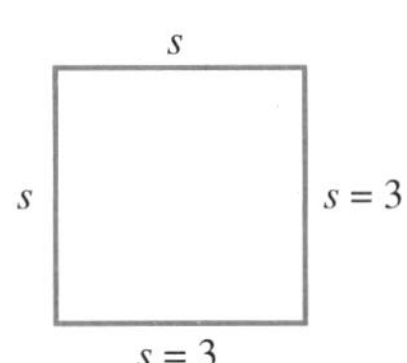

[10.2] $A = (3 \text{ in.})^2$
$= (3 \text{ in.})(3 \text{ in.})$
$= 9 \text{ in.}^2$
$P = 4(3 \text{ in.}) = 12 \text{ in.}$

[10.3] **Triangle**

- area: $A = \frac{1}{2}bh$
- perimeter: $P = s_1 + s_2 + s_3$

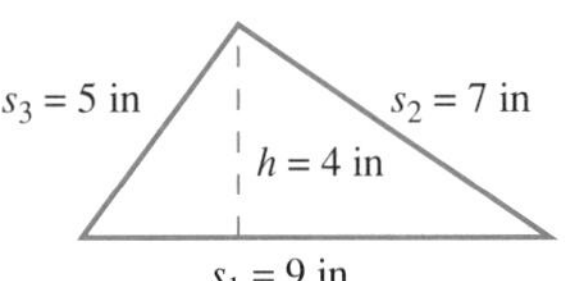

[10.3] $A = \frac{1}{2}(9 \text{ in.})(4 \text{ in.})$
$= 18 \text{ in.}^2$
$P = 9 \text{ in.} + 7 \text{ in.} + 5 \text{ in.}$
$= 21 \text{ in.}$

[10.4] **Circle**

- area: $A = \pi r^2$
- circumference: $C = \pi d = 2\pi r$

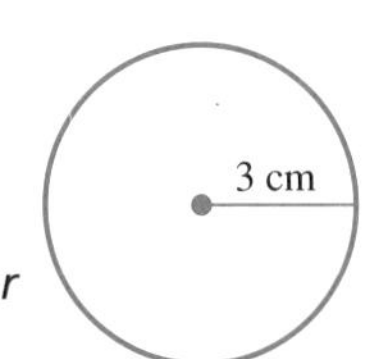

[10.4] $A = (3.14)(3 \text{ cm})$
$= (3.14)(3 \text{ cm})(3 \text{ cm})$
$= 28.26 \text{ cm}^2$
$C = 2(3.14)(3 \text{ cm})$
$= 18.84 \text{ cm}$

[10.5] **Rectangular solid**
Volume: $V = lwh$

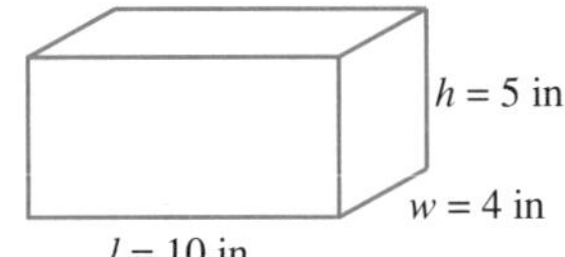

[10.5] $V = (10 \text{ in.})(4 \text{ in.})(5 \text{ in.})$
$= 200 \text{ in.}^3$

[10.5] **Cube**
Volume: $V = s^3$

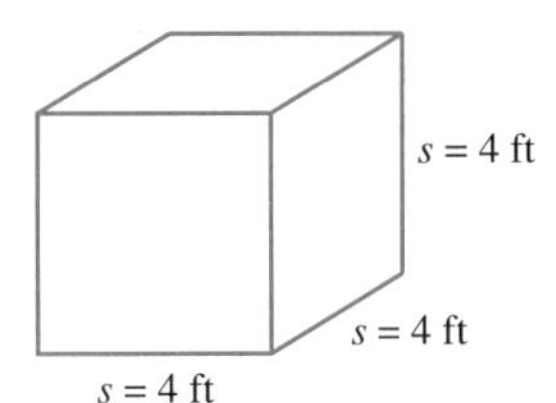

[10.5] $V = (4 \text{ ft})^3$
$= (4 \text{ ft})(4 \text{ ft})(4 \text{ ft})$
$= 64 \text{ ft}^3$

EXAMPLES

[10.6] **Sphere**
Volume: $V = \frac{4}{3}\pi r^3$

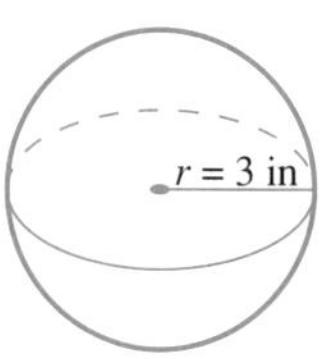

[10.6] $V = \frac{4}{3}(3.14)(3 \text{ in.})^3$
$= \frac{4}{3}(3.14)(3 \text{ in.})(3 \text{ in.})(3 \text{ in.})$
$= 113.04 \text{ in.}^3$

[10.6] **Cylinder**
Volume: $V = \pi r^2 h$

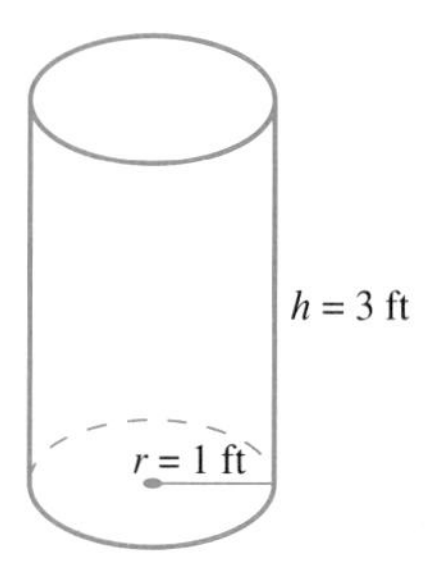

[10.6] $V = (3.14)(1 \text{ ft})^2(3 \text{ ft})$
$= (3.14)(1 \text{ ft})(1 \text{ ft})(3 \text{ ft})$
$= 9.42 \text{ ft}^3$

REVIEW EXERCISES—CHAPTER 10

The numbers in brackets indicate the section in the text in which similar problems are presented.

[10.1] Match the geometric figure with the correct name.

1.

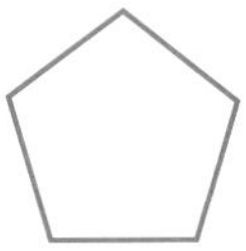

2.

3.

4.

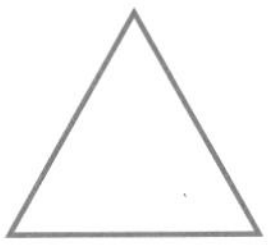

5.

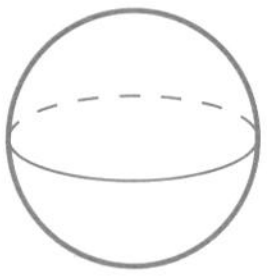

6.

7.

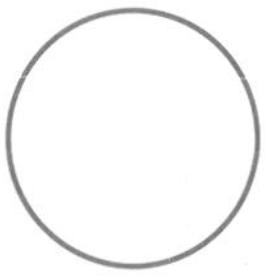

8.

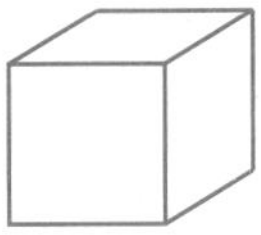

9.

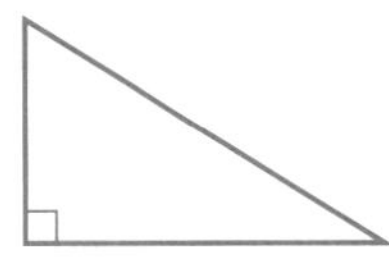

10.

11.

12.

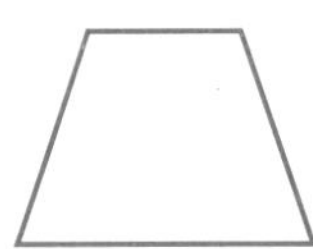

- Triangle
- Cylinder
- Trapezoid
- Right triangle
- Sphere
- Cube
- Equilateral triangle
- Parallelogram
- Circle
- Pentagon
- Rectangle
- Square

▼ [10.2]

13. Find the area and perimeter of a rectangle that measures 8 ft by 4 ft.

▼ [10.2]

14. Find the area and perimeter of a square that measures 10 in. on an edge.

▼ [10.2]

15. Find the area and perimeter of Figure 10.40.

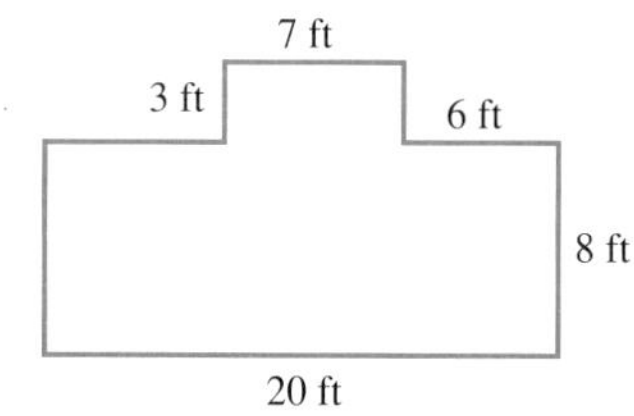

FIGURE 10.40.

▼ [10.3]

16. Find the area of a triangle with a base of 15 cm and a height of 6 cm.

▼ [10.3]

17. Find the area and perimeter of Figure 10.41.

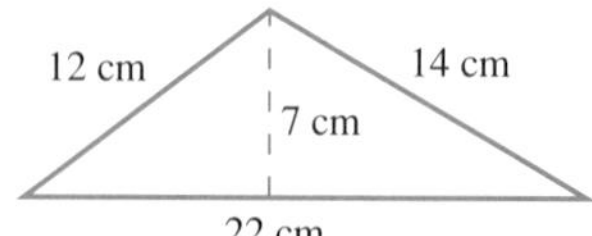

FIGURE 10.41.

▼ [10.2]–[10.3]

18. Find the area of Figure 10.42.

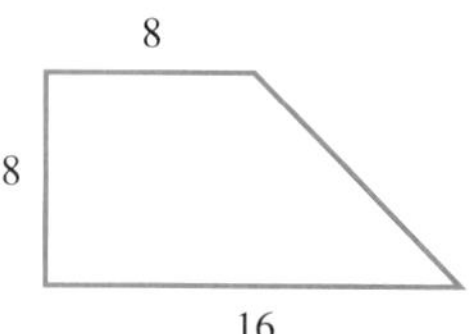

FIGURE 10.42.

▼ [10.4]

19. Find the area and circumference of a circle with a radius of 6 in.

▼ [10.4]

20. What is the area and circumference of a circular flower garden that has a diameter of 15 ft?

▼ [10.5]

21. Find the volume of Figure 10.43.

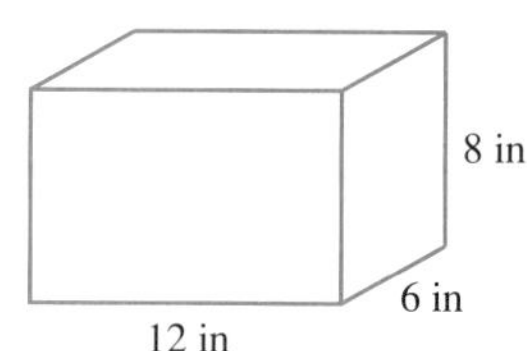

FIGURE 10.43.

▼ [10.5]

22. An open-top box is 18 cm long, 6 cm deep, and 7 cm high. What is its volume and surface area?

▼ [10.5]

23. Find the volume of a cube that measures 20 cm on each edge.

▼ [10.6]

24. Find the volume of a 12-in. sphere.

▼ [10.6]

25. A five-gallon pail in the shape of a right circular cylinder has a base with diameter 14 in. and is 20 in. tall. What is its volume?

▼ [10.6]

26. Find the volume of the hemisphere shown in Figure 10.44.

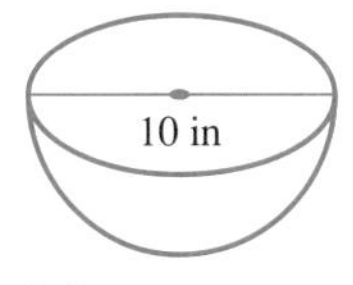

FIGURE 10.44.

CRITICAL THINKING EXERCISES—CHAPTER 10

Tell whether each of the following is true or false and give reasons why.

1. If the length and width of a rectangle are both doubled, the perimeter will be doubled.

2. If the length and width of a rectangle are both doubled, the area will be doubled.

3. If the radius of a circle is doubled, the circumference will be doubled.

4. If the radius of a circle is doubled, the area will be doubled.

5. In a triangle, if you double the length of the base and cut the altitude in half, the area will be the same.

▼ Find the error in each of the following:

6. A 16-in.-diameter pizza contains twice as much as an 8-in. diameter pizza.

7. Every square is a rectangle, therefore every rectangle must be a square.

8. The circumference of a circle is the amount of area inside of it.

9. Mr. Jardine ordered 168 ft of baseboard molding to put around the edge of his 14-ft-by-12-ft room.

10. Mrs. Amelio ordered 144 cu. ft of sand to fill her child's sandbox, which measures 6 ft long, 4 ft wide, and 6 in. high.

NAME ▮▮▮▮▮ CLASS

ACHIEVEMENT TEST—CHAPTER 10

1. Find (a) the area and (b) the perimeter of a rectangle that measures 6 m by 5 m.

1. (a) ______ (b) ______

2. Find (a) the area and (b) the perimeter of a square with each edge equal to 16 cm.

2. (a) ______ (b) ______

3. Find (a) the area and (b) the perimeter of Figure 10.45.

3. (a) ______ (b) ______

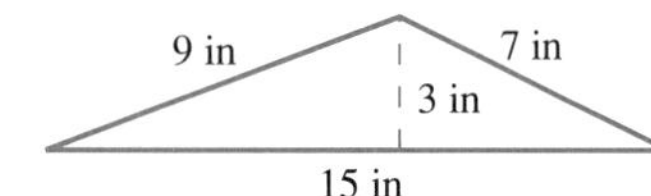

FIGURE 10.45.

4. Find (a) the area and (b) the circumference of a circle with a radius of 5 in.

4. (a) ______ (b) ______

5. Find (a) the area and (b) the circumference of the circle in Figure 10.46.

5. (a) ______ (b) ______

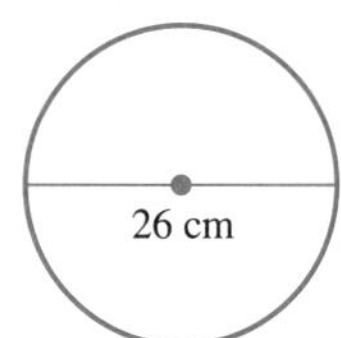

FIGURE 10.46.

6. Find the volume in cubic feet of a box that measures 2.5 ft high, 3.4 ft long, and 1.8 ft high.

6. ______

7. Find (a) the volume and (b) the surface area of a cube that measures 4 cm on each edge.

7. (a) ______ (b) ______

8. Find (a) the area and (b) the perimeter of Figure 10.47.

8. (a) ______

(b) ______

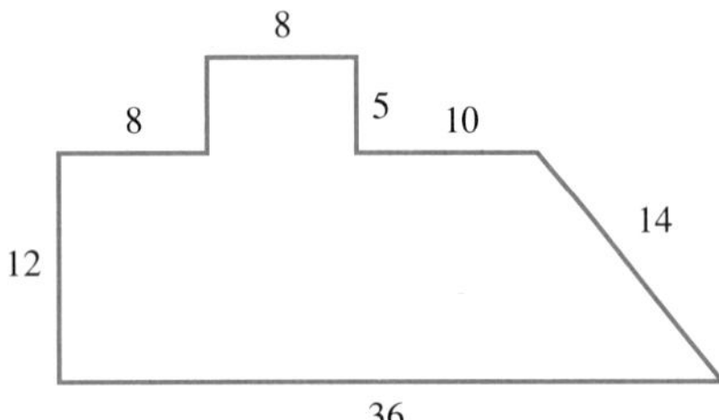

FIGURE 10.47.

9. Find (a) the volume and (b) the surface area of Figure 10.48.

9. (a) ______

(b) ______

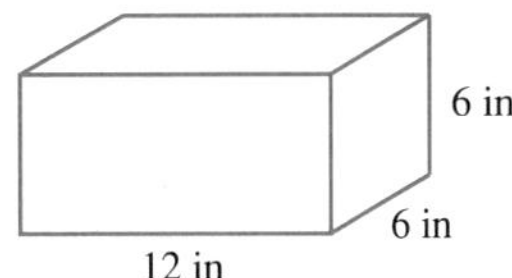

FIGURE 10.48.

10. Find the volume of a 10-in.-diameter spherical basketball.

10. ______

11. Find the volume of the cylinder in Figure 10.49.

11. ______

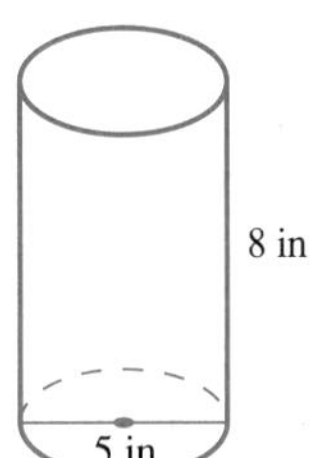

FIGURE 10.49.

CHAPTER 11

Measurement and Units

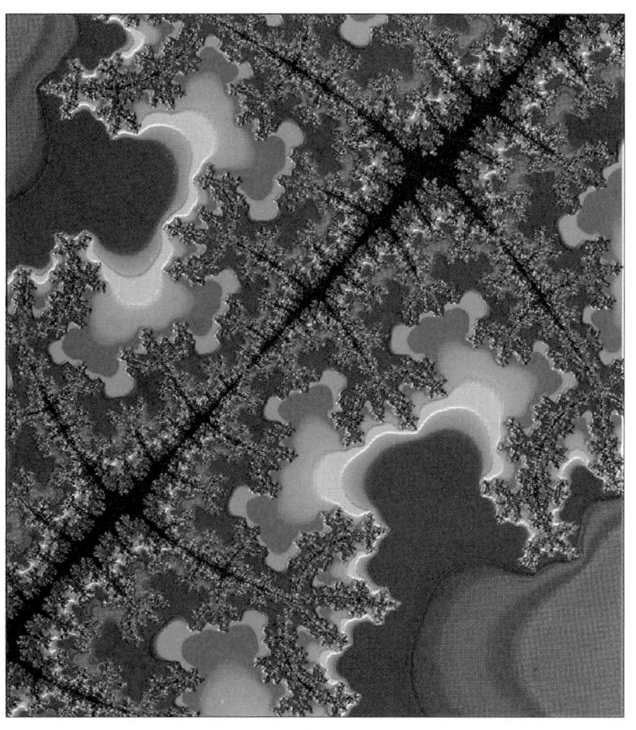

INTRODUCTION

In any practical situation, numbers are most often attached to a unit or a measurement of some kind: 3 children, 8 feet, 16 dollars, 10.8 gallons, and so on. The list is endless.

In this chapter, we will investigate the metric system and how to convert metric units to English units. We will also learn how to combine numbers that contain units when using the operations of arithmetic.

CHAPTER 11—NUMBER KNOWLEDGE

If you walk one mile, you're going to walk the same distance regardless of what units you use.

For example:

1 mile = 5,280 feet

1 mile = 63,360 inches

1 mile = 8 furlongs

1 mile = 160,934 centimeters

1 mile = 320 rods

1 mile = 1760 yards

1 mile = 1609 meters

The number of steps that you take is going to be the same regardless of the units that you use. A mile by any other name is still a mile, especially if you have to walk it.

11.1 The Metric System

The metric system or International System (SI) has become the dominant language of measurement in the world. The United States remains the only major industrialized country that still has not adopted the metric system as the national standard. However, the United States uses the metric system extensively in such fields as science and medicine.

The basic units in the metric system are the meter (length), gram (weight), and liter (volume or capacity). To give you an idea of their size, 1 meter equals approximately 39.37 inches, 28.35 grams equals approximately 1 ounce, and approximately 3.785 liters equals 1 gallon. Prefixes are added to the words meter, gram, and liter to form new units. Conversion from one unit to another in the metric system is easy because all the units use 10 as a base, which means that we need only move a decimal point to change from one unit to another.

For example, 1 millimeter means one thousandth of a meter, 1 centimeter means one hundredth of a meter, and 1 kilogram means one thousand grams. The following tables list the most commonly used metric units for length, weight, and volume, and their relationships to one another.

NAMES OF METRIC PREFIXES

Prefix	Means	Multiplier
milli	one thousandth of	0.001 times
centi	one hundredth of	0.01 times
deci	one tenth of	0.1 times
deka	ten times	10 times
hecto	one hundred times	100 times
kilo	one thousand times	1000 times

LENGTH

1 millimeter (1 mm)	= 0.001 meter
1 centimeter (1 cm)	= 0.01 meter
1 decimeter (1 dm)	= 0.1 meter
1 meter (1 m)	= standard unit of length
1 dekameter (1 dam)	= 10 meters
1 hectometer (1 hm)	= 100 meters
1 kilometer (1 km)	= 1000 meters

WEIGHT

1 milligram (1 mg)	= 0.001 gram
1 centigram (1 cg)	= 0.01 gram
1 decigram (1 dg)	= 0.1 gram
1 gram (1 g)	= standard unit of weight
1 dekagram (1 dag)	= 10 grams
1 hectogram (1 hg)	= 100 grams
1 kilogram (1 kg)	= 1000 grams

VOLUME OR CAPACITY

1 milliliter (1 mL)	= 0.001 liter
1 centiliter (1 cL)	= 0.01 liter
1 deciliter (1 dL)	= 0.1 liter
1 liter (1 L)	= standard unit of volume
1 dekaliter (1 daL)	= 10 liters
1 hectoliter (1 hL)	= 100 liters
1 kiloliter (1 kL)	= 1000 liters

Converting from one unit to another can now be accomplished by counting lines in the table and then moving the decimal point. Here's how:

EXAMPLE 1 Change 35 centimeters to hectometers.

SOLUTION In the table of lengths, count the number of lines from centimeters to hectometers. There are 4 lines, so we move the decimal point 4 places. If we are changing to a *larger* unit (counting *down* in the table), we move the decimal point to the *left*. If we are changing to a *smaller* unit (counting *up* in the table), we move the decimal point to the *right*. In our case, we are counting down 4 lines in the table, so we move the decimal point 4 places to the left and we are done.

35 centimeters = 0.0035 hectometer ▮▮

EXAMPLE 2 Change 12 dekaliters to deciliters.

SOLUTION We count up 2 rows in the volume table from dekaliters to deciliters, so we move the decimal point 2 places to the right. This gives us our answer:

12 daL = 1200 dL ▮▮

EXAMPLE 3 38.6 kilometers is equal to how many centimeters?

SOLUTION From km to cm is 5 lines up in the table, which means that we move the decimal point 5 places to the right, giving us

$$38.6 \text{ km} = 3\ 860\ 000 \text{ cm}$$

Notice that in the metric system, spaces are used rather than commas in numbers above 4 places.

EXAMPLE 4 How many milligrams are there in 4 grams?

SOLUTION We count from grams to milligrams (since we are changing from g to mg) 3 rows in an upward direction, so we move the decimal point to the right 3 places to obtain our answer:

$$4 \text{ g} = 4000 \text{ mg}$$

11.1 Exercises

1. Change 27 decigrams to centigrams.
2. Convert 77 liters to hectoliters.
3. How many grams are there in a kilogram?
4. How many centimeters are there in 7.04 hectometers?
5. Convert 1360 milliliters to liters.
6. Change 38.4 decimeters to kilometers.
7. Change 278 425 milligrams to kilograms.
8. Convert 812 centigrams to kilograms.
9. How many milligrams are there in 76 grams?
10. Convert 706.84 millimeters to kilometers.
11. If a car has a 76-liter gas tank, how many centiliters will it hold?
12. A line is 34 centimeters long. How many millimeters is this?
13. The distance from New York to Phoenix is approximately 4000 kilometers. Express this distance in decimeters.
14. Add 30.7 centimeters, 127 millimeters, and 0.375 meters. (***Hint:*** Change all terms to the same units and then add.)
15. Subtract 206 centiliters from 4 hectoliters.

11.2 Unit Conversion

In the last section we discovered how easy it is to convert from one metric unit to another simply by shifting a decimal point. In the *English* system, or *U.S. Customary* system, which is currently used in the United States, it is not nearly so easy to change from one unit to another. To change tons to ounces, for example, or inches to miles, requires longer calculations. Also, as long as *both* the English system and the metric system are used, we must be concerned with changing from units in one system to units in the other system, for example, changing from miles to kilometers. These conversions often involve changing units many times in one problem, which greatly increases the chance of making an error. We shall now investigate a method that minimizes the chance of error by handling unit changes in an orderly fashion.

Following are unit conversion tables that give common equivalent measurements in the English system, and between the English and metric systems.

LENGTH

English	English–Metric
12 inches (in.) = 1 foot (ft)	1 in = 2.54 cm
3 feet (ft) = 1 yard (yd)	39.37 in. = 1 m
$5\frac{1}{2}$ yd = 1 rod	1 mile = 1.6 kilometers (km)
5280 ft = 1 mile	

WEIGHT

English	English–Metric
16 ounces (oz) = 1 pound (lb)	1 oz = 28.35 g
2000 lb = 1 ton	2.2 lb = 1 kilogram (kg)

VOLUME

English	English–Metric
16 fluid ounces = 1 pint (pt)	1 qt = 0.946 liter
2 pt = 1 quart (qt)	1.057 qt = 1 liter
4 qt = 1 gallon (gal)	

TIME (FOR BOTH SYSTEMS)

60 seconds (sec) = 1 minute (min)
60 min = 1 hour (hr)
24 hr = 1 day
7 days = 1 week
365 days = 1 year (yr)

OTHER MEASURES

1 liter = 1,000 cubic centimeters (cm^3 or cc)
7.48 gal = 1 ft^3
1 ft^3 of water weighs 62.4 lb
1 cm^3 (or cc) of water weighs 1 g

The method for converting units is described in the following examples, so work through each of them carefully, paying attention to each detail.

EXAMPLE 1 Convert 3 meters to feet.

SOLUTION We start by looking in the conversion tables for some combination of unit conversions that will get us from meters to feet. We find that

$$1 \text{ meter} = 39.37 \text{ inches}$$
$$12 \text{ inches} = 1 \text{ foot}$$

In other words, we go from meters to inches and then from inches to feet.

Now, since 1 meter = 39.37 inches, the ratio $\frac{39.37 \text{ inches}}{1 \text{ meter}}$ has the value 1 (since the numerator is equal to the denominator). We multiply our original quantity by this ratio:

$$3 \cancel{\text{meters}} \times \frac{39.37 \text{ inches}}{1 \cancel{\text{meter}}} = 118.11 \text{ inches}$$

Note that meters divided by meters is 1 or, said another way, the meters "cancel." This leaves us with 3 × 39.37 inches = 118.11 inches. This quantity is still equal to 3 meters, because we obtained it by multiplying 3 meters by a ratio that was equal to 1.

We are being asked to convert to feet, so we must now convert 118.11 inches to feet. Since 12 inches = 1 foot, we have $\frac{1 \text{ foot}}{12 \text{ inches}} = 1$. Again we are multiplying by a ratio that is equal to 1, so our result is equal to the original quantity:

$$118.11 \cancel{\text{inches}} \times \frac{1 \text{ foot}}{12 \cancel{\text{inches}}}$$

This time the inches cancel, leaving us with

$$118.11 \times \frac{1 \text{ foot}}{12} = \frac{118.11 \text{ feet}}{12} = 9.84 \text{ feet}$$

Normally we do both steps at once, in the following way:

$$3 \cancel{\text{meters}} \times \frac{39.37 \cancel{\text{inches}}}{1 \cancel{\text{meter}}} \times \frac{1 \text{ foot}}{12 \cancel{\text{inches}}} = \frac{3 \times 39.37 \text{ feet}}{12} = 9.84 \text{ feet}$$ ❚❚

EXAMPLE 2 Convert 6 yards to inches.

SOLUTION From the conversion table for length, we find that we can go from yards to feet, and from feet to inches.

$$1 \text{ yard} = 3 \text{ feet}$$
$$1 \text{ foot} = 12 \text{ inches}$$

From these we form ratios, each equal to 1, in such a way that when our original quantity, 6 yards, is multiplied by these ratios, the units cancel to give us inches.

$$6 \cancel{\text{yards}} \times \frac{3 \cancel{\text{feet}}}{1 \cancel{\text{yard}}} \times \frac{12 \text{ inches}}{1 \cancel{\text{foot}}} = 6 \times 3 \times 12 \text{ inches} = 216 \text{ inches}$$

This method actually tells us when we have committed an error. Suppose, for example, that we try to use the ratio $\frac{1 \text{ foot}}{12 \text{ inches}}$ instead of $\frac{12 \text{ inches}}{1 \text{ foot}}$. We would then have

$$6 \cancel{\text{yards}} \times \frac{3 \text{ feet}}{1 \cancel{\text{yard}}} \times \frac{1 \text{ foot}}{12 \text{ inches}}$$

We find that the units will not cancel. This tells us to reverse the ratio and try again.

EXAMPLE 3 Change 4 gallons to liters.

SOLUTION From the conversion table we have

$$1 \text{ gallon} = 4 \text{ quarts}$$
$$1 \text{ quart} = 0.946 \text{ liter}$$

Therefore

$$4 \cancel{\text{gallons}} \times \frac{4 \cancel{\text{quarts}}}{1 \cancel{\text{gallon}}} \times \frac{0.946 \text{ liter}}{1 \cancel{\text{quart}}} = 4 \times 4 \times 0.946 \text{ liter} = 15.136 \text{ liters}$$

EXAMPLE 4 How many seconds are there in 1 week?

SOLUTION

$$1 \text{ week} = 7 \text{ days}$$
$$1 \text{ day} = 24 \text{ hours}$$
$$1 \text{ hour} = 60 \text{ minutes}$$
$$1 \text{ minute} = 60 \text{ seconds}$$

Therefore

$$1 \cancel{\text{week}} \times \frac{7 \cancel{\text{days}}}{1 \cancel{\text{week}}} \times \frac{24 \cancel{\text{hours}}}{1 \cancel{\text{day}}} \times \frac{60 \cancel{\text{minutes}}}{1 \cancel{\text{hour}}} \times \frac{60 \text{ seconds}}{1 \cancel{\text{minute}}}$$
$$= 1 \times 7 \times 24 \times 60 \times 60 \text{ seconds} = 604{,}800 \text{ seconds}$$

11.2 Exercises

▼ Round answers to two decimal places.

1. Change 70 miles to kilometers.

2. Six rods equals how many inches?

3. Two days is the same as how many seconds?

4. How many rods are there in 2 miles?

5. Convert 4.3 meters to feet.

6. Find how many fluid ounces there are in 6 gallons.

7. Change 1 rod to centimeters.

8. How many meters are equivalent to 17.4 rods?

9. How many minutes are there in 4.5 days?

10. Convert 9000 grams to pounds.

11. Convert 1 ton to grams.

12. Which is greater, the number of seconds in 1 week or the number of minutes in 1 year?

13. Mount Everest is 29,028 feet high. Approximately how many meters is this?

14. Approximately how many miles high is an airplane that is flying at an altitude of 11,000 feet?

15. The distance from New York to Flagstaff, Arizona, is about 2400 miles. Express this distance in kilometers.

11.3 Adding Denominate Numbers

A **denominate number** is any number that **measures** something. It always has some standard unit of measure as a part of it, for example 5 gallons, 26 feet, 7 years, and 56 dollars. Plain numbers that do not have any units connected to them, such as 3, −4, 7.8, and $\sqrt{3}$, are called **abstract** numbers.

In most of the activities that occupy our time every day, such as sports, driving, shopping, and cooking, and in most common occupations, the majority of numbers that we encounter have units attached to them. It is also true that many of the errors that people make occur because of incorrect conversion of units and not because of problems working with abstract numbers. Think of all the denominate numbers involved in the everyday occurrence of driving a car: you put so many gallons of gasoline in the tank, so many pounds per square inch of air pressure in the tires, drive at a certain number of miles per hour, and so on.

When adding denominate numbers, only those numbers that have the same units—called **like units**—can be combined. That is, yards are added to yards, feet to feet, inches to inches, and so on. In every case, we simplify our answer by converting to the largest unit possible. For example, an answer like 5 yards 7 feet should be changed to 7 yards 1 foot, because 7 feet = 2 yards + 1 foot.

EXAMPLE 1 Add 3 yd 2 ft 4 in. to 1 yd 1 ft 11 in.

SOLUTION It is generally best to write the denominate numbers under one another, with the like units in vertical columns. Then we can add by starting with the smallest unit:

3 yd	2 ft	4 in.
+ 1 yd	1 ft	11 in.
1 yd	1 ft	
	~~4 ft~~	~~15 in.~~
	1 yd	1 ft
5 yd	1 ft	3 in.

Since 15 in. = 1 ft + 3 in., we add the 1 ft in with the ft column, giving us 4 ft = 1 yd + 1 ft. We add the 1 yd to the yd column. Our answer in simplest form is 5 yd 1 ft 3 in.

EXAMPLE 2 Add 2 weeks 3 days 14 hours 22 minutes; 1 week 4 days 38 minutes; and 4 weeks 17 hours 46 minutes.

SOLUTION We place like units in vertical columns, putting in zeros for missing units, and then we add, starting with the smallest unit. We work from right to left:

2 weeks	3 days	14 hr	22 min
1 week	4 days	0 hr	38 min
+ 4 weeks	0 days	17 hr	46 min
1 week	1 day	1 hr	
	~~8 days~~	~~32 hr~~	~~106 min~~
	1 week	1 day	1 hr
8 weeks	1 day	8 hr	46 min

We simplify by converting the sum of smaller units to the next larger unit when necessary.

EXAMPLE 3 Add:

	3 lb	5 oz
	4 lb	6 oz
	9 lb	12 oz
+	5 lb	3 oz
	1 lb	
	←	~~26 oz~~
		1 lb
	22 lb	10 oz

11.3 Exercises

Add the following denominate numbers. Be sure that your answer is in simplest form.

1.

	6 gal	2 qt	1 pt
	5 gal	1 qt	1 pt
+	2 gal	3 qt	0 pt

2.

	5 lb	6 oz
	3 lb	2 oz
+	1 lb	12 oz

3.

	2 yd	1 ft	11 in.
		2 ft	9 in.
+	1 yd		2 in.

4.

	1 mile	2429 ft
	1 mile	3726 ft
+	4 mile	911 ft

5.

	3 days	6 hr	15 min	12 sec
	1 day	12 hr	50 min	56 sec
+		5 hr	6 min	14 sec

6.

	7 tons	1136 lb
	9 tons	213 lb
+	1 ton	1460 lb

7.

	7 gal		
	2 gal	1 qt	1 pt
+	1 gal	2 qt	1 pt

8.

	36 hr	14 min
	5 hr	17 min
		51 min
	1 hr	16 min
+	2 hr	48 min

11.4 Subtracting Denominate Numbers

As in adding, when we subtract denominate numbers, we can subtract only like units. We place the like units under one another and subtract. The only time we encounter any problem at all is when we find it necessary to borrow, as illustrated in the following example.

EXAMPLE 1 Subtract 3 yd 2 ft from 5 yd 1 ft.

SOLUTION We place the like units under one another.

$\overset{4}{\cancel{5}}$ yd	$\overset{4}{\cancel{1}}$ ft
−3 yd	2 ft

When we try to subtract, we immediately hit a roadblock, since 1 ft is less than 2 ft. We borrow 1 yd from the 5 yd, write it as 3 ft, and add it to the 1 ft that is already there. The subtraction is now accomplished easily:

4 yd	4 ft
−3 yd	2 ft
1 yd	2 ft

EXAMPLE 2 Subtract 3 gal 3 qt from 9 gal 2 qt.

SOLUTION

$\overset{8}{\cancel{9}}$ gal	$\overset{6}{\cancel{2}}$ qt
−3 gal	3 qt
5 gal	3 qt

The 1 gal that we borrow equals 4 qt, which we then add to the 2 qt already there to make 6 qt.

EXAMPLE 3 Subtract:

$\overset{4}{\cancel{5}}$ days	$\overset{27}{\overset{3}{\cancel{4}}}$ hr	$\overset{83}{\cancel{23}}$ min
−1 day	7 hr	31 min
3 days	20 hr	52 min

EXAMPLE 4 Subtract:

$\overset{4}{\cancel{5}}$ yd	$\overset{2}{\overset{3}{\cancel{0}}}$ ft	$\overset{15}{\cancel{3}}$ in.
−4 yd	2 ft	9 in.
0 yd	0 ft	6 in.

SOLUTION In this case, when we try to borrow from the ft column, we find 0 ft, so we must borrow 1 yd, which we write as 3 ft. Now we are able to borrow 1 ft (leaving 2 ft), which is equal to 12 in., this, when added to the 3 in. already there, gives a total of 15 in. Subtracting gives us 6 in., and both of the other columns yield zeros.

11.4 Exercises

Perform the following subtractions:

1.

	4 gal	3 qt	2 pt
−	1 gal	1 qt	2 pt

2.

	6 hr	20 min
−	2 hr	42 min

3.

	4 yd	1 ft	9 in.
−	1 yd	2 ft	11 in.

4.

	113 min	16 sec
−	12 min	18 sec

5.

	7 gal		
−	2 gal	3 qt	1 pt

6.

	7 miles	4584 ft
−	7 miles	3160 ft

7.

	5 tons	715 lb
−	2 tons	1240 lb

8.

	6 yr	23 weeks	4 days
−	3 yr	46 weeks	6 days

9. A scheduled flight from New York to Los Angeles takes 6 hours 21 minutes, and the return trip takes 7 hours 4 minutes. What is the difference in time between the two flights?

10. Find the difference between 6 gal 1 qt 1 pt and 1 gal 3 qt 1 pt.

11.5 Multiplying Denominate Numbers

Suppose you have a pile of boards, all the same width and length, and you want to know the total length of the boards in the pile. One way to find out would be to measure each one separately and then add all of these numbers together. Of course, the easier way to accomplish the same result would be to measure one board, count the number of boards in the pile, and multiply the two numbers together to obtain the total length of the boards in the pile.

To multiply a denominate number by an abstract number:

Multiply each part of the denominate number by the abstract number and simplify the answer.

EXAMPLE 1 Multiply 4 gal 2 qt 1 pt by 5.

SOLUTION

4 gal	2 qt	1 pt
		× 5
20 gal	10 qt	~~5~~ pt
3 gal	2 qt	1 pt
	~~12~~ qt	
	0	
23 gal	0 qt	1 pt

Simplify the answer beginning with the smallest unit.

EXAMPLE 2 Multiply 6 times 2 days 7 hr 8 min.

SOLUTION

2 days	7 hr	8 min
		× 6
12 days	~~42~~ hr	48 min
1 day	18	
13 days	18 hr	48 min

Since 48 min is less than 1 hr, it cannot be simplified further. Moving to the next larger unit, 42 hr = 1 day + 18 hr.

EXAMPLE 3 Multiply 12 times 2 yd 2 ft.

2 yd	2 ft
	× 12
24 yd	~~24~~ ft
8 yd	0 ft
32 yd	

Under certain conditions it is possible to multiply two denominate numbers together. The general rule is that it is okay only if the answer makes sense. If a room measures 10 ft wide by 12 ft long and we want to know the area of the floor, we multiply the length by the width:

$$A = 12 \text{ ft} \times 10 \text{ ft} = 120 \text{ ft}^2$$

We have multiplied two denominate numbers, and our answer has meaning.

On the other hand, if we multiply 4 hr × 7 hr, we get 28 square hours, which makes no sense.

11.5 Exercises

Multiply and simplify in each of the following problems.

1. 6 × (4 gal 3 qt 1 pt)

2. (5 yd 2 ft 9 in.) × 12

3. (14)(4 hr 8 min 3 sec)

4. (4)(6 miles 1740 ft)

5. 5 lb 7 oz
   ×15

6. 7 tons 1725 lb
   × 8

7. 4 yd 2 ft
   × 11

8. 23 dollars 75 cents
   × 5

9. 5 yr 37 weeks 4 days
   × 8

10. Twelve boards, each of length 3 yd 1 ft 8 in., lie in a pile. What is the total length of all the boards?

11.6 Dividing Denominate Numbers

To divide a denominate number by an abstract number:

Divide the largest unit by the abstract number first. If there is a remainder, convert it to smaller units and add it to those units already there. Next, divide the next smaller unit by the abstract number, converting any remainder and adding it to the next smaller unit. Repeat this process until each unit has been divided by the abstract number.

EXAMPLE 1 Divide 5 gal 2 qt 1 pt by 3.

SOLUTION First we divide the largest unit, 5 gal, by 3:

$$\begin{array}{r} 1\text{ gal; remainder} = 2\text{ gal} \\ 3\overline{)5\text{ gal}} \\ \underline{3} \\ 2 \end{array}$$

We convert the 2 gal remainder to 8 qt and add it to the 2 qt already there, giving a total of 10 qt. Now we divide 10 qt

$$\begin{array}{r} 3\text{ qt; remainder} = 1\text{ qt} \\ 3\overline{)10\text{ qt}} \\ \underline{9} \\ 1 \end{array}$$

We convert the 1 qt to 2 pt and add it to the 1 pt given in the original problem, giving us 3 pt. Next we divide the 3 pt by 3.

$$\begin{array}{r} 1\text{ pt; no remainder} \\ 3\overline{)3\text{ pt}} \\ \underline{3} \\ 0 \end{array}$$

Finally, we collect all the quotients to obtain the final answer:

1 gal 3 qt 1 pt

As in any division problem, we can check our work by use of multiplication.

Check

$$\begin{array}{rrr} 1\text{ gal} & 3\text{ qt} & 1\text{ pt} \\ & & \times 3 \\ \hline 3\text{ gal} & 9\text{ qt} & \not{3}\text{ pt} \\ 2\text{ gal} & \underline{1\text{ qt}} & 1 \\ & \not{1}\not{0}\text{ qt} & \\ & 2 & \\ \hline 5\text{ gal} & 2\text{ qt} & 1\text{ pt} \end{array}$$

EXAMPLE 2 Divide 311 yd 2 ft by 5.

SOLUTION

$$\begin{array}{r} 62\text{ yd; remainder} = 1\text{ yd} = 3\text{ ft} \\ 5\overline{)311\text{ yd}} \\ \underline{30} \\ 11 \\ \underline{10} \\ 1 \end{array}$$

We add the remainder of 1 yd = 3 ft to the 2 ft already given to obtain 5 ft. We divide again:

$$\begin{array}{r} 1\text{ ft; no remainder} \\ 5\overline{)5\text{ ft}} \\ \underline{5} \\ 0 \end{array}$$

We collect the quotients, which yields 62 yd 1 ft.

EXAMPLE 3 Divide 30 hr 43 min 35 sec by 7.

SOLUTION

$$\begin{array}{r} 4\text{ hr; remainder} = 2\text{ hr} = 120\text{ min} \\ 7\overline{)30\text{ hr}} \\ \underline{28} \\ 2 \end{array}$$

Since 120 min + 43 min = 163 min, we now divide 7 into 163 min:

$$\begin{array}{r} 23\text{ min; remainder} = 2\text{ min} = 120\text{ sec} \\ 7\overline{)163\text{ min}} \\ \underline{14} \\ 23 \\ \underline{21} \\ 2 \end{array}$$

Adding seconds, we get 120 sec + 35 sec = 155 sec. We again divide:

$$\begin{array}{r} 22\text{ sec; remainder} = \tfrac{1}{7}\text{ sec} \\ 7\overline{)155\text{ sec}} \\ \underline{14} \\ 15 \\ \underline{14} \\ 1\text{ sec remainder} \end{array}$$

Note that when we divide the remainder, 1 sec, by 7, we obtain $\frac{1}{7}$ sec, so the last division yields $22\frac{1}{7}$ sec. Collecting the quotients gives the final answer: 4 hr 23 min $22\frac{1}{7}$ sec.

11.6 Exercises

Divide in each of the following problems.

1. (5 hr 20 min) ÷ 4

2. (7 yd 2 ft 3 in.) ÷ 5

3. (15 weeks 4 days 14 hr) ÷ 8

4. (36 gal 3 pt) ÷ 4

5. (11 miles 614 ft) ÷ 9

6. (375 yd 10 in.) ÷ 14

7. (26 lb 13 oz) ÷ 12

8. (12 rods 7 yd) ÷ 4

9. (346 yd 1 ft) ÷ 6

10. (8 yr 24 weeks 6 days 5 hr) ÷ 5

11. (124 gal 1 pt) ÷ 17

12. A strip of aluminum 8 yd 2 ft 9 in. is to be divided equally into five pieces. How long will each piece be?

11.7 More About Measurement

In a previous section we cancelled units just as though they were numbers or algebraic quantities. We continue this mode of thinking now as we consider problems of area and volume. For example, if we want to know the area of a room that measures 12 ft long by 9 ft wide, we use the formula for the area of a rectangle:

$$\text{area} = \text{length} \times \text{width}$$
$$A = (12\text{ ft})(9\text{ ft}) = 108\text{ ft}^2$$

We write ft^2 since $(\text{ft})(\text{ft}) = \text{ft}^2$, just as $(5)(5) = 5^2$ or $x \cdot x = x^2$. We think of 1 ft^2 as having the same meaning as 1 square foot, a unit of area. In a similar fashion, 1 ft^3 is the same as 1 cubic foot, which is a unit of volume.

Changing units involving area or volume involves a procedure similar to the one that we have already learned. In each case we use an appropriate table of conversion.

EXAMPLE 1 Change 2 square yards to square feet.

SOLUTION Using the notation we just discussed, we want to change 2 yd^2 to ft^2. From the conversion table of units, we find that 1 yd = 3 ft. Now we square both sides of the equation, which yields

$$1\text{ yd}^2 = 9\text{ ft}^2$$

From this equation we learn that the ratio $\dfrac{9\text{ ft}^2}{1\text{ yd}^2}$ is equal to 1. Multiplying our original quantity by this ratio gives us

$$2\,\cancel{\text{yd}^2} \times \frac{9\text{ ft}^2}{1\,\cancel{\text{yd}^2}} = 2 \times 9\text{ ft}^2 = 18\text{ ft}^2$$

As before, yd^2 cancels with yd^2, leaving us with the desired units, ft^2. ▮▮

EXAMPLE 2 How many square inches are there in 1 square yard?

SOLUTION From the conversion table we find that

$$1\text{ yd} = 3\text{ ft}$$

and

$$1\text{ ft} = 12\text{ in.}$$

Since we are dealing with square units in this problem, we square both sides of each equation, which yields

$$1\text{ yd}^2 = 9\text{ ft}^2$$

and

$$1\text{ ft}^2 = 144\text{ in.}^2$$

We form ratios equal to 1 in such a way that the units cancel in a desirable manner. This gives us

$$1\,\cancel{\text{yd}^2} \times \frac{9\,\cancel{\text{ft}^2}}{1\,\cancel{\text{yd}^2}} \times \frac{144\text{ in}^2}{1\,\cancel{\text{ft}^2}} = 1 \times 9 \times 144\text{ in.}^2 = 1296\text{ in.}^2$$ ▮▮

EXAMPLE 3 An area of 6 square inches is equal to how many square centimeters?

SOLUTION We know that 1 in. = 2.54 cm. Squaring both sides of the equation gives us

$$1 \text{ in.}^2 = 6.45 \text{ cm}^2 \qquad \text{(approximately)}$$

so

$$6 \cancel{\text{in.}^2} \times \frac{6.45 \text{ cm}^2}{1 \cancel{\text{in.}^2}} = 38.7 \text{ cm}^2 \qquad \text{(approximately)}$$

EXAMPLE 4 How many cubic inches are there in 4 cubic feet?

SOLUTION Since 1 ft = 12 in. and we are dealing with cubic units this time, we cube both sides of the equation, yielding 1 ft^3 = 1728 in.^{3} Now we multiply:

$$4 \cancel{\text{ft}^3} \times \frac{1728 \text{ in.}^3}{1 \cancel{\text{ft}^3}} = 6912 \text{ in.}^3$$

11.7 Exercises

1. Convert 3 square meters to square centimeters.

2. Change 716 square feet to square yards.

3. Convert 20 square meters to square yards.

4. Change 1 square foot to square centimeters.

5. How many square rods are there in 1 square mile?

6. How many cubic centimeters are there in 2 gallons?

7. A volume of 2 cubic inches contains how many cubic centimeters?

8. One square rod is equivalent to how many square inches?

9. Convert 0.5 square meters to square millimeters.

10. How many square feet (to the nearest tenth) are there in 500 square inches?

11. How many square yards of carpeting are needed to cover a floor 10 feet wide by 12 feet long?

12. How many cubic inches are contained in 1 gallon?

13. If you are building a concrete sidewalk that measures 70 cubic feet, how many cubic yards of cement should you order?

14. Which is greater, 10 square meters or 12 square yards?

15. The weight of a cubic foot of water is approximately 62.4 pounds. From this information, determine how much a cubic inch of water weighs.

11.8 Still More Measurements

In certain cases, measurements will have more than one unit: for example, miles per hour, or pounds per square inch. In each of these the word **per** means **divided by.** Miles per hour means miles divided by hours, or $\frac{\text{miles}}{\text{hr}}$. Similarly, pounds per square inch is written $\frac{\text{lb}}{\text{in.}^2}$. We are often required to make changes from one set of such units to another.

EXAMPLE 1 Convert 30 miles per hour to feet per second.

SOLUTION Since **per** means **divided by,** we want to change 30 $\frac{\text{miles}}{\text{hr}}$ to $\frac{\text{ft}}{\text{sec}}$. This involves changing the unit of length, miles, to feet, and also changing the unit of time, hours, to seconds. We find each of these conversions in the appropriate conversion tables.

$$1 \text{ mile} = 5280 \text{ feet}$$
$$1 \text{ hour} = 60 \text{ minutes}$$

and

$$1 \text{ minute} = 60 \text{ seconds}$$

Now we form ratios from these equations and multiply in such a way as to produce the desired units, $\frac{\text{ft}}{\text{sec}}$:

$$\frac{30 \text{ miles}}{\text{hr}} \times \frac{5280 \text{ ft}}{1 \text{ mile}} \times \frac{1 \text{ hr}}{60 \text{ min}} \times \frac{1 \text{ min}}{60 \text{ sec}} = \frac{30 \times 5280 \text{ ft}}{60 \times 60 \text{ sec}}$$

$$= 44 \frac{\text{ft}}{\text{sec}}$$

EXAMPLE 2 A tire holds 32 pounds per square inch of air pressure. How many grams per square centimeter is this?

SOLUTION We must change pounds to grams and also square inches to square centimeters.

$$1 \text{ lb} = 16 \text{ oz}$$

$$1 \text{ oz} = 28.35 \text{ g}$$

and

$$1 \text{ in.} = 2.54 \text{ cm}$$

Squaring yields $1 \text{ in.}^2 = 6.45 \text{ cm}^2$. Forming ratios and multiplying then gives us

$$32 \frac{\text{lb}}{\text{in.}^2} \times \frac{16 \text{ oz}}{1 \text{ lb}} \times \frac{28.35 \text{ g}}{1 \text{ oz}} \times \frac{1 \text{ in.}^2}{6.45 \text{ cm}^2} = \frac{32 \times 16 \times 28.35 \text{ g}}{6.45 \text{ cm}^2}$$

$$= 2250 \frac{\text{g}}{\text{cm}^2} \quad \text{(approximately)}$$

11.8 Exercises

1. Change 70 kilometers/hour to meters/second.

2. Ten meters/second is how much in miles/hour?

3. Convert 20 pounds/square foot to kilograms/square meter.

4. The speed of sound in air is approximately 1100 feet/second. How fast is this in miles/hour?

5. One atmosphere of pressure is 14.7 pounds/square inch. Express this in kilograms/square centimeter.

6. If a man runs 100 meters in 10 seconds (the world record is a little faster than that), how fast does he run in miles/hour?

7. Acceleration due to gravity of a freely falling body is 32 feet/second^2. What is this acceleration in meters per second^2?

8. In France, a speed sign reads 100, meaning 100 kilometers/hour. How many miles/hour is this?

SUMMARY—CHAPTER 11

The numbers in brackets refer to the section in the text in which each concept is presented.

[11.1] To convert from one metric unit to another: Count the number of lines in the conversion table between the units being considered and move the decimal point that number of places. If you count *down* in the table, move the decimal point to the left. If you count *up* in the table, move the decimal point to the right.

EXAMPLES

[11.1] Change 6.2 g to mg.
Count *up* 3 lines in the table, from grams to milligrams, and move the decimal point 3 places to the *right*.

$$6.2 \text{ g} = 6200 \text{ mg}$$

[11.2] Conversions that involve English units are accomplished by multiplying the quantity to be converted by ratios, each equal to 1, in such a way that all units cancel except those to which you are converting.

[11.2] Change 6 in. to cm.

$$6 \cancel{\text{in.}} \times \frac{2.54 \text{ cm}}{1 \cancel{\text{in.}}} = 15.24 \text{ cm}$$

[11.3] A **denominate number** is one that measures something. An **abstract number** is a plain number with no units.

[11.3] To add denominate numbers: Add like units and simplify the answer, if possible, by converting each unit to the next largest unit.

[11.3]

6 lb	9 oz
+ 4 lb	10 oz
1 lb	
	19 oz
	1 lb
11 lb	3 oz

[11.4] To subtract denominate numbers: Subtract like units; if necessary, borrow 1 from the next larger unit and convert it to the units needed.

[11.4]

7	20
~~8~~ ft	~~8~~ in.
− 2 ft	10 in.
5 ft	10 in.

[11.5] To multiply a denominate number by an abstract number: Multiply each part of the denominate number by the abstract number and simplify the answer.

[11.5]

4 ft	8 in.
	× 4
16 ft	~~32~~ in.
2 ft	8 in.
18 ft	8 in.

[11.6] To divide a denominate number by an abstract number: Divide the largest unit by the abstract number first, then convert any remainder to the next smaller unit and add it to the units already there. Repeat this process until each unit has been divided by the abstract number.

[11.6] Divide: 6 hr 40 min ÷ 5.

$$\frac{6 \text{ hr}}{5} = 1 \text{ hr; remainder} = 60 \text{ min}$$

$$40 \text{ min} + 60 \text{ min} = 100 \text{ min}$$

$$\frac{100 \text{ min}}{5} = 20 \text{ min}$$

The answer is 1 hr 20 min.

EXAMPLES

[11.7] **To change area and volume from one set of units to another:** Multiply by ratios, each equal to 1, so that all of the units cancel except the required ones.

[11.7] Change 3 in.^{3} to cm^3.

$$1 \text{ in.} = 2.54 \text{ cm}$$

so

$$1 \text{ in.}^3 = 16.39 \text{ cm}^3 \quad \text{(rounded)}$$

$$3 \cancel{\text{in.}^3} \times \frac{16.39 \text{ cm}^3}{1 \cancel{\text{in.}^3}} = 49.17 \text{ cm}^3$$

[11.8] **To change measurements involving more than one unit:** Convert each unit to the appropriate new unit using the rules previously stated for unit conversion.

[11.8] Change $10 \dfrac{\text{miles}}{\text{hr}}$ to $\dfrac{\text{ft}}{\text{sec}}$.

$$10 \frac{\cancel{\text{miles}}}{\cancel{\text{hr}}} \times \frac{5280 \text{ ft}}{1 \cancel{\text{mile}}} \times \frac{1 \cancel{\text{hr}}}{60 \cancel{\text{min}}} \times \frac{1 \cancel{\text{min}}}{60 \text{ sec}}$$

$$= 14.67 \frac{\text{ft}}{\text{sec}}$$

REVIEW EXERCISES—CHAPTER 11

The numbers in brackets refer to the section in which similar problems are presented.

[11.1]

1. Change 31.2 hectometers to centimeters.

2. Convert 96 deciliters to liters.

3. How many dekagrams are equivalent to 7625 mg?

[11.2]

4. How many ounces are equal to 3.6 kilograms?

5. If you weigh 157 pounds, how much will a scale in Rome indicate when you step on it? (It reads in kilograms.)

6. In track competition there is a half-mile race and also a 1000-meter race. Which one is longer and by how many feet?

7. How many centimeters are there in a foot?

8. Three liters of water weigh how many grams?

9. The speed of light is 186,000 miles/sec. A *light-year* is the distance that light travels in one year at this rate of speed. Express this distance in miles.

[11.3]

10. Add and simplify:

	3 yd	2 ft	4 in.
	1 yd	1 ft	9 in.
+		1 ft	2 in.

11. Add and simplify:

	8 week	4 days	3 hr	8 min
	1 week	6 days	23 hr	48 min
+	2 week	1 day	9 hr	56 min

[11.4]

12. Subtract:

	8 yd		
−	2 yd	2 ft	2 in.

13. Subtract:

	12 hr	15 min	8 sec
−	3 hr	18 min	26 sec

▼ [11.5]

14. Multiply and simplify:

8 miles	4720 ft
	× 12

▼ [11.6]

15. Divide 6 yd 1 ft 9 in by 4.

16. Divide 8 gal 3 qt 1 pt by 3.

▼ [11.7]

17. How many pounds does 20 cm^3 of water weigh?

18. A certain model Honda motorcycle has an engine with a displacement of 100 cm^3. What is this displacement in cubic inches?

19. How many cubic centimeters does a 2.5-liter container hold?

20. An automobile engine has a displacement of 4.2 liters. What is this displacement in cubic inches?

▼ [11.8]

21. An airplane flying at 525 mph is going how many feet per second?

22. An airplane flying at 600 mph is going how many meters per second?

23. The discharge pipe from a storage tank has a flow of 70 kiloliters per hour. What is this rate in gallons per minute?

24. Change 50 lb/in.^{2} to kg/cm^2.

25. The flow of natural gas in a pipe is 200 yd^3/min. Give this flow in cubic meters per minute.

NAME ▌▌▌▌▌ CLASS

ACHIEVEMENT TEST—CHAPTER 11

1. Change 327 deciliters to kiloliters. 1. ________

2. Convert 1285 millimeters to hectometers. 2. ________

3. Convert 15 miles to kilometers. 3. ________

4. Five kilograms equal how many pounds? 4. ________

5. Add and simplify: 5. ________

	4 gal	3 qt	1 pt
	2 gal		1 pt
+	1 gal	1 qt	

6. Subtract: 6. ________

	3 days	17 hr	4 min	26 sec
−	1 day	16 hr	9 min	42 sec

7. Multiply and simplify: 12 lb 6 oz × 14 7. ________

8. Divide 12 yd 2 ft 11 in. by 5. 8. ________

9. Convert 5 yd^2 to square meters.

9. ______

10. Which is greater, 3 m^2 or 25 ft^2? How much greater? Give your answer in square feet.

10. ______

______

11. Convert 60 mph to kilometers/hour.

11. ______

12. Convert 30 lb/ft^2 to kilograms/square meter.

12. ______

NAME ▌▌▌▌▌ CLASS

CUMULATIVE REVIEW—CHAPTERS 10 AND 11

▼ Name the geometric figures in Exercises 1–3.

1. 1. ____________________

2. 2. ____________________

3. 3. ____________________

4. Find the area and the perimeter of a rectangle that measures 3 ft by 4 ft. 4. ____________________

5. Find the area of the following figure. 5. ____________________

6

8

15

6. Find the area and circumference of a circle with a radius of 8 in. Round your answers to two decimal places. 6. ____________________

7. Find the volume of a box that measures 2 ft long, 1 ft wide, and $1\frac{1}{2}$ ft high. 7. ____________________

8. Convert 3 km to meters. 8. ____________________

9. How many inches are there in 5 m? 9. ____________________

10. Mt. Kilimanjaro is 19,340 ft high. How many meters is this?

10. ______

11. How many grams are there in a 5-lb bag of sugar?

11. ______

12. Add and simplify:

4 yd	2 ft	4 in.
2 yd	2 ft	6 in.
+ 1 yd	1 ft	8 in.

12. ______

13. Subtract:

9 hr	12 min	8 sec
− 3 hr	20 min	11 sec

13. ______

14. A Harley-Davidson Low-Rider motorcycle has an engine displacement of 80 $in.^3$ How many cubic centimeters is this?

14. ______

CHAPTER 12

Using a Calculator

INTRODUCTION

Almost everyone owns a hand-held calculator of some sort. They have become relatively inexpensive in recent years. They make an easy graduation gift from an aunt who has no idea what to buy an 18-year-old. Sometimes you can even get a free calculator as a promotional gift when subscribing to a magazine or opening a bank account.

Calculators usually come with an instruction booklet that most people plan to read but somehow never get around to. In this chapter you will be shown what some of the buttons on your calculator mean and we will unlock some of the secrets that your calculator holds. This will allow you to put your calculator to work for you in a meaningful way.

CHAPTER 12—NUMBER KNOWLEDGE

In the scheme of things, calculators in their present form have not been around very long at all, only about 20 years. In this time they have made it possible to solve many kinds of problems with ease. Before they became available, such problems were solvable only by people with a great deal of mathematical expertise.

Here are some of the many types of problems that a calculator can greatly help to solve:

- *Keys to saving energy.* Settle the arguments—how much does it cost if you leave a 100-watt light burning all night? Is it worth hanging clothes instead of putting them in the dryer?
- *Inflation.* What will a $1.29 loaf of bread cost in the year 2050 if inflation continues at an annual rate of 6%? How much money will I need to retire in the year 2025?
- *Permutations and combinations.* You like ten foods. Each day you pack three of them in your lunch. How many days can you go without eating the same lunch twice?
- *Puzzles and recreation.* How many times does your heart beat if you live to be 80? How much air do you breathe while you are alive? How many Saturday nights are left if you live to be 100? How much do I weigh on the moon? A steel-belted radial tire is guaranteed for 40,000 miles. If the tread is 1 cm thick and the diameter of the tire is 65 cm, how much rubber comes off in one revolution?

All of these problems can be solved without a calculator. However, all of them can be solved much more rapidly, with more ease and with less special mathematical expertise, *with* the aid of a calculator.

12.1 Different Types of Calculators

There are two major types of calculators: those that use *algebraic logic* and those that use *reverse Polish notation (RPN).* Calculators that use RPN circuitry are usually identified by the presence of an ENTER key and the absence of an = key. In our discussion we will assume that your calculator is an algebraic logic calculator. If your calculator is the RPN type, the best help will probably be the instruction book provided with the calculator.

Order of Operations

Some less expensive calculators do not obey the correct order of operations. An easy experiment will tell you whether or not your calculator has the correct order of operations built into it. Key in the following sequence on your calculator:

3 + 4 X 5 =

If the answer in the display is 23, then your calculator will multiply before it adds, which is the correct order of operations. If the answer in the display is 35, then the calculator added the 3 + 4 before multiplying by 5, which is not the proper order of operations. In this chapter we will assume that your calculator automatically uses the correct order of operations. If your calculator is not of this type, then you must be sure that you enter the operations in the correct order.

Number of Digits

Not all calculators show the same number of digits in the display. If your display shows fewer digits than indicated in our work that follows, round our answer to the same number of places shown in your display. The answers should then be the same.

12.2 One-Step Operations on the Calculator

Sooner or later everyone presses a wrong button on the calculator, so it is important to estimate your answer whenever possible.

EXAMPLE 1 Add: $184.7 + 409.8$.

SOLUTION

Estimate: $200 + 400 = 600$

Calculator: 184.7 [+] 409.8 [=] 594.5

As you can see, our estimate is close to the answer in the display on the calculator.

EXAMPLE 2 Multiply: 46.31×72.08.

SOLUTION

Estimate: $50 \times 70 = 3500$

Calculator: 46.31 [×] 72.08 [=] 3338.0248

Some answers are not so easy to estimate, so be extra careful not to press any wrong buttons.

EXAMPLE 3 Evaluate the following:

	PROBLEM	KEY-IN	DISPLAY
(a)	$38.26 - 14.56$	38.26 [−] 14.56 [=]	23.7
(b)	$(15.96)^2$	15.96 [x^2]	254.7216
(c)	$279.4 \div 36$	279.4 [÷] 36 [=]	7.7611111
(d)	$\frac{348.91}{6.53}$	348.91 [÷] 6.53 [=]	53.431853
(e)	$91.6 - 115.8$	91.6 [−] 115.8 [=]	−24.2
(f)	$(-14.8)(9.4)$	14.8 [±] [×] 9.4 [=]	−139.12

The [±] key changes the sign of any number, so we use it to obtain −14.8. It is pressed *after* the number is keyed in.

Clearing An Incorrect Entry

Most calculators allow you to clear the last entry without having to start the problem over from the beginning. You may need to read your instruction booklet or experiment with the keys.

If your calculator has these keys	To clear the last entry	To clear the entire problem
CE/C	CE/C once	CE/C twice
C and AC	C	AC
C and CE	CE	C

Study the following examples carefully to see how to change incorrect entries. Try them on your own calculator.

EXAMPLE 4 Correct a number that has been incorrectly entered by using the clear-entry key.

Multiply: 224×18

SOLUTION

224 × 17 CE/C 18 = 4032

Incorrect entry (17) — Correct entry (18)

EXAMPLE 5 To correct an operation that has been entered by mistake, simply re-enter the correct operation.

Multiply: 224×18

SOLUTION

224 + × 18 = 4032

Incorrect operation (+) — Correct Operation (×)

EXAMPLE 6 To correct an error made early in the problem, you must clear the entire problem.

Multiply: 224×18

SOLUTION

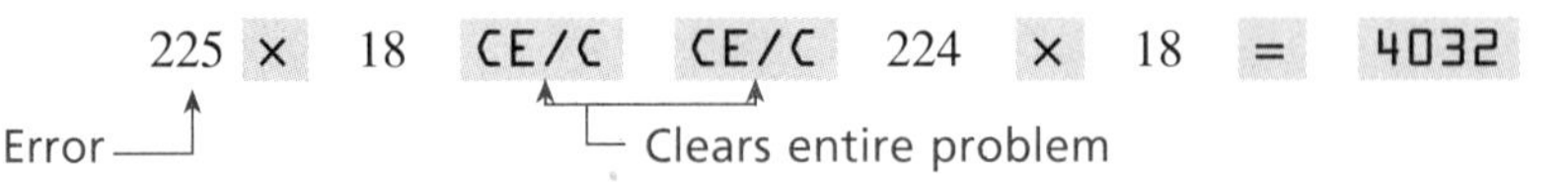
225 × 18 CE/C CE/C 224 × 18 = 4032

Error (225) — Clears entire problem (CE/C CE/C)

12.2 Exercises

▼ Do the following problems on your calculator. Also estimate your answer where applicable.

1. 3,264 + 5,986
2. 6,158 + 9,946
3. 16.236 + 18.91
4. 13.66 + 33.104
5. 13,069 − 5,998
6. 68,982 − 53,114
7. 19,216 − 39,427
8. 8,436 − 26,555
9. 3.201 − 14.823
10. 7.426 − 13.009
11. 68 × 43
12. 91 × 86
13. 9.63 × 21.6
14. 13.87 × 92.4
15. 268.5 × 199.1
16. 183.9 × 696.4
17. 7214 ÷ 13
18. 1921 ÷ 46
19. 23.6 ÷ 72.4
20. 18.8 ÷ 38.4
21. $\frac{248.9}{17.1}$
22. $\frac{311.6}{28.2}$
23. $\frac{19.3}{36.5}$
24. $\frac{11.1}{84.6}$
25. $(-62)(34.8)$
26. $(-26)(41.7)$
27. $(-32)(-447)$
28. $(-94)(-188)$
29. $\sqrt{1066}$
30. $\sqrt{342}$
31. $\sqrt{857}$
32. $\sqrt{921}$
33. $\sqrt{8850}$
34. $\sqrt{4127}$
35. $\sqrt{21.6}$
36. $\sqrt{9.86}$
37. 21^2
38. 16^2
39. 268^2
40. 139^2
41. $(31.42)^2$
42. $(63.78)^2$
43. $(0.4)^2$
44. $(0.6)^2$

12.3 Multi-Step Operations

Memory

The **memory** in a calculator allows it to store a number for use later on. Some common memory keys are listed below. If your calculator has a different key than those listed, check your instruction booklet.

Key	What it does
MIN or STO	Puts the displayed number into the memory.
MR	Recalls the number in memory for your use.
CM or AC	Erases the number in the memory.
M+	Adds the displayed number to the number in the memory.
M−	Subtracts the displayed number from the number in the memory.
RM	Pressed once, it recalls the number from the memory for your use.
CM or MRC	Pressed twice, it clears the number from the memory.

Experiment with the memory keys on your calculator and then use your calculator to follow along with the examples.

EXAMPLE 1 Find $\dfrac{62.4}{17 + 14.7}$.

SOLUTION Since we are dividing 62.4 by the *entire* denominator, we evaluate the denominator first.

17 + 14.7 =	31.7 STO	Store 31.7 in the memory.
62.4 ÷ RCL =	1.9684543	Divide 62.4 by 31.7, which is in the memory. ▮▮

EXAMPLE 2 Mrs. Serio went to the grocery store with $20.00 and bought two bottles of shampoo at $2.79 each, two dozen eggs at $1.23 per dozen, and six cans of tomatoes at $1.29 per can. Use your calculator to keep a running record of how much money she has left after buying each item.

SOLUTION

20 STO		Place $20 in storage.
2 × 2.79 M−	5.58	The cost of the shampoo is $5.58.
MRC 14.42		There is $14.42 left in memory.
2 × 1.23 M−	2.46	The cost of the eggs is $2.46.
MCR 11.96		There is $11.96 left of the original $20.
6 × 1.29 M−	7.74	The tomatoes cost $7.74
MCR 4.22		The change from the $20 will be $4.22. ▮▮

Order of Operations and Parentheses

You must be especially careful when more than one operation is involved in a calculation. Regardless of the type of calculator you use, if you are careful about which operation you do first, you will obtain the correct result. Always treat what appears inside parentheses as a single unit.

EXAMPLE 3 Find (16 + 23) × 14.

SOLUTION

16 + 23 =	39	To add first, key in = .
× 14 =	546	Now multiply. ▮▮

EXAMPLE 4 Find 4^3.

SOLUTION

4 y^x 3 =	64	y^x raises numbers to powers. ▮▮

EXAMPLE 5 Find $\sqrt{79 - 46}$.

SOLUTION

79 [−] 46 [=]	33	Evaluate 79 − 46 first.
[√]	5.7445626	Find the square root.

Note: If you key in the sequence 79 [−] 46 [√] [=], the result is 72.21767. Can you see why an incorrect result is obtained? Since the calculator performed the square root before it subtracted, we obtained $79 - \sqrt{46}$, which is not what we wanted to do.

EXAMPLE 6 $42 \div 11 - (156 - 118)$

SOLUTION We work inside the parentheses first and store the result in the memory.

156 [−] 118 [=]	38 [MIN]	38 is in the memory.
42 [÷] 11 [=]	3.18181818	Perform the division.
[−] [MR] [=]	−34.181818	Subtract what is in the memory.

The result is −34.181818.

12.3 Exercises

Perform the following calculations.

1. $16 \times 24 + 56$
2. $17 \times 35 + 18$
3. $32 \times 18 - 72$
4. $46 \times 6 - 112$
5. $\sqrt{76 + 44}$
6. $\sqrt{36 + 58}$
7. $\sqrt{36 - 12}$
8. $\sqrt{114 - 75}$
9. 6^3
10. 11^3
11. 5^4
12. 6^4
13. 9^5
14. 8^6
15. $(3.2)^3$
16. $(6.1)^3$
17. $(9.46)^4$
18. $(3.86)^4$
19. $(14 - 9) \div (36 - 8)$
20. $(15 - 4) \div (56 - 19)$
21. $(36 + 4) \times (46 - 32)$
22. $(52 + 18) \times (76 - 52)$
23. $72.4 \div (13 + 52)$
24. $46.1 \div (75 - 8)$
25. $(26.4 - 7.6) \div 9.7$
26. $(58.1 - 13.7) \div 5.8$
27. $\frac{(14)(36)}{(75)(21)}$
28. $\frac{(24)(85)}{(13)(76)}$
29. $\frac{(23)(19)(52)}{(71)(43)}$
30. $\frac{(15)(77)(18)}{(52)(12)}$
31. $15(4^3)$
32. $17(5^4)$
33. $(-6.3)(2.38)$
34. $(-9.1)(5.61)$
35. $(-3.4)^3$
36. $(-8.7)^3$
37. $\frac{(14.7)(6.8)^2}{15.2}$
38. $\frac{(42.1)(3.7)^3}{42.8}$
39. $\frac{24\pi}{16}$
40. $\frac{32\pi}{18}$

12.4 The Calculator and Scientific Notation

Large Numbers

Most calculators can display only numbers that are 8 or perhaps 10 digits long. What happens if we need to make calculations involving larger numbers?

Scientific calculators accomplish this by using powers of 10. For example, if we multiply 35,000 by 67,500 on a scientific calculator, the result in the display will look like this:

2.3625 09

What this means is 2.3625×10^9. Multiplying 2.3625 by 10^9 will move the decimal point nine places to the right.

2.3625 09 = 2,362,500,000.

Move 9 places to the right.

If the display shows a negative number on the right, we move the decimal point to the left. Study the following example.

EXAMPLE 1

	DISPLAY	ORDINARY NOTATION
(a)	3.264 07	32,640,000
	Move the decimal point seven places to the right.	
(b)	4.15 −04	0.000415
	Move the decimal point four places to the left.	

EXAMPLE 2 Evaluate 26^6.

SOLUTION

$26^6 = 26$ y^x 6 = 3.0891578 08

To change our answer to ordinary notation, we move the decimal point eight places to the right.

3.0891578 08 = 308915780

EXAMPLE 3 Multiply: (26.4)(3659)(4000).

SOLUTION

26.4 × 3659 × 4000 = 3.863904 08

This is written 386,390,400 in ordinary notation.

EXAMPLE 4 Evaluate $(0.045)^6$.

SOLUTION

0.045 y^x 6 = 8.3038 −09

The negative sign means that we move the decimal point nine places to the left.

8.3038 −09 = 0.0000000083038

EXAMPLE 5 Evaluate $\frac{(65{,}000)(900)}{(0.0009)}$.

SOLUTION 65000 × 900 ÷ 0.0009 = 6.5 10 = 65,000,000,000

12.4 Exercises

The following numbers appear in your calculator display. Write them in ordinary notation.

1. 1.6356 06
2. 2.4391 08
3. 7.432961 08
4. 9.551257 09
5. 6.421 −08
6. 9.523 −08
7. 8.881 −11
8. 1.008 −11
9. 4.9 11
10. 6.8 12

Evaluate each of the following on a scientific calculator. Write the answer in ordinary notation.

11. 26,000 × 24,000
12. 390,000 × 20,000
13. (2,850)(3,800)(4,200)
14. (1,600)(2,3800)(9,450)
15. 860,000 × 760,000
16. 925,000 × 180,000
17. 125,000 ÷ 0.0006
18. 350,000 ÷ 0.0017
19. (0.0075)(0.00238)(0.003)
20. (0.0236)(0.0042)(0.0005)
21. $(6.35)^{10}$
22. $(7.41)^{11}$
23. 720^3
24. 885^3
25. $(0.0041)^3$
26. $(0.0035)^3$
27. $\frac{(23{,}000)(850)}{(0.0007)}$
28. $\frac{(67{,}500)(3{,}000)}{(0.0065)}$
29. $\frac{1}{(38{,}000)(450)}$
30. $\frac{1}{(78{,}500)(300)}$

12.5 Number Patterns—Fun with Your Calculator

Patterns occur in mathematics all of the time and illustrate the concept of *inductive reasoning*.

EXAMPLE 1 Using your calculator, evaluate each line. See if you can find a pattern and then evaluate the last line without using your calculator by writing the next number in the pattern.

$$1^2 = 1$$
$$11^2 = 121$$
$$111^2 = 12321$$
$$1111^2 = 1234321$$
$$11111^2 =$$

Your answer should be 123454321, which you can check on your calculator.

12.5 Exercises

▼ Perform the first four given calculations on your calculator, find a pattern, and write the last number without using your calculator.

1. $9 \cdot 9$
$99 \cdot 89$
$999 \cdot 889$
$9999 \cdot 8889$
$99999 \cdot 88889$

2. $55 \cdot 15$
$555 \cdot 15$
$5555 \cdot 15$
$55555 \cdot 15$
$555555 \cdot 15$

3. $4 \cdot 6$
$44 \cdot 6$
$444 \cdot 6$
$4444 \cdot 6$
$44444 \cdot 6$

4. 9^2
99^2
999^2
9999^2
99999^2

5. 3^2
33^2
333^2
3333^2
33333^2

6. $4 \cdot 6$
$4 \cdot 66$
$4 \cdot 666$
$4 \cdot 6666$
$4 \cdot 66666$

7. $3 \cdot 37$
$6 \cdot 37$
$9 \cdot 37$
$12 \cdot 37$
$15 \cdot 37$

8. $9 \cdot 9 + 19$
$98 \cdot 9 + 18$
$987 \cdot 9 + 17$
$9876 \cdot 9 + 16$
$98765 \cdot 9 + 15$

9. $0.04 \cdot 9.9$
$0.004 \cdot 99.9$
$0.0004 \cdot 999.9$
$0.00004 \cdot 9999.9$
$0.000004 \cdot 99999.9$

10. $3 \cdot 26$
$3 \cdot 226$
$3 \cdot 2226$
$3 \cdot 22226$
$3 \cdot 222226$

SUMMARY—CHAPTER 12

The numbers in brackets refer to the section in the text in which each concept is presented.

		EXAMPLE
[12.2]	To avoid errors, *estimate* your answer first whenever possible.	[12.2] Add: 31.6 + 88.7. Estimate: 30 + 90 = 120. Calculator: 31.6 [+] 88.7 [=] 120.3
[12.2]	The [±] key changes the sign of a number. It is pressed *after* keying in the number.	[12.2] To enter −13.6, enter 13.6 and press [±]. The display reads −13.6
[12.2]	The **clear entry** button ([CE/C] or [C] or [CE]) clears the last entry only and allows you to correct a number entered in error.	[12.2] Find 62 × 38. 62 [×] 37 [C] 38 [=] 2356 (37: error)
[12.3]	The **memory** allows you to store a number for future use.	[12.3] Find 116 − (13 + 58). Evaluate 13 + 58 and store it in memory. 13 [+] 58 [=] 71 [MIN] 116 [−] [MR] [=] 45
[12.3]	[M+] adds the number in the display to the number in the memory.	
[12.3]	[M−] subtracts the number in the display from the number in the memory.	
[12.3]	[y^x] raises a number to a power.	[12.3] Find 4^6. 4 [y^x] 6 [=] 4096
[12.4]	Numbers that do not fit in the display appear in **scientific notation.**	[12.4] 3.065 08 = 306,500,000 2.5 −06 = 0.0000025

REVIEW EXERCISES—CHAPTER 12

The numbers in brackets refer to the section in the text in which similar problems are presented.

Evaluate each of the following on your calculator and also estimate the answer.

[12.2]

1. $3{,}886 + 9{,}142$
2. $13.85 + 30.46$
3. $8{,}285 - 1{,}850$
4. 1.856×327.9
5. $7258 \div 695$
6. -85×76
7. $(-49)(-712)$
8. $\sqrt{3047}$
9. $\sqrt{12.95}$
10. $(38.4)^2$

Use your calculator to evaluate each of the following:

[12.3]

11. $23 \times 18 + 56$
12. $59 \times 12 - 132$
13. $\sqrt{785 - 429}$
14. 8^3
15. 4^6
16. $(7.46)^4$
17. $(36 - 29) \div (52 - 38)$
18. $17.4 \div (13.4 + 19.8)$
19. $\dfrac{(32)(85)}{(76)(4)}$
20. $6.4(6^3)$
21. $(-4.2)^3$
22. $\dfrac{(-6.4)(7.8)^2}{6.9}$

Evaluate the following on your calculator. Give your answer in ordinary notation.

[12.4]

23. $875{,}000 \times 1{,}200$
24. $(7{,}650)(3{,}000)(8{,}500)$
25. $(0.0006)(0.0035)(0.0075)$
26. $(0.0046)^4$
27. $(8.56)^{11}$
28. $\dfrac{(67{,}000)(3{,}600)}{(0.064)}$

NAME ▮▮▮▮▮ CLASS

ACHIEVEMENT TEST—CHAPTER 12

In Exercises 1-8, estimate the answer, then evaluate the problem on your calculator.

1. $7{,}846 + 13{,}062$ 1. ______ ______

2. $65.8 + 91.7$ 2. ______ ______

3. $(189.1)(312.4)$ 3. ______ ______

4. $7592 \div 811$ 4. ______ ______

5. -624×39 5. ______ ______

6. $\sqrt{5016}$ 6. ______ ______

7. $\sqrt{15.25}$ 7. ______ ______

8. $(21.4)^2$ 8. ______ ______

Evaluate each of the following on your calculator.

9. $46 \times 32 + 23$ 9. ______

10. $21 \times 76 - 225$ 10. ______

11. $\sqrt{785 - 521}$ 11. ______

12. 3^5 12. ______

13. $(8.45)^4$ 13. ______

14. $75.8 \div (52.1 - 18.7)$

15. $(-5.7)^5$

16. -26.3×73.6

17. $\dfrac{(2.01)(3.2)^2}{15.2}$

▼ Evaluate Exercises 18–22 on your calculator. Also give your answer in ordinary notation.

18. $445,000 \times 2,900$

19. $(0.003)(0.00045)(0.009)$

20. $(0.0032)^3$

21. $(29.52)^7$

22. $\dfrac{(81,500)(9^4)}{0.058}$

14. ______

15. ______

16. ______

17. ______

18. Display ______

Ordinary ______

19. Display ______

Ordinary ______

20. Display ______

Ordinary ______

21. Display ______

Ordinary ______

22. Display ______

Ordinary ______

Chapter 1

Exercise 1.1

1. ten thousands **3.** ones **5.** thousands
7. hundreds **9.** hundred millions **11.** thousands
13. tens **15.** ones **17.** tens
19. thousands **21.** fifty-two
23. five thousand, three hundred twenty-eight
25. seven hundred two
27. three hundred eight thousand
29. sixty billion, six hundred six million, sixty thousand, six hundred six
31. four thousand one
33. 36 **35.** 74,997 **37.** 70,000,000
39. 70,007,000 **41.** 74,060,000,460 **43.** 600,000
45. 20 + 6 **47.** 500 + 8 **49.** 20,000 + 30 + 8
51. 20,000,000 + 6,000,000 + 300,000 + 8,000 + 100 + 20 + 4
53. 30,000,000 + 30,000 + 1,000 + 100 + 10 + 8
55. hundreds **57.** hundreds **59.** ten thousands
61. The whole numbers are the counting numbers together with zero: 0, 1, 2, 3, 4, ...

Exercise 1.2

1. 700 **3.** 400 **5.** 900
7. 1,000 **9.** 4,400 **11.** 6,200
13. 4,900 **15.** 4,000 **17.** 5,300
19. 7,821,000 **21.** 7,900

	Ten	Hundred	Thousand
23. 3,852	3,850	3,900	4,000
25. 23,676	23,680	23,700	24,000
27. 3,006	3,010	3,000	3,000
29. 7,111	7,110	7,100	7,000
31. 42,999	43,000	43,000	43,000

33. 16,800,000 **35.** $229,000,000 **37.** 210,000
39. 600,000 **41.** $1,700 **43.** $1,300

Exercise 1.3

1. 78 **3.** 37 **5.** 469
7. 889 **9.** 91 **11.** 125
13. 1,014 **15.** 1,154 **17.** 12,175
19. 11,372 **21.** 108,904 **23.** 98,169
25. 181 **27.** 258 **29.** 1,502
31. 951 **33.** 14,865 **35.** 17,573
37. 48,306 **39.** 52,664 **41.** 76,117
43. 11,070 **45.** 113,224 **47.** 1,169 gal
49. $221 **51.** $5,830
53. Any number larger than 9 has a digit in the next place-value column, so it can be written in or "carried" into the next column.

Exercise 1.4

1. 501 **3.** 572 **5.** 4,113
7. 108 **9.** 9 **11.** 2,944
13. 959 **15.** 356 **17.** 4,502
19. 4,879 **21.** 55,049 **23.** 2,776 votes
25. $108
27. profit: $36,000; net profit: $25,000
29. $198

Exercise 1.5

1. 60 **3.** 400 **5.** 3,000
7. 350 **9.** 7,500 **11.** 73,800
13. 2,120 **15.** 16,583 **17.** 21,312
19. 438,594 **21.** 24,720,645 **23.** 2,472,448
25. 4,410 **27.** 3,993,750 **29.** 4,022,480
31. 476 students **33.** $18,480 **35.** 559,680 ft
37. 736 miles **39.** $15,964 **41.** $54.27
43. The cyclist is 1100 ft ahead of the jogger.
45. Factors are numbers that are multiplied together.
47. When you multiply two numbers together, say 8 times 5, it is the same as adding 5 eight times.

Exercise 1.6

1. 1,537 R 1 **3.** 636 R 4 **5.** 193,064 R 1
7. 3,755 R 6 **9.** 1,666 R 4 **11.** 7,640
13. 140,030 R 1 **15.** 80,256 R 3 **17.** 154 R 4
19. 606 R 1 **21.** 2,750 R 1 **23.** 0
25. undefined **27.** 0 **29.** undefined
31. undefined **33.** undefined **35.** undefined

Exercise 1.7

1. 36 R 7 **3.** 140 R 24 **5.** 378 R 11
7. 95 R 429 **9.** 1,511 R 10 **11.** 136 R 169
13. 27 R 34 **15.** 79 R 385 **17.** 124
19. 83 R 17 **21.** 108 R 16 **23.** 85 R 56
25. 172 R 13 **27.** 6 R 703 **29.** 38 hr
31. 17 min **33.** $398/week **35.** 367 bales
37. 16 glasses
39. Sometimes one number does not divide another evenly.

Exercise 1.8

1. 9 **3.** 81 **5.** 216
7. 64 **9.** 1,296 **11.** 6,859
13. 0 **15.** 1 **17.** 1
19. undefined **21.** 1 **23.** 1

25. 1 **27.** 100 **29.** 1,000,000
31. 10,000,000,000
33. $10^{10} = 10{,}000{,}000{,}000$; $10^6 = 1{,}000{,}000$; $10^5 = 100{,}000$
35. **(a)** 256 **(b)** 16,384
37. 81 **39.** 100 **41.** 9
43. 1 **45.** 1 **47.** 0
49. 1,000 **51.** 64
53. 10^{100}; because it's much shorter and therefore saves time.

Exercise 1.9

1. 34 **3.** 19 **5.** 8
7. 75 **9.** 1 **11.** 14
13. $\frac{3}{4}$ **15.** 28 **17.** 39
19. 2 **21.** 42 **23.** 4
25. 4 **27.** 19 **29.** 2
31. 2 **33.** 3 **35.** 2
37. $131.80 **39.** 13 **41.** 8
43. 16 **45.** 4 **47.** 0
49. 39 **51.** 17 **53.** 5
55. 5 **57.** 5 **59.** 3

Exercise 1.10

1. 5; commutative property of addition
3. 3; commutative property of multiplication
5. 6; multiplicative identity
7. 8; additive identity
9. m; commutative property of addition
11. x; commutative property of addition
13. commutative property of addition
15. commutative property of multiplication
17. multiplicative identity
19. additive identity
21. multiplicative identity
23. commutative property of multiplication
25. associative property of addition
27. associative property of multiplication
29. distributive property of multiplication over addition
31. The commutative property of addition means that if you change the order of two numbers in an addition problem, the sum does not change.
33. The associative property of multiplication means that if we multiply three numbers together, it makes no difference which two we multiply first.
35. Subtraction is not commutative for whole numbers because if the order of subtraction is reversed, a different answer (not a whole number) is obtained.

Review Exercises—Chapter 1

1. thousands **3.** 52,609 **5.** 24,700
7. 10,382 **9.** 69,120 **11.** 3,776 R 6
13. 0 **15.** undefined **17.** 27 mpg
19. 27 **21.** 2
23. associative property of addition
25. 787,745 **27.** 8,760 hr **29.** 78
31. 815

Critical Thinking Exercises—Chapter 1

1. False: if the two numbers are large enough, they will add to 10,000 or more.
3. False: division is not commutative.
5. False: 0^0 is undefined.
7. Always do division before addition. The correct answer is 7.
9. Always do multiplication before subtraction. The correct answer is 24.
11. 45, 46, 47, 48, 49, 50, 51, 52, 53, 54

Achievement Test—Chapter 1

1. 3,436,533
2. five thousand, eighty-four
3. 62,084 **4.** 5,000 + 200 + 80 **5.** 61,950
6. 62,000 **7.** 62,000 **8.** 8,260
9. $9,693 **10.** 132,912 **11.** 13,403
12. undefined **13.** 0 **14.** undefined
15. $175 **16.** 247 passengers **17.** 1
18. 256 **19.** 4
20. The associative property of multiplication

Chapter 2

Exercise 2.1

1. 2, 3 **3.** none **5.** none
7. 2, 3, 5 **9.** 3 **11.** none
13. composite **15.** composite **17.** composite
19. prime **21.** composite **23.** prime
25. $2 \cdot 3^2$ **27.** $2 \cdot 7^2$ **29.** 2^5
31. $2^2 \cdot 3^2 \cdot 5$ **33.** $2^3 \cdot 5^2$ **35.** $2 \cdot 23$
37. $7 \cdot 11$
39. $12 = 2^2 \cdot 3$; $355 = 5 \cdot 71$
41. $212 = 2^2 \cdot 53$; $32 = 2^5$
43. 2 **45.** 3, 5 **47.** prime
49. 2, 3 **51.** 3 **53.** 2, 5
55. composite **57.** prime **59.** composite
61. prime **63.** composite **65.** composite
67. composite **69.** composite
71. No, because the two prime numbers will be factors of the product
73. yes

Exercise 2.2

1. $\frac{6}{8}$ **3.** $\frac{2}{18}$ **5.** $\frac{1}{6}$
7. $\frac{3}{6}$ **9.** $\frac{3}{10}$
11.

$\frac{4}{8}$

$\frac{1}{2}$

13.

$\frac{2}{6}$

$\frac{4}{12}$

15.

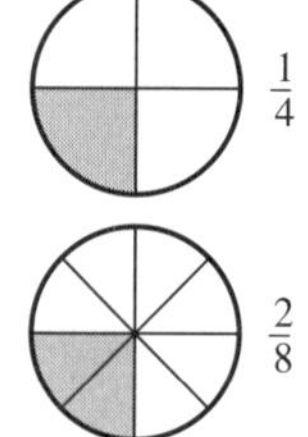

$\frac{1}{4}$

$\frac{2}{8}$

17. $\frac{2}{4}, \frac{3}{6}, \frac{4}{8}, \ldots$ **19.** $\frac{6}{16}, \frac{9}{24}, \frac{12}{32}, \ldots$ **21.** $\frac{2}{8}, \frac{3}{12}, \frac{4}{16}, \ldots$
23. $\frac{12}{22}, \frac{18}{33}, \frac{24}{44}, \ldots$ **25.** $\frac{16}{14}, \frac{24}{21}, \frac{32}{28}, \ldots$ **27.** $\frac{6}{20}, \frac{9}{30}, \frac{12}{40}, \ldots$
29. $\frac{32}{10}, \frac{48}{15}, \frac{64}{20}, \ldots$ **31.** $\frac{4}{6}, \ldots$ **33.** $\frac{2}{16}, \ldots$
35. $\frac{4}{14}, \ldots$ **37.** $\frac{20}{14}, \ldots$ **39.** $\frac{16}{18}, \ldots$
41. $\frac{8}{30}, \ldots$

Exercise 2.3

1. $\frac{1}{5}$ **3.** $\frac{5}{7}$
5. cannot be reduced **7.** $\frac{1}{7}$
9. $\frac{6}{7}$ **11.** $\frac{1}{3}$
13. $\frac{3}{5}$ **15.** $\frac{4}{1}$ or 4
17. $\frac{16}{9}$ **19.** $\frac{1}{3}$
21. $\frac{3}{4}$ **23.** cannot be reduced
25. $\frac{9}{4}$ **27.** cannot be reduced
29. $\frac{2}{7}$ **31.** $\frac{4}{3}$
33. cannot be reduced **35.** $\frac{3}{2}$
37. $\frac{12}{18}$ would normally be reduced to $\frac{2}{3}$.
39. $\frac{25}{32}$ cannot be reduced **41.** $\frac{1}{3}$ hr
43. $\frac{1}{3}$ days **45.** $\frac{3}{4}$
47. $\frac{1}{2}$ **49.** $\frac{1}{3}$
51. $\frac{1}{7}$ **53.** $\frac{3}{2}$
55. $\frac{2}{3}$ **57.** cannot be reduced
59. 2
61. The fraction $\frac{5}{6}$ cannot be reduced. The 3s cannot be cancelled because they are not *factors* of both the numerator and denominator.
63. See the rule for reducing fractions to lowest terms.

Exercise 2.4

1. $\frac{1}{4}$ **3.** $\frac{2}{21}$ **5.** $\frac{1}{12}$
7. $\frac{1}{6}$ **9.** $\frac{4}{9}$ **11.** $\frac{15}{32}$
13. $\frac{1}{4}$ **15.** $\frac{5}{9}$ **17.** $\frac{20}{3}$
19. 10 **21.** $\frac{1}{10}$ **23.** $\frac{3}{8}$
25. 2 **27.** $\frac{5}{2}$ **29.** $\frac{1}{12}$
31. $\frac{5}{64}$ **33.** 1 **35.** $\frac{3}{4}$
37. $\frac{3}{4}$ **39.** 162 boys **41.** $\frac{3}{8}$ of a cup
43. $\frac{1}{4}$, 2 slices **45.** 3640 students

Exercise 2.5

1. 3 **3.** $\frac{5}{6}$ **5.** $\frac{27}{64}$
7. $\frac{4}{25}$ **9.** $\frac{7}{8}$ **11.** 4
13. 27 **15.** $\frac{1}{16}$ **17.** $\frac{1}{8}$
19. 6 **21.** $\frac{3}{5}$ **23.** 16
25. 20 pieces **27.** $\frac{5}{128}$ **29.** 6 times

Exercise 2.6

1. $\frac{6}{8} = \frac{3}{4}$ **3.** $\frac{6}{9} = \frac{2}{3}$ **5.** $\frac{2}{3}$
7. $\frac{4}{4} = 1$ **9.** $\frac{3}{28}$ **11.** $\frac{12}{12} = 1$
13. $\frac{10}{11}$ **15.** $\frac{11}{16}$ **17.** $\frac{0}{8} = 0$
19. $\frac{15}{8}$ **21.** $\frac{15}{3} = \frac{5}{1} = 5$ **23.** $\frac{42}{45} = \frac{14}{15}$
25. $\frac{12}{4} = 3$ **27.** $\frac{5}{8}$

Exercise 2.7

1. $2 \cdot 3 \cdot 7 = 42$ **3.** $2 \cdot 2 \cdot 2 \cdot 3 \cdot 3 = 72$
5. $2 \cdot 2 \cdot 2 \cdot 2 \cdot 3 = 48$ **7.** $2 \cdot 2 \cdot 3 \cdot 3 \cdot 7 = 252$
9. $2 \cdot 3 \cdot 3 \cdot 5 = 90$ **11.** $2 \cdot 2 \cdot 2 \cdot 3 \cdot 3 \cdot 5 = 360$
13. $2 \cdot 3 \cdot 5 \cdot 11 = 330$ **15.** $2 \cdot 2 \cdot 2 \cdot 2 \cdot 3 \cdot 3 = 144$

17. 20 **19.** 10 **21.** 6
23. 12 **25.** 40 **27.** $\frac{5}{6}$
29. $\frac{25}{45} = \frac{5}{9}$ **31.** $\frac{47}{48}$ **33.** $\frac{77}{144}$
35. $\frac{105}{112} = \frac{15}{16}$ **37.** $\frac{3}{16}$ **39.** $\frac{13}{20}$
41. $\frac{53}{48}$ **43.** $\frac{87}{84} = \frac{29}{28}$ **45.** $\frac{23}{42}$
47. $\frac{47}{60}$ **49.** $\frac{7}{12}$ **51.** $\frac{3}{8}$ cup
53. $\frac{5}{8}$ teaspoon
55. The LCD is equal to the product of all of the denominators when there are no prime factors common to any of them. The denominators are called *relatively prime.*

Review Exercises—Chapter 2

1. 2, 3 **3.** numerator, denominator
5. $\frac{6}{42}, \frac{9}{63}, \frac{12}{84}$ are typical **7.** $\frac{3}{20}$
9. $\frac{31}{40}$ **11.** $\frac{16}{15}$
13. $\frac{59}{48}$ **15.** $\frac{23}{72}$
17. $\frac{5}{32}$ **19.** $\frac{1}{5}$
21. $\frac{8}{9}$ **23.** $\frac{1}{10}$
25. $\frac{3}{8}$ of a cup **27.** 6 times
29. $\frac{2}{12} = \frac{1}{6}$
31. $\frac{5}{6}$ of James' day is spent eating, sleeping, and socializing, and $\frac{1}{6}$ of his day is spent studying.

Critical Thinking Exercises—Chapter 2

1. True: both reduce to $\frac{1}{3}$.
3. False: when zero occurs in the denominator, the fraction is not defined.
5. False: denominators are not *added* together when adding fractions. The correct result is $\frac{7}{5}$.
7. The result is not in lowest terms. The correct answer is $\frac{\cancel{3}}{\cancel{4}} \cdot \frac{\cancel{4}}{\cancel{9}} = \frac{1}{3}$.
9. $\frac{8}{24} = \frac{\cancel{2} \cdot \cancel{2} \cdot \cancel{2}}{\cancel{2} \cdot \cancel{2} \cdot \cancel{2} \cdot 3} = \frac{1}{3}$, not 0.

Achievement Test—Chapter 2

1. 3, 5 **2.** $2^2 \cdot 3^2 \cdot 5$
3. denominator, numerator **4.** **(a)** $\frac{2}{5}$ **(b)** $\frac{3}{8}$
5. $\frac{20}{30}, \frac{30}{45}, \frac{40}{60}, \ldots$ **6.** **(a)** $\frac{9}{20}$ **(b)** cannot be reduced
7. **(a)** 21 **(b)** 22 **8.** 48 **9.** $\frac{1}{2}$
10. $\frac{5}{24}$ **11.** $\frac{7}{12}$ **12.** $\frac{4}{11}$
13. $\frac{41}{48}$ **14.** $\frac{25}{32}$ **15.** $\frac{5}{24}$
16. $\frac{8}{12} = \frac{2}{3}$ **17.** $\frac{3}{8}$ of a cup

Chapter 3

Exercise 3.1

1. $1\frac{3}{8}, \frac{11}{8}$ **3.** $4\frac{1}{4}, \frac{17}{4}$ **5.** $2\frac{2}{3}, \frac{8}{3}$
7. $3, \frac{12}{4}$ **9.** $\frac{11}{6}$ **11.** $\frac{23}{16}$
13. $\frac{44}{7}$ **15.** $\frac{58}{5}$ **17.** $\frac{50}{3}$
19. $\frac{269}{22}$ **21.** 1 **23.** $19\frac{1}{3}$
25. $3\frac{1}{9}$ **27.** $2\frac{11}{15}$ **29.** $36\frac{1}{5}$
31. $7\frac{13}{17}$ **33.** $\frac{73}{12}$ **35.** 7 times; $\frac{7}{4}$
37. $\frac{249}{8}$ **39.** $\frac{16}{5}$ **41.** $\frac{13}{4}$
43. $\frac{9}{5}$ **45.** $\frac{27}{4}$ **47.** $3\frac{1}{4}$
49. $4\frac{1}{2}$ **51.** $\frac{23}{3}$ **53.** $4\frac{3}{4}$
55. $\frac{48}{5}$

57. A mixed number is a combination of a whole number and a fraction.

Exercise 3.2

1. $\frac{77}{8}$ or $9\frac{5}{8}$ **3.** $\frac{123}{224}$ **5.** $\frac{171}{7}$ or $24\frac{3}{7}$
7. $\frac{204}{125}$ or $1\frac{79}{125}$ **9.** 8 **11.** $\frac{3}{28}$
13. 66 **15.** $\frac{5}{27}$ **17.** 6
19. $\frac{2}{3}$ **21.** $\frac{175}{16}$ or $10\frac{15}{16}$ **23.** 162
25. $\frac{77}{8}$ or $9\frac{5}{8}$ **27.** 4 **29.** $\frac{26}{5}$ or $5\frac{1}{5}$
31. 2 **33.** 12 **35.** $\frac{7}{18}$
37. 1 **39.** $\frac{15}{11}$ or $1\frac{4}{11}$ **41.** 24 yd
43. 4 shelves **45.** $\frac{5}{4}$ or $1\frac{1}{4}$ cups
47. No: $\left(3\frac{1}{2}\right)^2 = 12\frac{1}{4}$ but $3^2 + \left(\frac{1}{2}\right)^2 = 9 + \frac{1}{4} = 9\frac{1}{4}$.

Exercise 3.3

1. $5\frac{1}{10}$ **3.** $10\frac{11}{16}$ **5.** $16\frac{11}{18}$
7. $11\frac{25}{48}$ **9.** $12\frac{23}{40}$ **11.** $12\frac{19}{24}$
13. $21\frac{3}{4}$ **15.** $11\frac{11}{12}$ **17.** $14\frac{13}{24}$
19. $16\frac{11}{18}$ **21.** $52\frac{3}{14}$ **23.** $37\frac{3}{20}$
25. $40\frac{11}{15}$ **27.** $306\frac{1}{2}$ **29.** $35\frac{1}{4}$
31. $103\frac{3}{8}$ **33.** $14\frac{5}{16}$ in. **35.** $2\frac{29}{32}$ in.
37. $10\frac{1}{4}$ hr **39.** 11 hr **41.** 39 in.

Exercise 3.4

1. $2\frac{1}{2}$ **3.** $2\frac{5}{8}$ **5.** $5\frac{1}{4}$
7. $5\frac{3}{4}$ **9.** $11\frac{13}{24}$ **11.** $2\frac{1}{3}$
13. $4\frac{7}{12}$ **15.** $3\frac{13}{40}$ **17.** $7\frac{1}{3}$
19. $12\frac{2}{5}$ **21.** $4\frac{11}{16}$ **23.** $4\frac{31}{48}$
25. $4\frac{25}{48}$ **27.** $35\frac{1}{2}$ **29.** 18 hr
31. $2\frac{3}{4}$ more hours **33.** $5\frac{1}{2}$ more hours
35. If the whole-number parts are small, changing the mixed numbers to improper fractions and then subtracting can be easier. If the whole-number parts are large, it is better to leave as mixed numbers and subtract.

Exercise 3.5

1. 4 **3.** 2 **5.** $\frac{7}{2}$ or $3\frac{1}{2}$
7. $\frac{1}{16}$ **9.** $\frac{32}{5}$ or $6\frac{2}{5}$ **11.** $\frac{17}{2}$ or $8\frac{1}{2}$
13. $\frac{8}{15}$ **15.** $\frac{4}{27}$ **17.** $\frac{50}{53}$
19. $\frac{28}{477}$ **21.** $\frac{136}{177}$ **23.** $\frac{3}{26}$
25. $\frac{69}{80}$ **27.** $\frac{512}{945}$ **29.** $\frac{68}{151}$
31. $5\frac{85}{114}$
33. Then the correct *L*owest *C*ommon *D*enominator has not been found.

Review Exercises—Chapter 3

1. $\frac{40}{7}$ **3.** $\frac{113}{11}$ **5.** $9\frac{4}{5}$
7. $2\frac{1}{4}$; $\frac{9}{4}$ **9.** 34 **11.** $1\frac{13}{32}$
13. $24\frac{2}{5}$ **15.** $4\frac{8}{15}$ **17.** $47\frac{11}{16}$
19. $\frac{4}{3}$ or $1\frac{1}{3}$ **21.** $\frac{88}{105}$ **23.** $5\frac{1}{4}$ cups
25. $6\frac{1}{4}$ more hours

Critical Thinking Exercises—Chapter 3

1. False: $6\frac{2}{3} = 6 + \frac{2}{3}$ or $\frac{20}{3}$
3. True
5. False: $5\frac{1}{3} \cdot \frac{3}{4} = \frac{\overset{4}{\cancel{16}}}{\underset{1}{\cancel{3}}} \cdot \frac{\overset{1}{\cancel{3}}}{\underset{1}{\cancel{4}}} = \frac{4}{1} = 4$
7. $3\frac{1}{4} \div \frac{2}{3} = \frac{13}{4} \cdot \frac{3}{2} = \frac{39}{8}$ or $4\frac{7}{8}$
9. Never add the denominators; an LCD must be found: $2\frac{1}{3} + 1\frac{3}{4} = \frac{7}{3} + \frac{7}{4} = \frac{28}{12} + \frac{21}{12} = \frac{49}{12}$ or $4\frac{1}{12}$
11. $\left(2\frac{1}{2}\right)^2 = \left(\frac{5}{2}\right)^2 = \frac{25}{4}$ or $6\frac{1}{4}$

Achievement Test—Chapter 3

1. $\frac{70}{9}$ **2.** $7\frac{3}{8}$ **3.** $16\frac{2}{7}$
4. $\frac{4}{3}$ or $1\frac{1}{3}$ **5.** 66 **6.** 48
7. $38\frac{4}{7}$ **8.** $5\frac{3}{10}$ **9.** $60\frac{23}{36}$
10. $10\frac{2}{3}$ **11.** $3\frac{7}{8}$ yd **12.** $21\frac{3}{8}$ pages

Cumulative Review—Chapters 1, 2, and 3

1. hundreds place **3.** 24,000 **5.** 50,693
7. 28,800 seconds **9.** 139 R 14 **11.** 1
13. 43,963
15. $2 \cdot 2 \cdot 3 \cdot 3 \cdot 5$ or $2^2 \cdot 3^2 \cdot 5$
17. 3 **19.** $\frac{47}{30}$ or $1\frac{17}{30}$ **21.** $\frac{45}{8}$
23. 2 **25.** $95\frac{9}{16}$ **27.** $12\frac{3}{8}$

Chapter 4

Exercise 4.1

1. **(a)** $\frac{358}{1000}$ **(b)** three hundred fifty-eight thousandths **(c)** $\frac{3}{10} + \frac{5}{100} + \frac{8}{1000}$
3. **(a)** $\frac{12}{1000}$ **(b)** twelve thousandths **(c)** $\frac{0}{10} + \frac{1}{100} + \frac{2}{1000}$
5. **(a)** $\frac{4}{10}$ **(b)** four tenths **(c)** $\frac{4}{10}$
7. **(a)** $\frac{9{,}999}{10{,}000}$ **(b)** nine thousand nine hundred ninety-nine ten-thousandths **(c)** $\frac{9}{10} + \frac{9}{100} + \frac{9}{1000} + \frac{9}{10{,}000}$
9. **(a)** $\frac{364}{1000}$ **(b)** three hundred sixty-four thousandths **(c)** $\frac{3}{10} + \frac{6}{100} + \frac{4}{1000}$
11. **(a)** $3\frac{7}{10}$ **(b)** three and seven tenths
13. **(a)** $17\frac{22}{1000}$ **(b)** seventeen and twenty-two thousandths
15. **(a)** $47\frac{234}{1000}$ **(b)** forty-seven and two hundred thirty-four thousandths
17. 9.021 **19.** 0.14 **21.** 80.012
23. 600.09 **25.** tens **27.** tenths
29. thousandths **31.** hundredths **33.** thousandths
35. hundreds
37. two and fifty-four hundredths
39. three and seven hundred eighty-four thousandths
41. eight and one hundred sixty-two thousandths
43.

Frank Field 167
201 Windy Blvd.
Chicago, IL March 15, 19 94
Pay To The Order Of Sullivan's Dept. Store $ 301.46
Three hundred one and 46/100 Dollars
Memo Frank Field

45.

Eva Robbins 341
26 Thurston Ave., Apt. 4A
Oneonta, NY July 6, 19 93
Pay To The Order Of Oneonta Community College $ 684.12
Six hundred eighty-four and 12/100 Dollars
Memo Eva Robbins

47.

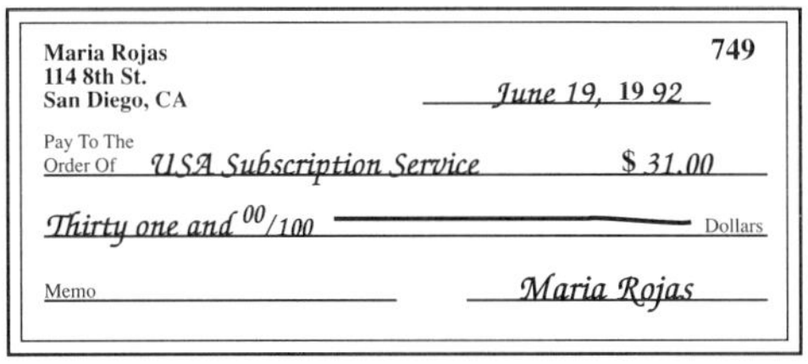

Maria Rojas
114 8th St.
San Diego, CA
749
June 19, 19 92
Pay To The Order Of USA Subscription Service $ 31.00
Thirty one and 00/100 Dollars
Memo Maria Rojas

49.

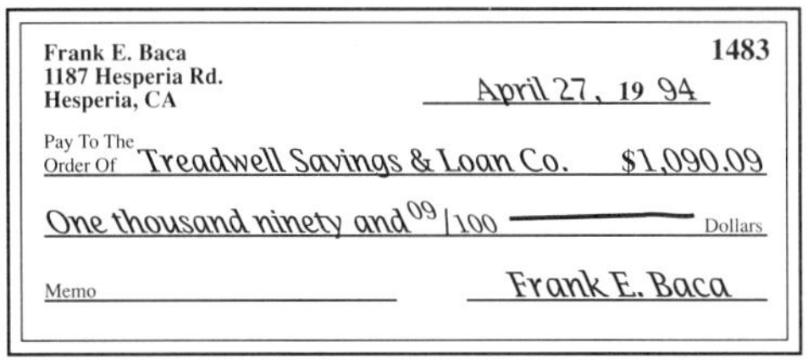

Frank E. Baca
1187 Hesperia Rd.
Hesperia, CA
1483
April 27, 19 94
Pay To The Order Of Treadwell Savings & Loan Co. $1,090.09
One thousand ninety and 09/100 Dollars
Memo Frank E. Baca

Exercise 4.2

	Number	Round to the nearest:		
		Thousandth	Hundredth	Tenth
1.	0.4614	0.461	0.46	0.5
3.	4.58263	4.583	4.58	4.6
5.	62.5198	62.520	62.52	62.5
7.	0.0984	0.098	0.10	0.1
9.	3.967	3.967	3.97	4.0
11.	39.9926	39.993	39.99	40.0

13. 322 **15.** 17
17. 8.6154, 8.5999, 8.6001 **19.** 1.01
21. 1.01
23. Both 0.91 and 1.09 are the same distance from 1.
25. 0.003, 0.09, 0.901, 0.95, 1.0, 1.001, 1.05
27. 2.3 kg **29.** 3.14 **31.** 136 mph

Exercise 4.3

1. 8.07 **3.** 7.984 **5.** 34.380
7. 21.321 **9.** 223.52834 **11.** 104.1504
13. 116.518 **15.** 53.78 **17.** 146.56
19. 71.128 **21.** 29.658 **23.** 26.6
25. 10.23 **27.** 21.825 **29.** 20.54
31. 27.171 **33.** 80.74 **35.** 24.791
37. 1.562 **39.** $295.15 **41.** $3.10
43. $51.78 **45.** $46.55 **47.** $3.65
49. $3.70; three one-dollar bills, two quarters, and two dimes
51. so that your change would include a nickel instead of four pennies

Exercise 4.4

1. 3.6 **3.** 0.48 **5.** 3.48
7. 0.2891 **9.** 1.41814 **11.** 0.05994
13. 0.00904 **15.** 0.39 **17.** 5.5
19. 0.0035 **21.** 0.03027 **23.** 0.02106
25. 24.9622 **27.** 0.02639
29. The decimal point moves one place to the right.
31. When multiplying by 0.1, the decimal point moves one place to the left; for 0.01, the decimal point moves two places to the left; and for 0.001, the decimal point moves three places to the left.
33. $20.50 **35.** $12.64 **37.** $2.94
39. $61.44 **41.** 834 lb **43.** $13.84
45. $0.97

Exercise 4.5

1. 2.435 **3.** 0.36 **5.** 0.0426
7. 0.0036498 **9.** 4.6 **11.** 0.36
13. 0.06 **15.** 0.009 **17.** 0.00031
19. 2.8 **21.** 3.6 **23.** 0.24
25. 2.6 **27.** 4.2 **29.** 0.27
31. 0.64 **33.** 0.82 **35.** 0.65
37. 1.96 **39.** 2.13 **41.** 7.71
43. 18 hr **45.** 26.8 **47.** $150.62/month
49. 26.4 mpg **51.** 0.342 **53.** 48.4 mph
55. 32.5 pts/game

Exercise 4.6

1. 0.2 **3.** 0.6 **5.** 0.38
7. 0.29 **9.** 0.22 **11.** 0.44
13. 0.41 **15.** 6.25 **17.** 8.67
19. 23.31 **21.** $\frac{3}{10}$ **23.** $\frac{1}{2}$
25. $\frac{3}{5}$ **27.** $\frac{9}{20}$ **29.** $2\frac{2}{5}$
31. $6\frac{3}{5}$ **33.** $16\frac{1}{4}$ **35.** $\frac{7}{200}$
37. $\frac{79}{125}$ **39.** $\frac{9}{2000}$ **41.** $13\frac{4}{125}$
43. $125\frac{1}{20}$ **45.** $12\frac{4}{5}$ **47.** 3.625
49. $30\frac{12}{25}$

Review Exercises—Chapter 4

1. **(a)** $\frac{13}{50}$ **(b)** twenty-six hundredths **(c)** $\frac{2}{10} + \frac{6}{100}$
3. 16.024
5. **(a)** 13.7 **(b)** 13.68 **(c)** 14
7. 21.774 **9.** 11.673 **11.** 9.796
13. 82.784 **15.** 0.82784 **17.** 1.2
19. 0.12 **21.** 5.38 **23.** $\frac{23}{100}$
25. $27\frac{11}{20}$ **27.** $5.45/hr

Critical Thinking Exercises—Chapter 4

1. False: the denominator must be divided into the numerator.
3. False: $3.061 = 3 + \frac{0}{10} + \frac{6}{100} + \frac{1}{1000}$
5. True: $10.54 = 10\frac{54}{100} = 10\frac{27}{50}$
7.
```
       1.5
2.3.)3.4.5
     2 3
     1 1 5
     1 1 5
         0
```
Before dividing, the decimal point must be moved the same number of places in the divisor and the dividend, making the divisor a whole number.

9. 0.027 is twenty-seven thousandths because the "7" of 27 is in the thousandths place.

Achievement Test—Chapter 4

1. seventy-three thousandths
2. $4 + \frac{2}{10} + \frac{7}{100}$
3. twenty-six and seven hundredths
4. 21.014 **5.** tenths
6. **(a)** 12.6 **(b)** 12.61 **(c)** 13

7. 16.579 **8.** 17.94 **9.** 13.255
10. 0.036144 **11.** 3.427 **12.** 5.6
13. **(a)** 1.08 **(b)** 4.21 **14.** **(a)** 0.6 **(b)** 5.6 **15.** $\frac{77}{200}$
16. $35\frac{3}{50}$ **17.** $67.00 **18.** $434.86

Chapter 5

Exercise 5.1

1. $\frac{3}{5}$ **3.** $\frac{3}{1}$ **5.** $\frac{7}{5}$
7. $\frac{1}{4}$ **9.** $\frac{1}{5}$ **11.** $\frac{3}{1}$
13. $\frac{3}{4}$ **15.** $\frac{3}{500}$ **17.** $\frac{17}{6}$
19. $\frac{9}{1}$ **21.** $\frac{7}{12}$ **23.** $\frac{5}{9}$
25. $\frac{4}{9}$ **27.** $\frac{153}{16}$ **29.** $\frac{1}{3}$
31. $\frac{1}{4}$ **33.** $\frac{1}{50}$ **35.** $\frac{5}{1}$
37. $\frac{1}{4}$ **39.** $\frac{5}{24}$ **41.** $\frac{2}{5}$
43. $\frac{1}{10}$ **45.** $\frac{2}{5}$

Exercise 5.2

1. 26 mpg **3.** 13 pts/game
5. 48 mph **7.** $\frac{1}{3}$ ton/acre
9. 34 words/min **11.** 5.2 gal/hr
13. 2.2 ft/sec **15.** 31.7 words/min
17. 241.7 sq. ft/gal **19.** 3.1 g/liter
21. 63.2 words/min **23.** 7.3 points/game
25. Brand A is 14.8¢ per ounce, and Brand B is 14.4¢ per ounce. Therefore, Brand B is the better buy.
27. In a *ratio,* either there are no units or, if there are, they are the same (or can be made the same), while a *rate* is the comparison of two quantities, each of a different type.
29. Unit pricing gives the consumer a basis on which to compare prices of different sizes or brands of an item. By comparing prices using unit pricing, a smart shopper can save money.

Exercise 5.3

1. = **3.** ≠ **5.** =
7. = **9.** ≠ **11.** ≠
13. = **15.** = **17.** yes
19. yes

Exercise 5.4

1. $x = 6$ **3.** $w = 50$ **5.** $m = 50$
7. $s = 330$ **9.** $k = 56$ **11.** $m = 28$
13. $y = \frac{30}{7}$ **15.** $x = \frac{7}{12}$

Exercise 5.5

1. 10 hr **3.** $7\frac{1}{2}$ cups
5. 7,000 sq ft **7.** 39 ft by 30 ft
9. approximately 2,583 families **11.** approximately 16.67 mg
13. $222.50 **15.** approximately 5.4 days
17. 16.33 in. **19.** 6.66 g
21. $1.33 **23.** $2,176.47
25. 126 items

Review Exercises—Chapter 5

1. $\frac{5}{2}$
3. **(a)** $9.00/hr **(b)** 42 words/min **(c)** 45.67 mph
5. **(a)** $x = 12$ **(b)** $y = 21$ **(c)** $t = 2.45$
7. 9 hr

Critical Thinking Exercises—Chapter 5

1. False: $\frac{3\text{ feet}}{12\text{ yards}} = \frac{3\text{ feet}}{36\text{ feet}} = \frac{1}{12}$
3. False: the ratio of strikeouts to home runs is $\frac{2{,}459}{534}$.
5. False: the unit price is $\frac{89\text{ cents}}{10\text{ oz}} = 8.9¢/\text{oz}$.
7. $\frac{6}{8} \neq \frac{12}{9}$ because $6 \cdot 9 \neq 8 \cdot 12$.
9. $\frac{\$420}{\$92} = \frac{x}{\$736}$
$92x = 309{,}120$
$x = \$3{,}360$

Achievement Test—Chapter 5

1. $\frac{9}{16}$
2. **(a)** $\frac{2}{5}$ **(b)** $\frac{1}{4}$
3. **(a)** $16/sq yd **(b)** 26.7 mpg
4. **(a)** = **(b)** ≠ **(c)** =
5. 24 ft by 22 ft
6. $1,023.08
7. 384 miles
8. The unit prices are 6.25¢/oz for the 12-oz size and 6.19¢/oz for the 16-oz size, so the 16-oz bottle is the best value.

Chapter 6

Exercise 6.1

1. 0.56 **3.** 0.49 **5.** 0.046
7. 0.076 **9.** 0.0005 **11.** 0.0032
13. 0.235 **15.** 0.00125 **17.** 3.4
19. 1.39 **21.** 31% **23.** 85%
25. 125% **27.** 204% **29.** 60%
31. 70% **33.** 6.3% **35.** 0.91%
37. 3.25% **39.** 87.6%
41. "Percent" means per one hundred.
43. You move the decimal point two places to the right when changing from a decimal to a percent because you are *multiplying* by 100.

Exercise 6.2

1. 80% **3.** 62.5% **5.** 37.5%
7. 16% **9.** 31.3% **11.** 150%
13. 160% **15.** 170% **17.** $\frac{3}{4}$
19. $\frac{2}{5}$ **21.** $\frac{9}{100}$ **23.** $\frac{2}{25}$
25. $\frac{1}{8}$ **27.** $\frac{23}{40}$ **29.** $1\frac{3}{4}$
31. $1\frac{4}{5}$ **33.** $\frac{3}{1000}$ **35.** $\frac{9}{10{,}000}$

	Percent	Decimal	Fraction
37.	25%	0.25	$\frac{1}{4}$
39.	32%	0.32	$\frac{8}{25}$
41.	17.5%	0.175	$\frac{7}{40}$
43.	11.6%	0.116	$\frac{29}{250}$
45.	0.008%	0.00008	$\frac{1}{12{,}500}$
47.	100%	1	$\frac{6}{6}$
49.	0.1%	0.001	$\frac{1}{1000}$

51. 0.22 **53.** 16.7% **55.** **(a)** 12.5% **(b)** $\frac{5}{12}$

Exercise 6.3

1. $A = 50$ 3. $A = 3.6$ 5. $P = 20\%$
7. $P = 58.33\%$ 9. $P = 400\%$ 11. $P = 250\%$
13. $A = 135$

Exercise 6.4

1. $P = 33.3\%$ 3. $A = 0.000632$ 5. $A = 128.128$
7. $P = 16.67\%$ 9. $A = 19.78$ 11. $W = 180$
13. $W = 18$ 15. $W = 70$ 17. $36,400
19. $3.90 21. $345.60 23. $1,080
25. 488 students will finish
27. The coat is reduced 15%, which is more than 8.33%, the amount the suit is reduced.
29. 22 problems
31. $90 by both methods. The second method is generally easier because it requires less work.

Review Exercises—Chapter 6

1. 0.18 3. 0.0005 5. 3.12
7. 285% 9. 0.93% 11. 60%
13. 180% 15. $\frac{3}{5}$ 17. $\frac{29}{200}$

	Percent	Decimal	Fraction
19.	30%	0.3	$\frac{3}{10}$
21.	37.5%	0.375	$\frac{3}{8}$
23.	124%	1.24	$1\frac{6}{25}$

25. $P = 75\%$ 27. $W = 190$ 29. $3,800
31. 6%

Critical Thinking Exercises—Chapter 6

1. False: when changing from a percent to a decimal, the decimal point is moved two places to the *left.*
3. False: change $\frac{3}{4}$ to a decimal by dividing the denominator into the numerator to obtain 0.75. Then change 0.75 to 75% by moving the decimal point two places to the right.
5. False: 20 is 74% of what number? $\frac{20}{W} = \frac{74}{100}$, so $W = 27$; there were 27 questions on the test.
7. 0.4 written as a percent is 40%; the decimal point moves to the *right* in converting from decimals to percents.

Chapter 6—Achievement Test

1. 0.14 2. 0.035 3. 0.1425
4. 38% 5. 4.5% 6. 20%
7. 80% 8. 114.286% 9. 160%
10. $\frac{17}{20}$ 11. $\frac{51}{200}$ 12. $\frac{3}{5000}$
13. $A = 119$ 14. $P = 16.67\%$ 15. $W = 330$
16. $P = 18.9\%$ 17. 21 games
18. 160 lb originally; 128 lb now

Cumulative Review—Chapters 4, 5, and 6

1. $\frac{17}{50}$ 3. 51.9
5. 48.165 7. 2.3
9. $\frac{4}{15}$ 11. 18
13. 0.055 15. 18.75%
17. 108 19. 260

Chapter 7

Exercise 7.1

1. 6 3. 84 5. 156
7. 28 9. 18 11. 33
13. 8 15. bimodal: 16, 19 17. no modes
19. mean: 13; median: 15; mode: 15
21. mean: 22.17; median: 21.5; mode: 18
23. mean: 3.12; median: 3.2; mode: none
25. 81
27. mean: 32 sit-ups; median: 35 sit-ups; mode: 35 sit-ups
29. mean: 60.5 mph; median: 62 mph; mode: 62 mph
31. 3.07 33. 2.2 35. 2.07
37. 2.69 39. 3.23 41. 2.28
43. 0.22
45. Yes: the average of any group of numbers must always lie between the largest and the smallest ("central tendency").

Exercise 7.2

1. 88° 3. 95° 5. 7°
7. 15° 9. 667 people 11. 3,071 people
13. by 257 people 15. It increased by 753 people.
17. cornflakes 19. raisin bran 21. 40 calories
23. 580 mg 25. cornflakes

Exercise 7.3

1. education 3. 4,500 5. education
7. technology 9. Lake Superior 11. 1,120 ft
13. 1993 15. 15 in. 17. upward
19. Tuesday 21. between Tuesday and Wednesday
23. 9° 25. Abe; 40 27. 140
29. 15 more 31. $2,000,000
33. Centerville and Goshen
35. $16,000,000 37. Scotchtown

Exercise 7.4

1. room and board 3. travel, clothing, miscellaneous
5. $1,500 7. brand *D*
9. 100 11. brand *E*

Review Exercises—Chapter 7

1. 28 3. 16.5
5. 164 7. 3.27
9. **(a)** 7 **(b)** 12 **(c)** 18 **(d)** 20 **(e)** 50 **(f)** 40
11. **(a)** game 9; 90 points **(b)** game 4; 55 points **(c)** 20 more **(d)** upward
13. **(a)** private homes **(b)** 45% **(c)** yes **(d)** 56 **(e)** 32

Critical Thinking Exercises—Chapter 7

1. Julian must add all three grades together and divide the total by 3, which yields 78.3.
3. It raises the class average by 10 points.
5. no solution; cannot be done

Achievement Test—Chapter 7

1. 47 2. 39 3. 2.6
4. mean: 161 lb; median: 161 lb; mode: 159 lb and 160 lb (bimodal)
5. 2.94

6. **(a)** 9.1 **(b)** 2.4 more **(c)** 116
7. **(a)** 90 **(b)** 20 more **(c)** 280
8. **(a)** 58° F **(b)** 18° **(c)** 7:00 A.M. and 8:00 A.M. **(d)** 3:00 A.M. **(e)** 1:00 A.M. and 2:00 A.M.
9. **(a)** 50 **(b)** English and social sciences **(c)** 100
10. **(a)** Stair climber and tennis are equally popular. **(b)** 25% **(c)** 560

Chapter 8

Answer to Number Knowledge Problem for Chapter 8

No, it can't be done. Each domino covers exactly two squares, one black and one red. When the upper left and lower right squares were removed, both of them were black, so what remains is 32 red squares and only 30 black squares. Since each domino must cover one square of each color, the entire board cannot be covered. Two red squares will remain uncovered.

Exercise 8.1

1.

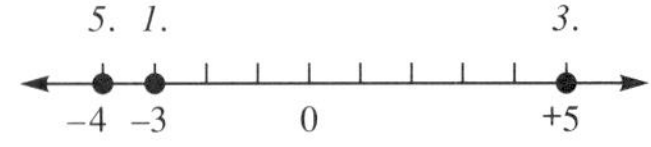

7. −5.00 **9.** +7,288 **11.** +1990
13. −15 **15.** −25 **17.** $-2\frac{1}{4}$
19. −36,198 **21.** −38.55

Exercise 8.2

1. 8 **3.** 4 **5.** 1462
7. 0 **9.** 6 **11.** 1.7
13. It is 15 miles from town *A* to town *B* and vice versa.
15. 10 **17.** 20 **19.** 21
21. 40 **23.** 11 **25.** 9

Exercise 8.3

1. true **3.** false **5.** true
7. false **9.** true **11.** false
13. false **15.** true **17.** false
19. > **21.** > **23.** <
25. < **27.** > **29.** >
31. > **33.** < **35.** >
37. the town at milepost 56

Exercise 8.4

1. 14 **3.** −3 **5.** 2
7. 7 **9.** −1 **11.** −6
13. −10 **15.** −8 **17.** −17
19. 13 **21.** −30 **23.** 31
25. 18 **27.** −22 **29.** 0
31. 50 **33.** 15 **35.** −121
37. −34 **39.** 66 **41.** 521
43. $-\frac{4}{8} = -\frac{1}{2}$ **45.** $-\frac{2}{7}$ **47.** −3 or 13
49. −3 **51.** 5 **53.** 15
55. −$9.55 (overdrawn) **57.** −$14.50 (overdrawn)
59. 20,602 ft

61.

−1	6	1
4	2	0
3	−2	5

63. 11 **65.** 6 **67.** 11
69. −10 **71.** −7 **73.** −7
75. 18 **77.** 0 **79.** 6
81. −5 **83.** −13 **85.** 6
87. −6 **89.** −15 **91.** 15
93. 2 **95.** −18 **97.** 0
99. −16 **101.** −10

Exercise 8.5

1. −2 **3.** −5 **5.** 9
7. −11 **9.** −19 **11.** −30
13. 0 **15.** 10 **17.** 17
19. 16 **21.** −4 **23.** 4
25. 1 **27.** $-8\frac{1}{2}$ **29.** $\frac{1}{7}$
31. 0 **33.** 33 **35.** 95
37. −744 **39.** $10\frac{3}{8}$
41. (−3) − (−5) = (−3) + (+5) = +2 2 units *right*

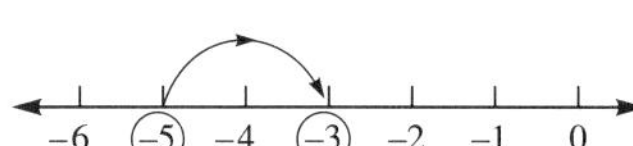

43. (−6) − (+5) = (−6) + (−5) = −11 11 units *left*

45. (+6) − (+6) = (+6) + (−6) = 0
To move from +6 to +6, you move zero units.

47. −2 **49.** 8 **51.** −4
53. 38° **55.** $68.97 **57.** 5
59. −4 **61.** 8 **63.** −11
65. −5 **67.** 0 **69.** 6
71. 11 **73.** 3 **75.** 0
77. 8 **79.** 5 **81.** −6
83. −20 **85.** 3 **87.** 9
89. 12 **91.** −2 **93.** 17
95. 6 **97.** −7 **99.** 16
101. 0

Exercise 8.6

1. −9 **3.** −29 **5.** −4 **7.** −11 **9.** −8 **11.** 13 **13.** 0 **15.** 8 **17.** 25 **19.** −2 **21.** 7.05 **23.** $-\frac{1}{10}$ **25.** **(a)** $-1\frac{1}{4}$ (net loss) **(b)** 63 **27.** $130.54

Exercise 8.7

1. 24 **3.** −28 **5.** −16 **7.** 66 **9.** −60 **11.** 132 **13.** −528 **15.** −98 **17.** −4.08 **19.** −54 **21.** −4 **23.** −38.7 **25.** −144 **27.** −120 **29.** −1 **31.** 17.64 **33.** 81 **35.** 1 **37.** 56 **39.** 4 **41.** −3 **43.** −$11,200 **45.** 48 **47.** −21 **49.** −27 **51.** 24 **53.** −70 **55.** 0 **57.** 36 **59.** −100 **61.** 250 **63.** 24 **65.** 8 **67.** 30 **69.** 0 **71.** −5

73. When multiplying signed numbers, each pair of negative numbers, when multiplied together, yields a positive answer. Therefore, if there is an even number of negative signs in the problem, the product will be positive. If there is an odd number of negative signs, each pair of negative factors yields a positive answer, but there will be one negative factor remaining that will make the final product negative.

Exercise 8.8

1. 3 **3.** 1 **5.** −7 **7.** −9 **9.** $-\frac{1}{6}$ **11.** 0 **13.** undefined **15.** 4 **17.** 4 **19.** −6 **21.** 8 **23.** −9 **25.** undefined **27.** 0 **29.** $\frac{3}{2}$ **31.** $-2\frac{1}{2}$ **33.** −0.4 **35.** undefined **37.** 6 **39.** 3 **41.** 21 **43.** −5 **45.** −64 **47.** $3.60 **49.** $18.30 **51.** 9 **53.** −6 **55.** −8 **57.** 6 **59.** 5 **61.** 0 **63.** undefined **65.** −7 **67.** undefined **69.** −2 **71.** −9 **73.** −2 **75.** −5 **77.** −5 **79.** 4

Exercise 8.9

1. True; commutative property of addition
3. False; subtraction is not associative.
5. True; associative property of addition
7. True; associative property of multiplication
9. False; division is not associative.
11. False; subtraction is not commutative.
13. True; commutative property of addition
15. $6(3 + 5) \stackrel{?}{=} 6 \cdot 3 + 6 \cdot 5$
$6(8) \stackrel{?}{=} 18 + 30$
$48 = 48$
17. $-8(4 + 7) \stackrel{?}{=} (-8)(4) + (-8)(7)$
$-8(11) \stackrel{?}{=} (-32) + (-56)$
$-88 = -88$
19. $4.3(1.2 + 0.3) \stackrel{?}{=} 4.3(1.2) + 4.3(0.3)$
$4.3(1.5) \stackrel{?}{=} 5.16 + 1.29$
$6.45 = 6.45$
21. (−7)(6)
23. 2 + (5 + 7)
25. Subtraction is not commutative because if you reverse the order, you obtain a different answer. For example: 10 − 2 = 8 but 2 − 10 = −8.

Exercise 8.10

1. 16 **3.** 4 **5.** 81 **7.** 64 **9.** 256 **11.** 1,000 **13.** 1 **15.** −1,000 **17.** $\frac{1}{4}$ **19.** 0.0009 **21.** $-\frac{8}{27}$ **23.** 1

25. Raising 10 to a whole-number exponent yields 1 followed by that number of zeros.

27. 9 **29.** 12 **31.** 3 **33.** −3 **35.** 5 **37.** −1 **39.** 4 **41.** −10 **43.** 7 **45.** $\frac{1}{3}$ is greater. **47.** positive **49.** the original fraction **51.** 25 **53.** 81 **55.** 4 **57.** 8 **59.** −8 **61.** 1,000 **63.** 6 **65.** 9 **67.** 4 **69.** −5 **71.** −3 **73.** 3 **75.** $\frac{1}{4}$ **77.** $\frac{1}{64}$ **79.** $\frac{1}{8}$

Exercise 8.11

1. 7 **3.** 1 **5.** 19 **7.** −7 **9.** 6 **11.** 5 **13.** 38 **15.** −9 **17.** −16 **19.** 100 **21.** −71

Exercise 8.12

1. −30 **3.** 13 **5.** 29 **7.** 15 **9.** −23 **11.** 42 **13.** −7 **15.** 15 **17.** 209 **19.** **(a)** −3 **(b)** 6 **21.** **(a)** 12 **(b)** −6 **23.** **(a)** 19 **(b)** 49 **25.** **(a)** 8 **(b)** 4 **27.** **(a)** −5 **(b)** −5 **29.** **(a)** −64 **(b)** 64 **31.** **(a)** 41 **(b)** 81 **33.** **(a)** 64 **(b)** 64

Review Exercises—Chapter 8

1.

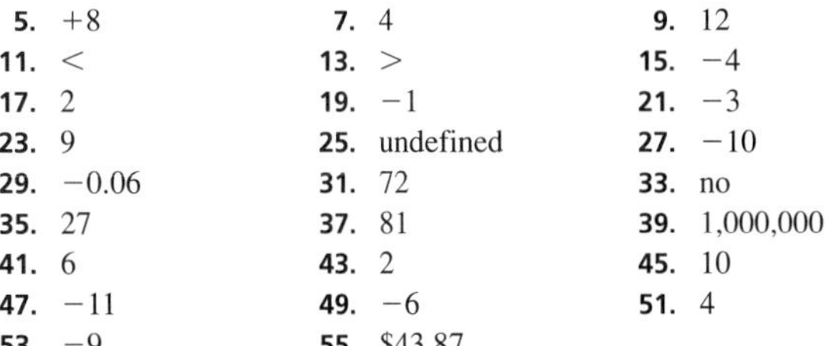

5. +8 **7.** 4 **9.** 12 **11.** < **13.** > **15.** −4 **17.** 2 **19.** −1 **21.** −3 **23.** 9 **25.** undefined **27.** −10 **29.** −0.06 **31.** 72 **33.** no **35.** 27 **37.** 81 **39.** 1,000,000 **41.** 6 **43.** 2 **45.** 10 **47.** −11 **49.** −6 **51.** 4 **53.** −9 **55.** $43.87

Critical Thinking Exercises—Chapter 8

1. True: −6 is to the *right* of −15 on the number line.
3. False: division by zero is *undefined.*
5. True: $-\sqrt[3]{-8} = -(-2) = 2$
7. −6 − 3 = −9
9. 0 ÷ (−4) = 0

11. $\frac{-8}{-8} = 1$

13. $18 \div 9 = 2$, whereas $9 \div 18 = \frac{1}{2}$; division is not commutative.

15. $|-6| + |+5| = 6 + 5 = 11$

17. $18 \div 4 + 5 = \frac{18}{4} + 5 = \frac{9}{2} + 5 = \frac{9}{2} + \frac{10}{2} = \frac{19}{2} = 9\frac{1}{2}$

19. $(-7)^2 \neq -49$

21. $(2 + 5)^2 = 7^2 = 49$; always perform operations inside the parentheses first.

Achievement Test—Chapter 8

1.

4. −24	**5.** +50	**6.** 0
7. 8	**8.** 13	**9.** <
10. >	**11.** <	**12.** −1
13. 1	**14.** 1	**15.** 20
16. 4	**17.** undefined	**18.** 0
19. −9	**20.** −4	**21.** −48
22. −3.16	**23.** $5\frac{2}{5}$	**24.** 64
25. 16	**26.** −8	**27.** 100,000
28. 3	**29.** 31	**30.** 9
31. 8	**32.** −8	**33.** $122.50
34. $168.00	**35.** $28.76	

Chapter 9

Exercise 9.1

1. 108 sq ft	**3.** $1,500	**5.** 176 miles
7. 48 sq units	**9.** 14.444 cm	**11.** 19
13. 6,400	**15.** 16	**17.** 900
19. 67.62	**21.** $16\frac{2}{3}$ sq. yd	**23.** 20 cu ft

25. 76 ft

27. **(a)** $21.08 **(b)** $6.08 **(c)** $181.12; no

Exercise 9.2

1. 18	**3.** −24	**5.** −8
7. 24	**9.** −4	**11.** −19
13. 9	**15.** 9	**17.** −94
19. −42	**21.** 58	**23.** 58
25. 2	**27.** −9	**29.** 25

Exercise 9.3

1. terms: $2x$, $3y$; coefficients: 2, 3; literal parts: x, y

3. terms: $-6x^3y$, $2x^2y^3$; coefficients: 6, 2; literal parts: x^3y, x^2y^3

5. terms: $7x^3y$, $-14xy^3$, $2z$, -5; coefficients: 7, −14, 2, −5; literal parts: x^3y, xy^3, z

7. terms: $-7xy^4$, $2x^2$, $-3xyz$; coefficients: −7, 2, −3; literal parts: xy^4, x^2, xyz

9. terms: $-11a^3$, $-14b^3$, $17ac^4$; coefficients: −11, −14, 17; literal parts: a^3, b^3, ac^4

11. $3x$

13. $5a$	**15.** $17x$
17. $3x$	**19.** $4a^2b$
21. $-2x^2y^2$	**23.** $-8x^3$
25. $10x^3 + 7x$	**27.** $14mn$
29. $8ab - 8b + 3$	**31.** cannot be combined
33. $9x^2y - 9$	**35.** $5m + 6n + 2$
37. $8a - 11b - 3c$	**39.** cannot be combined

Exercise 9.4

1. **(a)** yes **(b)** no **(c)** yes **(d)** no **(e)** yes

3. **(a)** no **(b)** yes **(c)** yes **(d)** yes **(e)** yes

Exercise 9.5

1. $x = 11$	**3.** $x = 11$	**5.** $x = -2$
7. $h = 10$	**9.** $x = -1$	**11.** $x = -9$
13. $x = -15$	**15.** $x = -2$	**17.** $x = 4.25$
19. $x = 7$	**21.** $x = 8.53$	**23.** $x = 0.3$
25. $y = -7$		

Exercise 9.6

1. $x = 4$	**3.** $x = 7$	**5.** $x = 8$
7. $y = -3$	**9.** $y = 9$	**11.** $y = 8$
13. $x = 32$	**15.** $x = 18$	**17.** $x = -24$
19. $x = 18$	**21.** $y = -56$	**23.** $y = 6$
25. $x = -\frac{1}{3}$	**27.** $x = 48$	**29.** $x = -6$
31. $x = -\frac{5}{14}$		

Exercise 9.7

1. $x = 2$	**3.** $x = 2$	**5.** $x = 1$
7. $x = 7$	**9.** $x = 3$	**11.** $x = 3$
13. $x = 8$	**15.** $y = -4$	**17.** $x = -\frac{14}{6}$ or $-\frac{7}{3}$
19. $x = -7$	**21.** $x = 2$	**23.** $x = 2$
25. $x = \frac{6}{11}$	**27.** $x = 3.1$	**29.** $x = 0$
31. $x = 10$	**33.** $x = -8$	**35.** $y = -12$
37. $x = -7$	**39.** $x = 8$	**41.** $x = 4$
43. $y = -3$	**45.** $y = 4$	**47.** $y = 0$
49. $x = 0$		

Exercise 9.8

1. $6 + x = 12$; $x = 6$	**3.** $17 = 4 + x$; $x = 13$
5. $\frac{x}{-4} = 5$; $x = -20$	**7.** $x = 8 + 5$; $x = 13$
9. $7x = 42$; $x = 6$	**11.** $2x + 6 = 25$; $x = \frac{19}{2}$ or $9\frac{1}{2}$
13. $6x + 4x = 30$; $x = 3$	**15.** $8 - x + 6 = 10$; $x = 4$
17. $6 + 2x = 22$; $x = 8$	**19.** $\frac{1}{4}x + 4 = 10$; $x = 24$
21. $21 - x = 2x$; $x = 7$	**23.** $6 + 4x = 6x - 4$; $x = 5$
25. $\frac{1}{3}(x + 6) = \frac{3}{4}x$; $x = \frac{24}{5}$ or $4\frac{4}{5}$	**27.** $\frac{1}{2}x - \frac{1}{3}x = 6$; $x = 36$
29. $40 = 2(w + 4) + 2w$; $w = 8$; $l = 12$	
31. $2x + x = 78$; 26 in., 52 in.	**33.** $1863 - (4x + 7) = 1776$; $x = 20$

Review Exercises—Chapter 9

1. 2040	**3.** 49
5. −2	**7.** 32

9. terms: $3a$, $2b$; coefficients: 3, 2; literal parts: a, b

11. terms: $-11m^2n$, $2mn^2$, -1; coefficients: −11, 2, −1; literal parts: m^2n, mn^2

13. $-2a - 5b$	**15.** cannot be combined
17. yes	**19.** no
21. no	**23.** $x = 11$
25. $x = -1$	**27.** $h = -12$
29. $y = -\frac{1}{13}$	**31.** $x = 6$
33. $x = 6$	**35.** $y = \frac{1}{3}$
37. $6 = x + 4$; $x = 2$	**39.** $x = 3x - 12$; $x = 6$

Critical Thinking Exercises—Chapter 9

1. False: to solve the equation $\frac{1}{3}x = 9$, we must *multiply* both sides of the equation by $\frac{3}{1}$ or 3.
3. False: to solve the equation $24x = 6$, we isolate the x by dividing both sides of the equation by 24, not 6.
5. False: when combining like terms, the literal part in the answer stays the same and we add the coefficients. Therefore $6x + 7x = 13x$.
7. We must subtract 4 from *both* sides to keep things balanced, so we have $x + 4 + (-4) = 12 + (-4)$ and $x = 8$.
9. The two terms $12a^2$ and 12 cannot be combined because they have different literal parts.

Achievement Test—Chapter 9

1. 33.49 2. 27 3. 35
4. -48
5. (a) $-2ab^3, 6a^2b^2, 2$ (b) $-2, 6, 2$ (c) ab^3, a^2b^2
6. cannot be combined 7. $-3a^2 - 7b^2$ 8. $2rst - 2r + 1$
9. no 10. yes 11. $x = 11$
12. $x = -6$ 13. $x = 5$ 14. $x = -7$
15. $x = 11$ 16. $x = 3$ 17. $x = 7$
18. $x = 4$ 19. $x = 14$
20. Johnstown 75, Freeport 87

Cumulative Review—Chapters 7, 8, 9

1. mean: 40.1; median: 39.5; mode: 35
3. 3.13 5. 1
7. 1 9. undefined
11. 100,000 13. 9
15. 52 sq. in. 17. $x = -22$
19. $x = 4$ 21. Hornets 95, Bulls 104

Chapter 10

Exercise 10.1

1. circle 3. parallelogram 5. square
7. rectangle 9. equilateral triangle

Exercise 10.2

1. $A = 12$ ft^2; $P = 14$ ft 3. $A = 196$ in.^{2}; $P = 56$ in.
5. $A = 16$ yd^2; $P = 16$ yd 7. \$280.00
9. \$354.62 for tile; \$79.20 for baseboard
11. a rectangle with all sides equal to 4

Exercise 10.3

1. $A = 110$ cm^2; $P = 52$ cm 3. $A = 3\frac{3}{4}$ ft^2; $P = 10$ ft
5. $A = 18$ ft^2; $P = 20$ ft 7. $h = 7$ in.

	Base	Height	Area
9.	15 in.	9 in.	67.5 in.^{2}
11.	9.6 m	1.2 m	5.76 m^2

Exercise 10.4

	Radius	Diameter	Circumference	Area
1.	2 ft	4 ft	12.6 ft	12.6 ft^2
3.	3 cm	6 cm	18.8 cm	28.3 cm^2
5.	1.4 in.	2.8 in.	8.8 in.	6.2 in.^{2}
7.	$4\frac{1}{2}$ ft	9 ft	28.3 ft	63.6 ft^2

9. 5 flats 11. $A = 8{,}826$ ft^2; $P = 388.4$ ft

Exercise 10.5

1. $V = 432$ in.^{3} 3. $V = 108$ ft^3 5. 4 in.
7. $V = 1{,}120$ ft^3 9. $V = 123\frac{3}{4}$ ft^3

Exercise 10.6

	Radius	Diameter	Volume
1.	3 in.	6 in.	113.0 in.^{3}
3.	5 cm	10 cm	523.3 cm^3
5.	3.5 ft	7 ft	179.5 ft^3

7. 113.04 ft^3 9. 261.7 in.^{3}

	Radius	Diameter	Height	Volume
11.	2 ft	4 ft	6 ft	75.4 ft^3
13.	3.5 in.	7 in.	9 in.	346.2 in.^{3}
15.	$3\frac{1}{3}$ m	$6\frac{2}{3}$ m	5 m	174.4 m^3

17. 904.3 cm^3 19. 14.7 ft^3

Review Exercises—Chapter 10

1. hexagon 3. rectangle
5. sphere 7. circle
9. right triangle 11. cylinder
13. $A = 32$ ft^2; $P = 24$ ft 15. $A = 181$ ft^2; $P = 62$ ft
17. $A = 77$ cm^2; $P = 48$ cm 19. $A = 113.04$ in.^{2}; $P = 37.68$ in.
21. 576 in.^{3} 23. 8,000 cm^3
25. 3,077.2 in.^{3}

Critical Thinking Exercises—Chapter 10

1. True: Consider a rectangle with perimeter $P = 2l + 2w$. Replacing l with $2l$ and w with $2w$ yields a new rectangle with perimeter $4l + 4w$, which is twice the original perimeter, since $2(2l + 2w) = 4l + 4w$.

3. True: Consider a circle with radius r; it has circumference $2\pi r$. Replacing r by $2r$ yields a new circle with circumference $2\pi(2r) = 4\pi r$; which is twice the circumference of the original circle.
5. True: Consider a triangle with base b and height h; it has area $\frac{1}{2}bh$. Replacing b by $2b$ and h by $\frac{1}{2}h$ yields a triangle with area $\frac{1}{2}(2b)(\frac{1}{2}h) = \frac{1}{2}bh$, which is the same as the area of the original triangle.
7. A rectangle has four right angles. Since a square also has four right angles, it is also a rectangle. However, a square must have four sides of equal length and a rectangle can have a different length and width.
9. Mr. Jardine used the formula for area when he needed the formula for perimeter. He needed only 52 ft.

Achievement Test—Chapter 10

1. **(a)** $A = 30$ m^2; **(b)** $P = 22$ m
2. **(a)** $A = 256$ cm^2; **(b)** $P = 64$ cm
3. **(a)** $A = 22.5$ in.^{2}; **(b)** $P = 31$ in.
4. **(a)** $A = 78.5$ in.^{2}; **(b)** $C = 31.4$ in.
5. **(a)** $A = 530.7$ cm^2; **(b)** $C = 81.6$ cm
6. $V = 15.3$ ft^3
7. **(a)** $V = 64$ cm^3; **(b)** surface area = 96 cm^2
8. **(a)** $A = 412$ sq. units; **(b)** $P = 98$ units
9. **(a)** $V = 432$ in.^{3}; **(b)** $A = 360$ in.^{2}
10. $V = 523.3$ in.^{3}
11. $V = 157$ in.^{3}

Chapter 11

Exercise 11.1

1. 270 cg
3. 1000 g
5. 1.36 liters
7. 0.278425 kg
11. 76 000 cL
13. 40 000 000 dm
15. 3.9794 hL

Exercise 11.2

1. 112 km
3. 17,800 sec
5. 14.11 ft
7. 502.92 cm
9. 6480 min
11. 907 200 g
13. 8847.75 m
15. 3840 km

Exercise 11.3

1. 14 gal 3 qt
3. 4 yd 1 ft 10 in.
5. 5 days 12 min 22 sec
7. 11 gal

Exercise 11.4

1. 3 gal 2 qt
3. 2 yds 1 ft 10 in.
5. 4 gal 1 pt
7. 2 tons 1475 lb
9. 43 min

Exercise 11.5

1. 29 gal 1 qt
3. 57 hr 52 min 42 sec
5. 81 lb 9 oz
7. 51 yd 1 ft
9. 45 yr 40 weeks 4 days

Exercise 11.6

1. 1 hr 20 min
3. 1 week 6 days $16\frac{3}{4}$ hr
5. 1 mile $1241\frac{5}{9}$ ft
7. 2 lb $3\frac{3}{4}$ oz
9. 57 yd $2\frac{1}{6}$ ft
11. 7 gal $2\frac{7}{17}$ pt

Exercise 11.7

1. 30 000 cm^2
3. 23.9 yd^2
5. 102 400 rod^2
7. 32.78 cm^2
9. 500 000 mm^2
11. 13.3 yd^2
13. 2.59 yd^3
15. 0.036 lb

Exercise 11.8

1. 19.6 m/sec
3. 97.9 kg/m^2
5. 1.034 kg/cm^2
7. 9.75 m/sec^2

Review Exercises—Chapter 11

1. 312 000 cm
3. 0.7625 dag
5. 71.36 kg
7. 30.48 cm
9. 5,865,696,000,000 miles/yr
11. 12 weeks 5 days 12 hr 52 min
13. 8 hr 56 min 42 sec
15. 1 yd 1 ft $11\frac{1}{4}$ in.
17. 0.044 lb (rounded)
19. 2500 cc (cm^3)
21. 770 ft/sec
23. 308.32 gal/min
25. 152.91 m^3/min

Achievement Test—Chapter 11

1. 0.0327 kiloliters
2. 0.01285 hectometers
3. 24 km
4. 11 lb
5. 8 gal 1 qt
6. 2 days 54 min 44 sec
7. 173 lb 4 oz
8. 2 yd 1 ft $9\frac{2}{5}$ in.
9. 4.18 m^2
10. 3 square meters is greater, by 7.29 ft^2.
11. 96 km/hr
12. 146.78 kg/m^2

Cumulative Review—Chapters 10 and 11

1. trapezoid
3. triangle
5. $A = 84$ sq. units
7. $V = 3$ ft^3
9. 196.85 in.
11. 2272.7 g
13. 5 hr 51 min 57 sec

Chapter 12

Exercise 12.2

1. 9250
3. 35.146
5. 7071
7. −20,211
9. −11.622
11. 2924
13. 208.008
15. 53458.35
17. 554.92308
19. 0.3259668
21. 14.555556
23. 0.5287671
25. −2157.6
27. 14304
29. 32.649655
31. 29.274562
33. 94.074439
35. 4.64758
37. 441
39. 71,824
41. 987.2164
43. 0.16

Exercise 12.3

1. 440
3. 504
5. 10.954451
7. 4.8989795
9. 216
11. 625
13. 59,049
15. 32.768
17. 8008.7465
19. 0.1785714
21. 560
23. 1.1138462
25. 182.36
27. 0.32
29. 7.4431707
31. 960
33. −14.994
35. −39.304
37. 44.718947
39. 4.718947

Exercise 12.4

1. 1,635,600
3. 743,296,100
5. 0.00000006421
7. 0.00000000008881
9. 490,000,000,000
11. 624,000,000
13. 45,486,000,000
15. 653,600,000,000
17. 208,333,333

19. 0.00000005355 **21.** 106,595,100 **23.** 373,248,000
25. 0.000000068921 **27.** 27,928,571,000
29. 0.000000058479532

Exercise 12.5

1. 8,888,811,111 **3.** 266,664 **5.** 1,111,088,889
7. 555 **9.** 0.3999996

Review Exercises—Chapter 12

1. 13,028 **3.** 6,435 **5.** 10.443165
7. 34,888 **9.** 3.5986108 **11.** 470
13. 18.867962 **15.** 4096 **17.** 0.5
19. 8.9473684 **21.** −74.088 **23.** 1,050,000,000
25. 0.00000001575 **27.** 18,080,541,000

Achievement Test—Chapter 12

1. 20,908 **2.** 157.5 **3.** 59,074.84
4. 9.3612824 **5.** −24,336 **6.** 70.823725
7. 3.9051248 **8.** 457.96 **9.** 1,495
10. 1,371 **11.** 16.248077 **12.** 243
13. 5098.317 **14.** 2.2694611 **15.** −6,016.9206
16. −1,935.68 **17.** 1.3541053
18. 1.2905 09; 1,290,500,000 **19.** 1.215 −08; 0.00000001215
20. 3.2768 −08; 0.000000032768
21. 1.9535048 10; 19,535,048,000
22. 9.2193362 09; 9,219,336,200

Index

L

M

N

O

P